AF595436

Advanced Coatings for Corrosion Protection

Advanced Coatings for Corrosion Protection

Editor

Wolfram Fürbeth

MDPI • Basel • Beijing • Wuhan • Barcelona • Belgrade • Manchester • Tokyo • Cluj • Tianjin

Editor
Wolfram Fürbeth
Frankfurt am Main
Germany

Editorial Office
MDPI
St. Alban-Anlage 66
4052 Basel, Switzerland

This is a reprint of articles from the Special Issue published online in the open access journal *Materials* (ISSN 1996-1944) (available at: https://www.mdpi.com/journal/materials/special_issues/advanced_coatings).

For citation purposes, cite each article independently as indicated on the article page online and as indicated below:

LastName, A.A.; LastName, B.B.; LastName, C.C. Article Title. *Journal Name* **Year**, *Volume Number*, Page Range.

ISBN 978-3-03943-921-8 (Hbk)
ISBN 978-3-03943-922-5 (PDF)

Contents

About the Editor

Wolfram Fürbeth has a diploma in Chemistry from the University of Dortmund and received a Ph.D. in Materials Sciences from the University of Erlangen-Nuremberg in 1997. He joined DECHEMA as a research scientist in 1997 and has been head of the corrosion research group since 2005. In 2012, he obtained a habilitation in Materials Technology at RWTH Aachen Technical University. He served as Adjunct Professor at RWTH in 2018, and since March 2017, he has been a member of the Institute Directorate of the DECHEMA Research Institute. Since 2017, he has served as Head of the Scientific Advisory Board of the German Corrosion Society, as well as Chairman of the European Federation of Corrosion Science and Technology Advisory Committee since 2017 and Chairman of Working Party Coatings since 2012.

Editorial

Special Issue: Advanced Coatings for Corrosion Protection

Wolfram Fürbeth

DECHEMA Research Institute, Theodor-Heuss-Allee 25, 60486 Frankfurt am Main, Germany; wolfram.fuerbeth@dechema.de; Tel.: +49-69-7564-398

Received: 29 July 2020; Accepted: 30 July 2020; Published: 1 August 2020

Abstract: Corrosion is an important issue in many industrial fields. Among others, coatings are by far the most important technology for corrosion protection of metallic surfaces. The special issue "Advanced Coatings for Corrosion Protection" has been launched as a means to present recent developments on any type of advanced coatings for corrosion protection. Fifteen contributions have been collected on metallic, inorganic, polymeric and nanoparticle enhanced coatings providing corrosion protection as well as partly other functionalities.

Keywords: metallic coatings; anodizing layers; passivation; polymeric coatings; laser cladding; PVD; superhydrophobic coatings; composite coatings

---

Corrosion is an important issue in many industrial fields. It leads to high economic losses of 3–4% of the GDP of an industrialized country year by year. Adequate corrosion protection is therefore essential in many applications. Among others, coatings are by far the most important technology for corrosion protection of metallic surfaces.

In the very traditional field of coatings for corrosion protection in the last years a deeper understanding of mechanisms of the protective action and corrosion mechanisms of and below protective coatings has been gained. This was necessary due to upcoming environmental and health issues for some well-established compounds used in former coating systems, e.g., lead or chromates, which have been banned from industrial application. This lead as well to a large amount of research in the field of advanced coating systems for corrosion protection.

This situation is the case for all the different types of protective coatings that are typically used. Novel metallic coatings, e.g., novel zinc alloys are under development, as well as novel pretreatment systems or passivating chemicals avoiding the use of chromates. The upcoming chemical nanotechnology fosters the development of hybrid or inorganic sol-gel coatings, as well as of nanoparticles and nanocapsules to be used as fillers in coating systems. This has also led, in recent years, to the development of novel self-healing and smart coatings. Furthermore, nowadays, bio-based substances are becoming increasingly used for organic coatings. Last but not least, new anodizing processes have also been developed in the frame of an increased use of light metals for light weight construction.

The special issue "Advanced Coatings for Corrosion Protection" has been proposed as a means to present recent developments on any of these types of advanced coatings for corrosion protection. Thus, 15 contributions have been collected on metallic, inorganic, polymeric and nanoparticle enhanced coatings providing corrosion protection as well as partly other functionalities.

Among all of them, inorganic coatings stand out for the number of contributions being submitted to this special issue; however, these are of many different types and for different applications. The thinnest but often quite effective type of an inorganic coating may be a passivating oxide layer. The most commonly used passivating agent to develop such an oxide layer is nitric acid. Lara-Banda et al. investigated an environmentally friendly alternative for the passivation of 15-5 and 17-5PH stainless steels based on citric acid [1]. It could be shown that, for both types of steel, the passive layer formed in citric acid as passivating solution had very similar characteristics to that formed with nitric acid.

Much thicker oxide layers can be obtained by anodizing techniques especially on light metals. Besides the conventional anodizing treatment at rather low voltages plasma-electrolytic oxidation (PEO) or micro arc oxidation (MAO), leading to ceramic oxide layers, has become an increasingly important alternative, being the topic of three contributions in this special issue [2–4]. Yao et al. prepared different MAO coatings on the magnesium alloy AZ91D [2]. It was found that especially a brown coating doped with Cu is able to significantly reduce the corrosion of magnesium parts in marine environments. Magnesium alloys to be used as biodegradable implant materials are the background of the paper by Anawati et al. [3]. Therefore, they investigated not only the corrosion resistance of PEO coatings, but also their ability to form bone mineral apatite. It was concluded that the alloying element Ca should be limited to 1 wt% as the excess tended to degrade the corrosion resistance and apatite-forming ability of the PEO coating. Kyziol et al. investigated the influence of MAO coatings on the stress corrosion cracking susceptibility of an AlMg6 alloy [4]. The pores in the MAO coating were insignificant and of limited depth. Therefore, the coating could increase the corrosion resistance.

Furthermore, inorganic coatings may be obtained from metal carbides [5], nitrides [6] or borides [7]. Jiang et al. used carburizing by a spark plasma sintering technique to enhance the erosion-corrosion resistance of tungsten in flowing coolant water [5]. W-Cr-C clad tungsten showed a different corrosion behavior than bare tungsten. Ti2AlN coatings were obtained by physical vapor deposition on ferritic steels and submitted to oxidation at a temperature of 700 °C by Gröner et al. [6]. The oxide scale of α-alumina was able to reduce the permeability for hydrogen significantly. Hu et al. obtained boride cermet coatings on carbon steel by a laser cladding process to improve the corrosion and wear resistance [7].

Polymeric coatings are widely used in corrosion protection and several contributions deal with this type of coatings as well [8–12]. Aging of the coating in terms of chain scission and phase separation may change the protective properties with time, as shown by Che et al. for polyurea coatings in marine atmosphere [8]. A wide variety of polymeric and hybrid systems can be chosen for protective coatings. Miela et al. showed how materials analysis and molecular dynamics simulation may help to identify the best performing coating system for erosion and corrosion protection of hydraulic water valves [9]. Polymeric coatings may also be used to add further functionalities to a barrier-type protective coating. As such, superhydrophobic properties were generated by Ou et al. on a zinc coating [10] and by Zhang et al. on the aluminum alloy AA5083 [11], providing water-repelling and long-term corrosion resistant surfaces. A very classical application of organic coatings is the insulation of buried pipelines. The paper by Hong et al. addresses this application, focusing on the additional cathodic protection design for a pre-insulated pipeline in a district heating system using computational simulation [12].

Finally, metallic coatings are widely used as noble barrier layers or as sacrificial layers providing cathodic protection to the substrate. One of the latter has been described by Tong et al. [13]. They produced a ZnAl diffusion layer on carbon steel by a mechanical energy aided diffusion method and characterized its corrosion behavior. On the other hand, barrier-type coatings may be reinforced by incorporation of nanoparticles into the coating matrix. Xu et al. demonstrated that the mechanical properties and the corrosion resistance of an aluminum foam can be improved by the electrodeposition of a NiMo coating that has been reinforced by SiC/TiN nanoparticles [14]. Furthermore, Fu et al. studied the effect of doping a NiFeCoP coating with cerium dioxide nanoparticles [15]. With an increased concentration of nano-CeO_2 in the composite coating, its corrosion resistance increased as well.

Conflicts of Interest: The authors declare no conflict of interest.

References

1. Lara-Banda, M.; Gaona-Tiburcio, C.; Zambrano-Robledo, P.; Delgado-E, M.; Cabral-Miramontes, J.; Nieves-Mendoza, D.; Maldonado-Bandala, E.; Estupiñan-López, F.; Chacón-Nava, J.G.; Almeraya-Calderón, F. Alternative to Nitric Acid Passivation of 15-5 and 17-4PH Stainless Steel Using Electrochemical Techniques. *Materials* **2020**, *13*, 2836. [CrossRef] [PubMed]
2. Yao, Y.; Yang, W.; Liu, D.; Gao, W.; Chen, J. Preparation and Corrosion Behavior in Marine Environment of MAO Coatings on Magnesium Alloy. *Materials* **2020**, *13*, 345. [CrossRef] [PubMed]
3. Anawati, A.; Asoh, H.; Ono, S. Corrosion Resistance and Apatite-Forming Ability of Composite Coatings formed on Mg-Al-Zn-Ca Alloys. *Materials* **2019**, *12*, 2262. [CrossRef] [PubMed]
4. Kyziol, L.; Komarov, A. Influence of Micro-Arc Oxidation Coatings on Stress Corrosion of AlMg6 Alloy. *Materials* **2020**, *13*, 356. [CrossRef] [PubMed]
5. Jiang, Y.; Yang, J.; Xie, Z.; Fang, Q. Enhanced Erosion–Corrosion Resistance of Tungsten by Carburizing Using Spark Plasma Sintering Technique. *Materials* **2020**, *13*, 2719. [CrossRef] [PubMed]
6. Gröner, L.; Mengis, L.; Galetz, M.; Kirste, L.; Daum, P.; Wirth, M.; Meyer, F.; Fromm, A.; Blug, B.; Burmeister, F. Investigations of the Deuterium Permeability of As-Deposited and Oxidized Ti2AlN Coatings. *Materials* **2020**, *13*, 2085. [CrossRef] [PubMed]
7. Hu, Z.; Li, W.; Zhao, Y. The Effect of Laser Power on the Properties of M3B2-Type Boride-Based Cermet Coatings Prepared by Laser Cladding Synthesis. *Materials* **2020**, *13*, 1867. [CrossRef] [PubMed]
8. Che, K.; Lyu, P.; Wan, F.; Ma, M. Investigations on Aging Behavior and Mechanism of Polyurea Coating in Marine Atmosphere. *Materials* **2019**, *12*, 3636. [CrossRef] [PubMed]
9. Mlela, M.K.; Xu, H.; Sun, F.; Wang, H.; Madenge, G.D. Material Analysis and Molecular Dynamics Simulation for Cavitation Erosion and Corrosion Suppression in Water Hydraulic Valves. *Materials* **2020**, *13*, 453. [CrossRef] [PubMed]
10. Ou, J.; Zhu, W.; Xie, C.; Xue, M. Mechanically Robust and Repairable Superhydrophobic Zinc Coating via a Fast and Facile Method for Corrosion Resisting. *Materials* **2019**, *12*, 1779. [CrossRef] [PubMed]
11. Zhang, B.; Xu, W.; Zhu, Q.; Yuan, S.; Li, Y. Lotus-Inspired Multiscale Superhydrophobic AA5083 Resisting Surface Contamination and Marine Corrosion Attack. *Materials* **2019**, *12*, 1592. [CrossRef] [PubMed]
12. Hong, M.S.; So, Y.S.; Kim, J.G. Optimization of Cathodic Protection Design for Pre-Insulated Pipeline in District Heating System Using Computational Simulation. *Materials* **2019**, *12*, 1761. [CrossRef] [PubMed]
13. Tong, J.; Liang, Y.; Wei, S.; Su, H.; Wang, B.; Ren, Y.; Zhou, Y.; Sheng, Z. Microstructure and Corrosion Resistance of Zn-Al Diffusion Layer on 45 Steel Aided by Mechanical Energy. *Materials* **2019**, *12*, 3032. [CrossRef] [PubMed]
14. Xu, Y.; Ma, S.; Fan, M.; Zheng, H.; Chen, Y.; Song, X.; Hao, J. Mechanical and Corrosion Resistance Enhancement of Closed-Cell Aluminum Foams through Nano-Electrodeposited Composite Coatings. *Materials* **2019**, *12*, 3197. [CrossRef] [PubMed]
15. Fu, X.; Ma, W.; Duan, S.; Wang, Q.; Lin, J. Electrochemical Corrosion Behavior of Ni-Fe-Co-P Alloy Coating Containing Nano-CeO_2 Particles in NaCl Solution. *Materials* **2019**, *12*, 2614. [CrossRef] [PubMed]

Article

Alternative to Nitric Acid Passivation of 15-5 and 17-4PH Stainless Steel Using Electrochemical Techniques

María Lara-Banda [1], Citlalli Gaona-Tiburcio [1], Patricia Zambrano-Robledo [1], Marisol Delgado-E [1], José A. Cabral-Miramontes [1], Demetrio Nieves-Mendoza [2], Erick Maldonado-Bandala [2], Francisco Estupiñan-López [1], José G. Chacón-Nava [3] and Facundo Almeraya-Calderón [1,*]

1 Universidad Autonoma de Nuevo Leon, FIME—Centro de Investigación e Innovación en Ingeniería Aeronáutica (CIIIA), Av. Universidad s/n. Ciudad Universitaria, San Nicolás de los Garza, Nuevo León 66455, Mexico; marialarabanda@yahoo.com.mx (M.L.-B.); citlalli.gaonatbr@uanl.edu.mx (C.G.-T.); patricia.zambranor@uanl.edu.mx (P.Z.-R.); marisol__1706@hotmail.com (M.D.-E); jose.cabralmr@uanl.edu.mx (J.A.C.-M.); francisco.estupinanlp@uanl.edu.mx (F.E.-L.)

2 Facultad de Ingeniería Civil, Universidad Veracruzana, Xalapa, Veracruz 91000, Mexico; dnieves@uv.mx (D.N.-M.); eemalban@gmail.com (E.M.-B.)

3 Centro de Investigación en Materiales Avanzados (CIMAV), Miguel de Cervantes 120, Complejo Industrial Chihuahua, Chihuahua, Chih 31136, Mexico; jose.chacon@cimav.edu.mx

* Correspondence: falmeraya.uanl.ciiia@gmail.com

Received: 5 May 2020; Accepted: 8 June 2020; Published: 24 June 2020

Abstract: Increasingly stringent environmental regulations in different sectors of industry, especially the aeronautical sector, suggest the need for more investigations regarding the effect of environmentally friendly corrosion protective processes. Passivation is a finishing process that makes stainless steels more rust resistant, removing free iron from the steel surface resulting from machining operations. This results in the formation of a protective oxide layer that is less likely to react with the environment and cause corrosion. The most commonly used passivating agent is nitric acid. However, it is know that high levels of toxicity can be generated by using this agent. In this work, a study has been carried out into the electrochemical behavior of 15-5PH (precipitation hardening) and 17-4PH stainless steels passivated with (a) citric and (b) nitric acid solutions for 60 and 90 min at 49 °C, and subsequently exposed to an environment with chlorides. Two electrochemical techniques were used: electrochemical noise (EN) and potentiodynamic polarization curves (PPC) according to ASTM G199-09 and ASTM G5-13, respectively. The results obtained indicated that, for both types of steel, the passive layer formed in citric acid as passivating solution had very similar characteristics to that formed with nitric acid. Furthermore, after exposure to the chloride-containing solution and according with the localization index (LI) values obtained, the stainless steels passivated in citric acid showed a mixed type of corrosion, whereas the steels passivated in nitric acid showed localized corrosion. Overall, the results of the R_n values derived show very low and similar corrosion rates for the stainless steels passivated with both citric and nitric acid solutions.

Keywords: stainless steel; passivated; electrochemical noise; precipitation hardening

1. Introduction

Corrosion in the aeronautical industry remains a major problem that directly affects safety, economic, and logistical issues. Stainless steel alloys have found increasing application in aircraft components that require great strength but can handle the increased weight. The high corrosion and temperature resistances found in stainless steel in harsh environments make it suitable for a range of aircraft parts such as fasteners, actuators and landing gear components [1–3]. Passivation is a chemical process to remove surface contamination, i.e., small particles of iron-containing shop dirt and iron particles from cutting tools that can act as initiation sites for corrosion. This process also can remove sulfides exposed on the surface of free-machining stainless alloys. In other words, by chemically removing free contaminants from the surface of stainless steel, the passivation process adds a thin oxide layer. More chromium available from a clean surface means a thicker chromium oxide layer at the top of the stainless steel surface. Moreover, this chemically non-reactive surface means more protection against corrosion [4–10].

Precipitation hardening (PH) stainless steels (SS) are a family of corrosion resistant alloys some of which can be heat treated to provide tensile strengths of 850 to 1700 MPa and yield strengths of 520 MPa to over 1500 MPa. These alloys contain 11–18% chromium, 3–4% nickel, and smaller counts of additional metals, including aluminum, niobium, molybdenum, titanium and tungsten. Nevertheless, chromium is the alloying element responsible for the formation of the passive film [11–14]. The family of precipitation hardening stainless steels can be divided into three main types—low carbon martensitic, semi-austenitic and austenitic. These stainless steels are widely used in aerospace structural applications due to its good corrosion resistance and high strength and toughness obtained by the formation of precipitates from age-hardening treatments. Previous investigations on aeronautical-aerospace sector has shown that 15-5PH and 17-4PH steels have good corrosion resistance regarding other stainless steels [15–20].

Back in 1997, the specification QQ-P-35 for passivation of stainless steel parts was withdrawn, and replaced by specification SAE-QQ-P-35, also withdrawn in 2005. The latter was replaced by specification ASTM A967-17. This indicates that both, citric and nitric acid can be used as passivating agents for stainless steels. To be effective, the nitric acid must be highly concentrated. However, many questions has been done regarding the production of harmful to health toxic vapors generated by the use of nitric acid in passivation baths [21,22]. On the other hand, citric acid is a biodegradable alternative that does no generate hazardous waste. Although the citric acid benefits as a passivating agent are well-established, technical information about the passivation process is scarce [23,24]. In 2003, Boeing Company evaluate the use of citric acid as an alternative for steel passivation in the aeronautic industry [6]. In 2008, the National Aeronautics and Space Administration (NASA) began a research program focused on the evaluation of the use of nitric acid in the passivation process of welded parts, using the salt chamber technique [10]. Later, NASA evaluated the use of citric acid on specimens exposed under atmospheric corrosion conditions using adherence tests [21].

It is well know that aggressive ions, especially chloride ions Cl-, affect the protecting nature of the passive film on stainless steels causing its breakdown. This leads to localized attack, mainly pitting corrosion [25,26]. In the study of corrosion mechanisms, a number of electrochemical techniques such as potentiodynamic, potentiostatic, and galvanostatic polarization tests, electrochemical impedance (EIS) and electrochemical noise (EN) are widely used. For instance, the evaluation of important parameters such as passive range, pitting potentials, corrosion rates and transpassive regions are studied using potentiodynamic polarization curves (PPC). Bragaglia et al. [27] studied the potentiodynamic polarization behavior of passivated citric and nitric acid baths) and unpassivated AISI 304 stainless steel samples after 1 h in 3.5 wt. % NaCl solution. The passivation treatment largely increased the pitting potential, particularly in the case of nitric acid. After 24 h exposure, electrochemical behavior for the nitric acid and the citric acid passivated samples were almost identical.

Electrochemical noise is a technique that does not alter the natural state of the system, since no external disturbance is applied [28]. This technique reflect random or spontaneous events of current and/or potential fluctuations. Under open-circuit conditions, these fluctuations appear to be related to variations in the rates of anodic and cathodic reactions causing small transients as a result of stochastic processes such as breakdown and repassivation of passive films and formation and propagation of pits. The fluctuations of current between two nominally identical electrodes as well as their potential versus a reference electrode (three electrode system) are recorded as time series, and by using several methods to analyze noise data, an understanding of the corrosion process occurring can be determined. The EN data can be analyzed by several methods. Perhaps the most commonly used are those related to frequency domain (power density spectral or spectral analysis), time domain (statistical methods as skewness, kurtosis, localization index (LI), and the variation of signal amplitude with time) and time-frequency domains [29,30]. Suresh and Mudali [31] studied the corrosion of UNS S30403 stainless steel in 0.05 M ferric chloride ($FeCl_3$) by spectral, statistical, and wavelet methods to deduce the corrosion mechanism. They found a good correlation of roll-off slopes derived from power spectral analysis and statistical parameters such as standard deviation, localization index (LI), and kurtosis with pitting as the corrosion mechanism. These authors reported a localization index (LI) in the range from 0.7 and 1. LI values of 0.1 to 1 has been attributed to pitting corrosion and hence the mechanism of corrosion was attributed to pitting attack [X]. Ortiz Alonso et al. [32], studied the stress corrosion cracking (SCC) behavior of a supermartensitic stainless steel by EN. They found that the LI value increased during the straining of specimens (in the range from 0.1 to 1), indicating the presence of localized events such as pits or cracks regardless of the susceptibility of the steel to stress corrosion cracking. In spite of some of its drawbacks, other studies also have found a good relationship between the LI parameter and pitting corrosion [33,34]

The aim of the present work is the study of the electrochemical behavior of 15-5PH and 17-4PH stainless steels passivated in nitric and citric acid and exposed to a 5 wt. % NaCl aqueous solution by PPC and EN.

2. Materials and Methods

2.1. Materials and Samples Preparation

The materials used in this work were 15-5PH and 17-4PH stainless steels used in the as received condition. The chemical composition of these steels was obtained by atomic absorption spectrometry, see Table 1.

Table 1. Chemical composition of the used stainless steels (wt. %).

Stainless Steel	Elements										
	C	Mn	P	S	Si	Cr	Ni	Mo	Nb	Cu	Fe
15-5PH	0.024	0.817	0.007	0.004	1.569	14.410	3.937	0.383	0.308	3.558	Bal.
17-4PH	0.022	0.827	0.023	0.029	1.637	15.204	3.050	0.340	0.144	3.908	Bal.

Stainless steel samples were machined as cylindrical coupons, according to ASTM A380-17 [35]. The specimens were polished with SiC grit paper till 4000 grade, followed by ultrasonic cleaning in ethanol and deionized water for about 10 min each.

2.2. Passivation Process

The passivation process was carried out under the specification ASTM A967-17 [36]. Gaydos et al. [21] reported that extended passivation treatments give a better protection against corrosion for a series of stainless steels. In the present work, two passivation baths (a) nitric acid (20%v) and (b) citric acid (15%v) solutions were used. A constant temperature of 49 °C was maintained along the passivation process. Specimens were immersed in the solutions for 60 and 90 min. Table 2 show the passivation exposure conditions for each type of steel.

Table 2. Passivation at a temperature of 49 °C.

Stainless Steel	Citric Acid ($C_6H_8O_7$)		Nitric Acid (HNO_3)	
	Passivated Time (min)			
	60	90	60	90
15-5PH	X	X	X	X
17-4PH	X	X	X	X

2.3. *Electrochemical Techniques*

In order to assess the corrosion behavior of passivated specimens (exposed area 4.46 cm^2), two electrochemical techniques were used: EN and PPC. The electrolyte was a 5 wt. % NaCl aqueous solution and all tests were carried out at room temperature.

2.3.1. Electrochemical Noise (EN)

This technique was carried out under ASTM G199-09 standard [37]. The experimental setup for EN measurements is schematically depicted in Figure 1. Here, two nominally identical electrodes (passivated stainless steels) as working electrodes (WE1 and WE2) were connected to measure the electrochemical current noise (ECN), whereas the electrochemical potential noise (EPN) was measured by connecting one working electrode to a saturated calomel reference electrode.

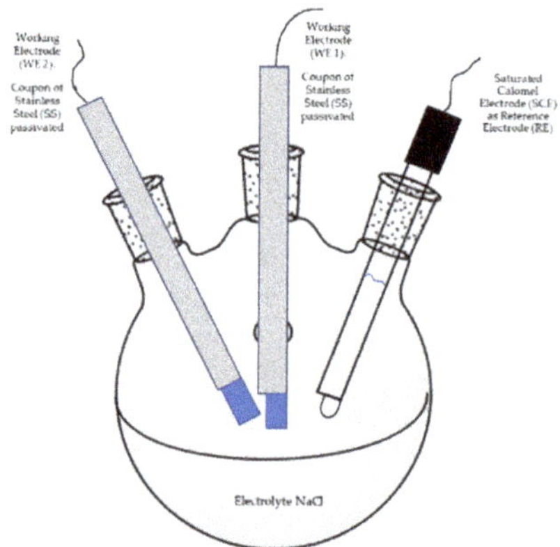

Figure 1. Experimental set up for electrochemical noise (EN) measurements.

The current and potential electrochemical noise was monitored as a function of time under open circuit condition for each particular electrode–electrolyte combination, using a Gill-AC (Alternating Current) potentiostat/galvanostat/ZRA (Zero Resistance Ammeter) from ACM Instruments. Electrochemical noise measurements started one after the open circuit potential stabilized (about 1 h after immersion in the electrolyte). Since the EN technique involves mostly non-stationary signals, trend removal was carried out. In each experiment, 1024 data were measured with a scanning speed of 1 data/s. The time series in current and potential were visually analyzed to interpret the signal transients and define the behavior of the frequency and amplitude of the fluctuations as a function of time. Resistance noise (*Rn*) data were obtained and used to calculate the corrosion rate according to Equation (1),

$$R_n = \frac{\sigma_E}{\sigma_I} \tag{1}$$

where σ_E is the standar deviation of potential noise, and σ_I is the standar deviation of current noise after trend removal. The LI, defined by Equation (2), is a parameter used to estimate, as a first approximation, the type of corrosion occurring in a given system [38–40]. LI values approaching zero, indicates uniform (general) corrosion; values in the range from 0.01 to 0.1 indicates mixed corrosion, whereas values from 0.1 to 1 correspond to pitting corrosion.

$$IL = \frac{\sigma_I}{I_{RMS}} \tag{2}$$

where I_{RMS} is the root mean square value of the corrosion current noise.

2.3.2. Potentiodynamic Polarization Curves (PPC)

This technique was carried out according to ASTM G5-13 [41] and ASTM G102-89 standards [42]. Here, a conventional three-electrode cell configuration was used, see Figure 2.

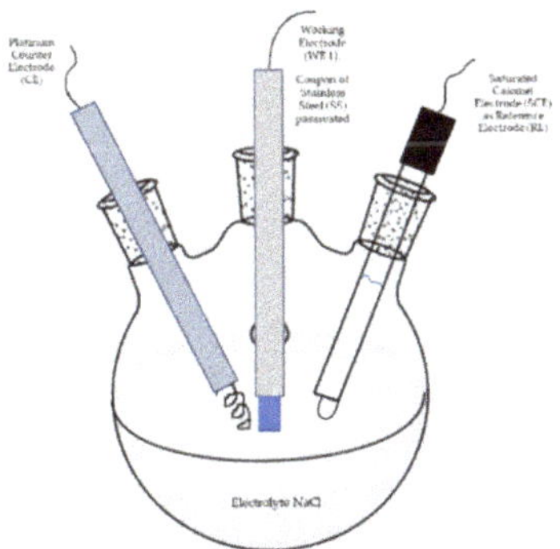

Figure 2. Conventional three-electrode cell configuration used in the potentiodynamic polarization curves (PPC) tests.

Potentiodynamic polarization curves were recorded in 5 wt. % NaCl aqueous solution at room temperature in a Gill-AC potentiostat/galvanostat from ACM Instruments. The potential scan was carried out from −1000 mV to +1200 mV, at a scan rate of 60 mV/min. A saturated calomel electrode (SCE) and a platinum wire were used as reference electrode and counter electrode, respectively. The working electrode (passivated sample) was hold for about 1 h at open circuit potential before tests.

3. Results

3.1. Electrochemical Noise

Figures 3 and 4 show the current and potential time series recorded for 15-5PH and 17-4PH stainless steel passivated in citric and nitric acid solutions at 60 and 90 min, respectively. Figure 2 shows that under passivation conditions at 60 and 90 min in citric acid, the passivated 15-5PH and 17-4PH stainless steel specimens did not present current fluctuations in time, this indicating that the specimens are in passive conditions; also, the potential noise signals remained constant without fluctuations in time (Figure 3a). The 17-4PH sample passivated for 60 min has higher current demand with low amplitude and high frequency transients, while the potential for this alloy has more active potentials (Figure 3d). For both types of stainless steel, the current-potential time series after 1000 s it has a tendency towards passivation.

Windowing analysis of electrochemical current noise between 0 and 200 s (Figure 3b) show no current increase for the 15-5PH samples passivated at 60 and 90 min. The 17-4PH steel passivated for 60 min, shows some transients of low amplitude and frequency, while for the 90 min passivation treatment only one anodic transient of high amplitude and low frequency was recorded 20 s after the start of the test. Another windowing analysis of current noise signal was performed between 900 and

1024 s (Figure 3c). For both types of stainless steel, irrespective of passivation conditions, no current fluctuations were observed. In some way, this behavior indicates stability of the passive layer.

For both types of stainless steel under passivation conditions, windowing analysis from 0 to 200 s and from 900 to 1024 s did not show frequency or amplitude transients, confirming the stability of potentials (Figure 3e,f). It is worth noting that the potentials of the 17-4PH samples are more negative than those recorded for the 15-5PH samples.

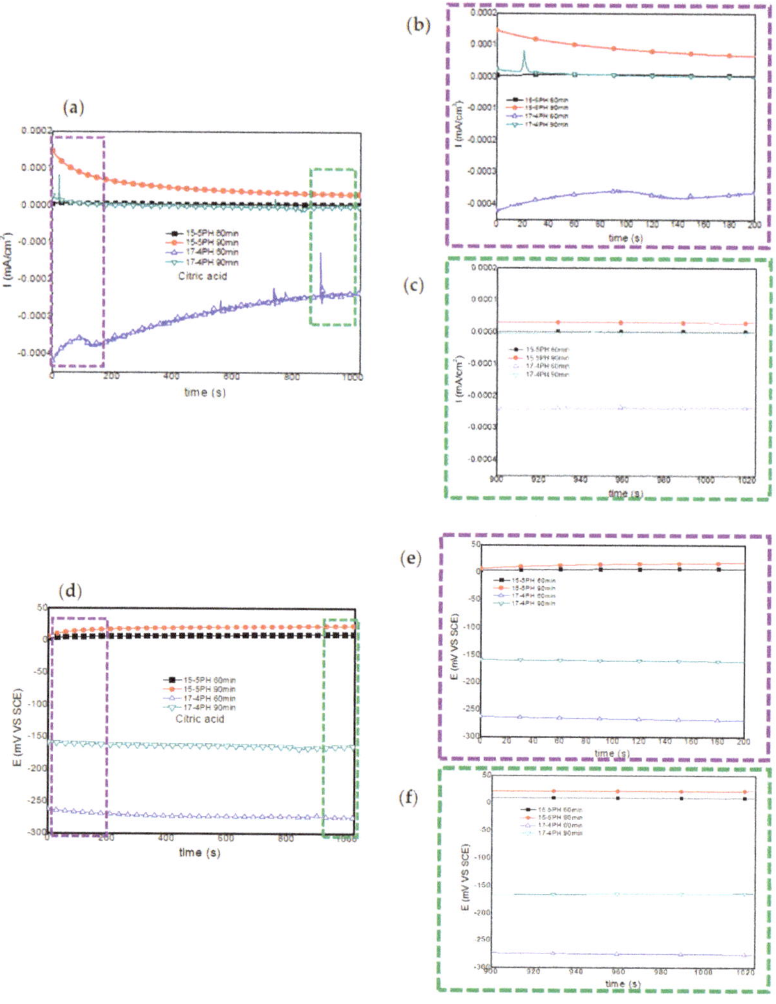

Figure 3. Electrochemical current and potential noise-time series for 15-5PH and 17-4PH samples passivated in citric acid at 49 °C, exposed in a 5 wt.% NaCl solution (**a**,**d**). Windowing of electrochemical current noise (ECN) from 0–200 and 900–1024 s (**b**,**c**); windowing of electrochemical potential noise (EPN) from 0–200 and 900–1024 s (**e**,**f**).

For the 15-5PH and 17-4PH stainless steels passivated in nitric acid, Figure 4 shows the current and potential noise time series recorded. The 17-4PH sample passivated for 90 min, show a decreases in current noise as a function of time; while the potential noise shifts to noble values, indicating stability of the passive layer. A similar behavior was observed for the 15-5PH samples passivated for 60 min.

The 15-5PH and 17-4PH samples passivated for 90 and 60 min show a small current demand during the first 300 and 700 s. Afterwards, no significant current or potential fluctuations (transients) were recorded, indicating stabilization of the passive layer (Figure 4a,d).

Windowing analysis of electrochemical current noise between 0 and 200 s (Figure 4b), shows a small increase in in current demand for the 15-5PH and 17-4PH samples passivated for 90 and 60 min, respectively (Figure 4b). From 900 to 1024 s, a windowing analysis of current noise did not show current transients (Figure 4c). Windowing analysis from 0 to 200 s and from 900 to 1024 s did not show frequency or amplitude transients, confirming the stability of potentials (Figure 4e,f). It is interesting to note that, irrespective of the time of passivation treatment, more noble potentials were attained by the 15-5PH stainless steel, in comparison with the 17-4PH steel.

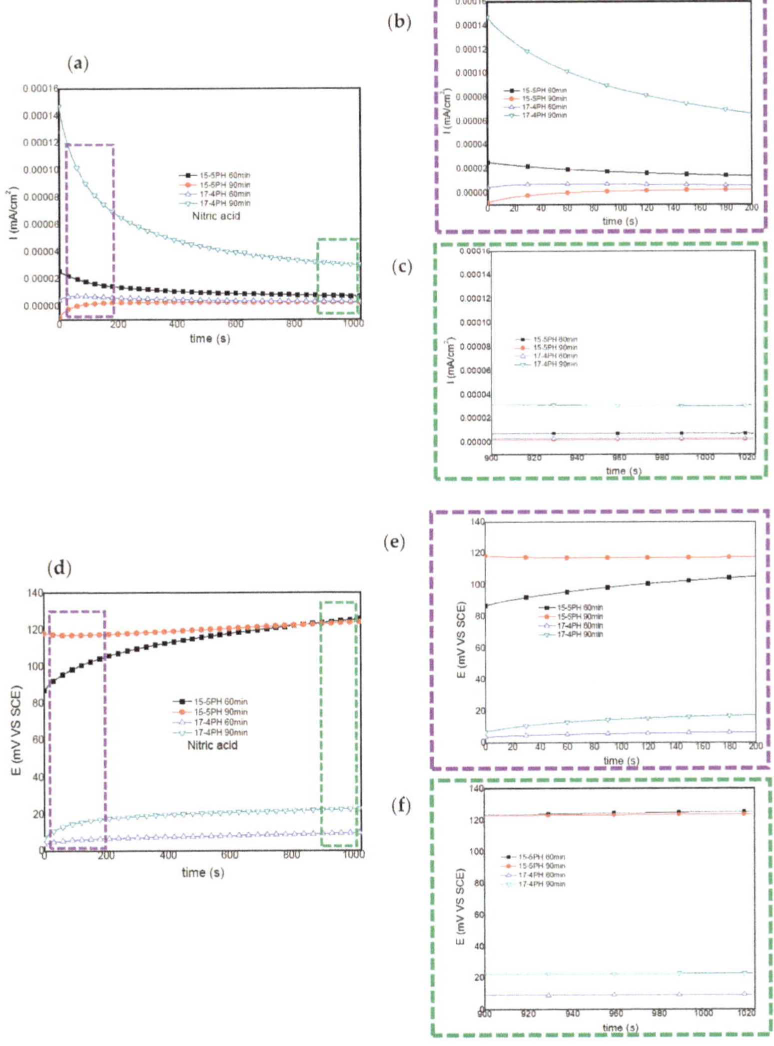

Figure 4. Electrochemical current and potential noise-time series for 15-5PH and 17-4PH samples passivated in nitric acid at 49 °C, exposed in a 5 wt. % NaCl solution (**a**,**d**). Windowing of ECN from 0–200 and 900–1024 s (**b**,**c**); windowing of EPN from 0–200 and 900–1024 s (**e**,**f**).

The EN parameters derived from the statistical analysis of current and potential time series measurements are shown in Table 3. The i_{corr} value obtained from noise resistance (Rn) for the samples passivated in citric acid is in the order of 10^{-4} (mA/cm^2), whereas for the samples passivated in nitric, i_{corr} values about 10^{-5} (mA/cm^2) were recorded. The very low values for i_{corr} obtained for both passivating agents indicate that citric acid could be a potential replacement for nitric acid as passivating agent. Information regarding the type of corrosion that could be occurring is given by the LI parameter. As can be seen from Table 3, the stainless steels passivated in citric acid solution mainly show a mixed corrosion type, whereas the stainless steels passivated in nitric acid solution the LI values indicates localized corrosion.

Table 3. Electrochemical noise parameters at various conditions in 5 wt. % NaCl at 49 °C.

Passivated Agent	Stainless Steel	Time (min)	Rn (Ω/cm^2)	i_{corr} (mA/cm^2)	LI	Corrosion Type
Citric acid	15-5PH	60	8.01×10^{-4}	6.49×10^{-4}	0.0862	Mixed
		90	5.00×10^{-5}	1.04×10^{-4}	0.0308	Mixed
	17-4PH	60	5.76×10^{-4}	4.51×10^{-4}	0.2492	Localized
		90	3.27×10^{-5}	1.59×10^{-4}	0.0900	Mixed
Nitric acid	15-5PH	60	2.35×10^{-6}	1.1×10^{-5}	0.1871	Localized
		90	1.51×10^{-6}	1.72×10^{-5}	0.1077	Localized
	17-4PH	60	1.03×10^{-6}	2.52×10^{-5}	0.1485	Localized
		90	1.34×10^{-6}	1.94×10^{-4}	0.1727	Localized

3.2. *Potenciodynamic Polarization*

The corrosion kinetic behaviour using potentiodynamic polarization can be observed through cathodic and anodic reactions in polarization curves. Corrosion rate in terms of penetration (mm/sec) is one of the main parameters obtained by potentiodynamic polarization curves, according to Faraday's law (Equation (3)) [40,42–44].

$$Corrosion\ rate = K_1 \frac{i_{corr}}{\delta}\ E.W \tag{3}$$

The potentiodynamic polarization curves obtained for the 15-5PH and 17-4PH stainless steels passivated for 60 min and 90 min in (a) citric acid and (b) nitric acid, and immersed in 5 wt. % NaCl solution are shown in Figure 5.The results for citric acid passivation (Figure 5a) show that the lower E_{corr} value was recorded for the 17-4PH sample passivated for 90 min, while the 15-5PH sample passivated for 90 min has the highest E_{corr}. Pitting potentials (E_{pitt}) were in the range from 42 mV up to 147 mV. This last value was recorded for the 15-5PH steel passivated for 90 min, this being the best treatment for nitric acid passivation, also corroborated by the lower corrosion rate obtained. For nitric acid passivation conditions, Figure 4b show that the E_{pitt} was largely improved, particularly for the 15-5PH steel passivated for 90 min, and also has the lower corrosion rate in this condition. The lower E_{pitt} value recorded was given by the 17-4PH passivated during 90 min, also giving the highest corrosion rate. On the whole, irrespective of the type of PH stainless steel used, the nitric acid passivation treatment largely increases the pitting potentials compared with the citric acid treatment.

The parameters (E_{corr}, E_{pitt}, i_{corr}, and corrosion rate (C.R.)) obtained from the polarization potentiodynamic curves are summarised in Table 4. Very low values of corrosion rate (within the same order of magnitude) were recorded for both 15-5PH and 17-4PH steels, irrespective of the passivation treatment conditions.

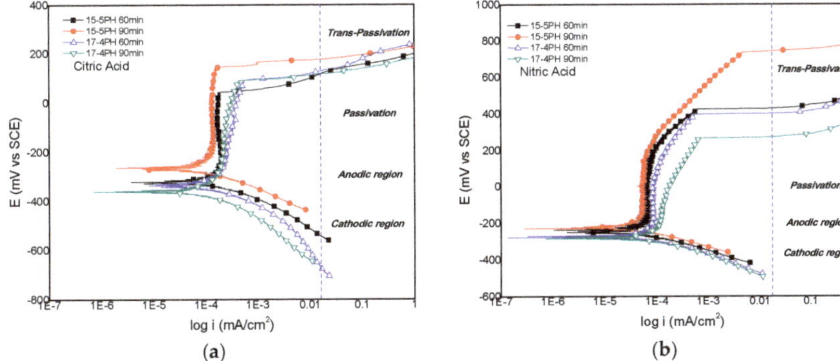

(**a**) (**b**)

Figure 5. Potentiodynamic polarization curves of 15-5PH and 17-4PH stainless steel passivated in (**a**) citric acid and (**b**) nitric acid, exposed in a 5 wt. % NaCl solution at 49 °C.

Table 4. Potentiodynamic polarization parameters in stainless steels passivated at 49 °C, in 5 wt. % NaCl.

Passivated Agent	Stainless Steel	Time (Min)	E_{corr} (mV)	E_{pit} (mV)	i_{corr} (mA/cm^2)	C. R. (mm/Year)
Citric Acid	15-5PH	60	−323	42	5.26×10^{-5}	5.54×10^{-7}
		90	−266	147	4.50×10^{-5}	4.75×10^{-7}
	17-4PH	60	−335	91	9.22×10^{-5}	9.64×10^{-7}
		90	−360	97	5.38×10^{-5}	5.63×10^{-7}
Nitric Acid	15-5PH	60	−228	467	2.16×10^{-5}	2.28×10^{-7}
		90	−228	765	2.27×10^{-5}	2.39×10^{-7}
	17-4PH	60	−271	439	3.51×10^{-5}	3.67×10^{-7}
		90	−279	323	4.41×10^{-5}	4.61×10^{-7}

4. Discussion

Several EN procedures correlating timed dependent fluctuation of current and potential during the corrosion process have been used to indicate the type of corrosion occurring. For instance, it is well recognized that the main source of electrochemical noise is the passive film breakdown process and repassivation process [45–49].

For the passivated 15-5PH and 17-4PH stainless steels in this work, the electrochemical potential time series recorded under nitric acid passivation show a passive region from 0 to 120 mV, whereas for citric acid passivation conditions, the passive region goes from −300 to 25 mV. Thus, passivation in nitric acid occurs at more positive (noble) potentials that in citric acid solutions. To some extent, this might indicate that the passive layer is more stable at more noble potentials [25]. Hence, higher corrosion resistance (*Rn*) values could be expected for passivation in nitric acid solutions [50]. To some extent, the results of *Rn* in Table 3 confirm this.

As a first approach, to assess the more likely type of corrosion occurring for the 15-5PH and 17-4PH stainless steels passivated in both citric acid and nitric acid solutions, the LI parameter was evaluated from the electrochemical noise data, and was found to be in the range from 0.03 to 0.249, see Table 3. From the LI values obtained for each passivating bath, the corrosion type occurring in citric acid passivation conditions can be attributed to mixed corrosion, whereas for nitric acid passivation conditions, the corrosion type could be attributed to pitting corrosion. LI has been used by several research groups for determining corrosion types under several conditions [51–56].

The use of LI to determine corrosion types has been the subject of many discussions among investigators on the data treatment and interpretations using LI [29,30,38,57–59]. Since the mean of the noise data (detrended) would be negligible, the standard deviation and root mean square current noise would converge to the same value and, hence, the LI evaluated from the data would be unity, irrespective of the corrosion type. Cottis [60] indicated that LI for identification of localization of corrosion is unduly influenced by the mean current and hence less reliable. In the present study, and as a first attempt, the LI parameter was estimated. Of course, it is recognized that in the study of stainless steels such as those in the present work, investigation of procedures based on the frequency domain and time-frequency domain deserves further attention.

The potentiodynamic polarization curves show passivation behavior for the 15-5PH and 17-4PH steels passivated in citric and nitric acid solutions. The passive zones on stainless steels are commonly made up of primary and secondary zones, which are formed before and after transpassivation, respectively. Potentials above E_{pitt} causes a rapid dissolution [61–63]. The passive zone involves the formation of iron and chromium oxide films [61,64,65]. Hence, selective dissolution on the surface of the alloy generates a surface enrichment of Cr^{3+} giving rise to $Cr(OH)_3$, as shown in Equation (4). Further dissolution of the hydroxide leads to the formation of a continuum layer of Cr_2O_3, according Equation (5) [66,67].

$$Cr^{3+} + 3OH^- \rightarrow Cr(OH)_3 + 3e^- \tag{4}$$

$$Cr(OH)_3 + Cr + 3OH^- \rightarrow Cr_2O_3 + 3H_2O + 3e^- \tag{5}$$

It has been argued that the anodic reactions during the film growth period are mainly from the oxidation of iron and chromium The following equations indicate the oxidation reactions of iron [68]:

$$3Fe + 8OH^- \rightarrow Fe_3O_4 + 4H_2O + 8e^- \tag{6}$$

$$2Fe_3O_4 + 2OH^- + 2H_2O \rightarrow 6FeOOH + 2e^- \tag{7}$$

$$2Fe_3O_4 + 2OH^- \rightarrow 3Fe_2O_3 + H_2O + 2e^- \tag{8}$$

For the nitric acid passivation treatment, the transpassive region is above 200 mV vs. ECS, whereas for citric acid passivation conditions, the transpassive region is above 50 mV vs. ECS. The passive film formed under nitric acid passivation conditions has higher E_{pitt} values, in comparison with the E_{pitt} values obtained under citric acid passivation conditions. This fact can be seen as a potential disadvantage for the citric acid treatment. Nevertheless, the corrosion rates obtained for both passivation treatments (Table 4) are very low and similar. Thus, for the PH stainless steel used in this work, citric acid passivation treatments can be as effective as nitric acid passivation treatments.

5. Conclusions

In this work, samples of 15-5PH and 17-4PH stainless steel were passivated in (a) citric acid and (b) nitric acid baths and exposed in a 5 wt. % NaCl solution. Their electrochemical behavior was studied by electrochemical noise and potentiodynamic polarization.

EN results show that, for citric solution passivation baths, the stabilization of the passive layer occurs at more active potentials compared to the stabilization potentials for nitric acid passivation baths. From noise resistance (*Rn*) data, very low corrosion rate values were derived for the PH stainless steels passivated in both (citric and nitric) passivating treatments.

Statistical evaluation of the time record was carried out and the localization index (LI) parameter was evaluated. According to the LI results, the PH stainless steels passivated in citric acid solution mainly show a mixed corrosion type, whereas LI values for the PH stainless steels passivated in nitric acid solution indicates localized corrosion.

In general, potentiodynamic polarization results indicated that, irrespective of the type of PH stainless steel used, the nitric acid passivation treatment largely increases the pitting potentials in comparison with the citric acid treatment. Also, for both passivation treatments, very low corrosion rate values (in the order of 10^{-7} mm/year) were recorded for both 15-5PH and 17-4PH steels.

On the whole, citric passivation treatments on PH stainless steels could be a green alternative route to the currently employed nitric passivation treatments.

Author Contributions: Conceptualization, M.L.-B., F.A.-C.; Methodology, P.Z.-R., M.D.-E, F.E.-L., E.M.-B.; Data Curation, C.G.-T., J.A.C.-M., and D.N.-M.; Formal analysis, J.G.C.-N., and F.A.-C.; Writing—Review and Editing, M.L.-B., J.G.C.-N., F.A.-C. All authors have read and agreed to the published version of the manuscript.

Funding: This research was funded by the Conacyt, Proyect "Estudio de las propiedades electroquímicas y mecanismos de crecimiento de la película de pasivación de aceros inoxidables endurecibles por precipitación en ambientes ácidos", con clave No. A1-S-8882 and UANL (Dirección de Investigación).

Acknowledgments: The authors acknowledge The Academic Body UANL—CA-316 "Deterioration and integrity of composite materials".

Conflicts of Interest: All authors of this article declare no conflicts of interest.

References

1. Siddiqui, T. *Aircraft Materials and Analysis*; McGraw Hill Education: New York, NY, USA, 2015; pp. 127–136.
2. ASM International. *ASM Handbook*; Corrosion Environments and Industries; Cramer, S.D., Covino, B.S., Jr., Eds.; ASM International: Materials Park, OH, USA, 2006; Volume 13, p. 27, Corrosion Environments and Industries.
3. Schorr, M.; Valdez, B.; Salinas, R.; Ramos, R.; Nedey, N.; Curiel, M. Corrosion control in military assets. *MRS Online Proc. Libr. Arch.* **2016**, *1815*. [CrossRef]
4. Lara-Banda, M.; Ortiz, D.; Gaoan-Tiburcio, C.; Zambrano, P.; Cabral-Miramontes, J.C.; Almeraya-Calderon, F. Citric Acid Passivation of 15-5PH and 17-4PH Stainless Steel Used in the Aeronautical Industry. In *International Materials Research Congress*; Springer: Cham, Switzerland, 2016; pp. 95–104.
5. Lopes, J.C. Material selection for aeronautical structural application. *Sci. Technol. Adv. Mat.* **2008**, *20*, 78–82.
6. Lewis, P.; Kolody, M. Alternative to Nitric Acid Passivation of Stainless Steel Alloys. In *Technology Evaluation for Environmental Risk Mitigation Compendium, Proceedings of the NASA Technology Evaluation for Environmental Risk Mitigation Principal Center (TEERM)*; Department od Defense (DoD) and NASA: Merritt Island, FL, USA, 2008.
7. Calle, L.M. Coatings on Earth and Beyond. In Proceedings of the Coatings Summit 2015, Cocoa Beach, FL, USA, 21–23 January 2015.
8. O'Laoire, C.; Tmmins, B.; Kremer, L.; Holmes, J.D.; Morris, M.A. Analysis of the acid passivation of stainless steel. *Anal. Lett.* **2006**, *39*, 2255–2271. [CrossRef]
9. Olsson, C.O.; Landolt, D. Passive films on stainless steels-chemistry, structure and growth. *Electrochim. Acta* **2003**, *48*, 1093–1104. [CrossRef]
10. Yasensky, D.; Reali, J.; Larson, C.; Carl, C. Citric acid passivation of stainless steel. In Proceedings of the Aircraft Airworthiness and Sustainment Conference, San Diego, CA, USA, 18–21 April 2009.
11. Lo, K.H.; Shek, C.H.; Lai, J.K.L. Recent developments in stainless steels. *Mater. Sci. Eng. R.* **2009**, *65*, 39–104. [CrossRef]
12. Cobb, H.M. *The History of Stainless Steel*; ASM International: Cleveland, OH, USA, 2010; pp. 189–192.
13. Ha, H.Y.; Jang, J.H.; Lee, T.H.; Won, C.; Lee, C.H.; Moon, J.; Lee, C.G. Investigation of the localized corrosion and passive behavior of type 304 stainless steels with 0.2–1.8 wt% B. *Materials* **2018**, *11*, 2097. [CrossRef] [PubMed]
14. Schmuki, P. From bacon to barriers: A review on the passivity of metal and alloys. *J. Solid State Electr.* **2002**, *6*, 145–164. [CrossRef]
15. Abdelshehid, M.; Mahmodieh, K.; Mori, K.; Chen, L.; Stoyanov, P.; Davlantes, D.; Foyos, J.; Ogren, J.; Clark, R., Jr.; Es-Said, O.S. On the correlation between fracture toughness and precipitation hardening heat treatments in 15-5PH stainless steel. *Eng. Fail. Anal.* **2007**, *14*, 626–631. [CrossRef]
16. Esfandiari, M.; Dong, H. The corrosion and corrosion–wear behaviour of plasma nitrided 17-4PH precipitation hardening stainless steel. *Surf. Coat. Technol.* **2007**, *202*, 466–478. [CrossRef]
17. Dong, H.; Esfandiari, M.; Li, X.Y. On the microstructure and phase identification of plasma nitrided 17-4PH precipitation hardening stainless steel. *Surf. Coat. Technol.* **2008**, *202*, 2969–2975. [CrossRef]
18. Hsiao, C.N.; Chiou, C.S.; Yang, J.R. Aging reactions in a 17-4 PH stainless steel. *Mater. Chem. Phys.* **2002**, *74*, 134–142. [CrossRef]
19. Gladman, T. Precipitation hardening in metals. *Mater. Sci. Technol.* **1999**, *15*, 30–36. [CrossRef]

20. Schade, C.; Stears, P.; Lawley, A.; Doherty, R. Precipitation Hardening PM Stainless Steels. *Adv. Powder Part.* **2006**, *1*, 7.
21. Gaydos, S.P. Passivation of aerospace stainless parts with citric acid solutions. *Plat. Surf. Finish.* **2003**, *90*, 20–25.
22. Shibata, T. Stochastic studies of passivity breakdown. *Corros. Sci.* **1990**, *31*, 413–423. [CrossRef]
23. Ashassi-Sorkhabi, H.; Seifzadeh, D.; Raghibi-Boroujeni, M. Analysis of electrochemical noise data in both time and frequency domains to evaluate the effect of ZnO nanopowder addition on the corrosion protection performance of epoxy coatings. *Arab. J. Chem.* **2016**, *9*, S1320–S1327. [CrossRef]
24. Isselin, J.; Kasada, R.; Kimura, A. Effects of aluminum on the corrosion behavior of 16% Cr ODS ferritic steels in a nitric acid solution. *J. Nucl. Sci. Techmol.* **2011**, *48*, 169–171. [CrossRef]
25. Estupiñán, F.H.; Almeraya, F.; Margulis, R.B.; Zamora, M.B.; Martínez, A.; Gaona, C. Transient analysis of electrochemical noise for 316 and duplex 2205 stainless steels under pitting corrosion. *Int. J. Electrochem. Sci.* **2011**, *6*, 1785–1796.
26. Upadhyay, N.; Pujar, M.G.; Sekhar, S.S.; Gopala, K.N.; Mallika, C.; Kamachi, M.U. Evaluation of the Effect of Molybdenum on the Pitting Corrosion Behavior of Austenitic Stainless Steels Using Electrochemical Noise Technique. *Corrosion* **2017**, *73*, 1320–1334. [CrossRef]
27. Bragaglia, M.; Cherubini, V.; Cacciotti, I.; Rinaldi, M.; Mori, S.; Soltani, P.; Nanni, F.; Kaciulis, S.; Montesperelli, G. Citric Acid Aerospace Stainless Steel Passivation: A Green Approach. In Proceedings of the CEAS Aerospace Europe Conference 2015, Delft, The Netherlands, 7–11 September 2015.
28. Huet, F. Electrochemical Noise Technique. In *Analytical Methods in Corrosion Science and Engineering*; Marcus, P., Florian, B., Eds.; Mansfeld CRC Taylor & Francis: Boca Raton, FL, USA, 2006; Chapter 14, pp. 507–570.
29. Suresh, G.U.; Kamachi, M.S. Electrochemical Noise Analysis of Pitting Corrosion of Type 304L Stainless Steel. *Corrosion* **2014**, *70*, 283–293. [CrossRef]
30. Homborg, A.M.; Cottis, R.A.; Mol, J.M.C. An integrated approach in the time, frequency and time-frequency domain for the identification of corrosion using electrochemical noise. *Electrochim. Acta* **2016**, *222*, 627–640. [CrossRef]
31. Nazarnezhad, B.M.; Neshati, J.; Hossein, S.M. Development of Time-Frequency Analysis in Electrochemical Noise for Detection of Pitting Corrosion. *Corrosion* **2019**, *75*, 183–191.
32. Ortiz, A.C.J.; Lucio–Garcia, M.A.; Hermoso-Diaz, I.A.; Chacon-Nava, J.G.; Martínez-Villafañe, A.; Gonzalez-Rodriguez, J.G. Detection of Sulfide Stress Cracking in a Supermartensitic Stainless Steel by Using Electrochemical Noise. *Int. J. Electrochem. Sci.* **2014**, *9*, 6717–6733.
33. Al-Zanki, I.A.; Gill, J.S.; Dawson, J.L. Electrochemical Noise Measurements on Mild Steel in 0.5 M Sulphuric Acid. *Mater. Sci. Forum* **1986**, *8*, 463–476. [CrossRef]
34. Cuevas-Arteaga, C.; Porcayo-Calderón, J. Electrochemical noise analysis in the frequency domain and determination of corrosion rates for SS-304 stainless steel. *Mater. Sci. Eng. A* **2006**, *435–436*, 439–446. [CrossRef]
35. *Standard Practice for Cleaning, Descaling and Passivation of Stainless-Steel Parts, Equipment, and Systems*; ASTM A380-17; ASTM International: West Conshohocken, PA, USA, 1999.
36. *Standard Specification for Chemical Passivation Treatments for Stainless Steel Parts*; ASTM A967-17; ASTM International: West Conshohocken, PA, USA, 1999.
37. *Standard Guide for Electrochemical Noise Measurement*; ASTM G199-09; ASTM International: West Conshohocken, PA, USA, 2009.
38. Mansfeld, F.; Sun, Z. Localization index obtained from electrochemical noise analysis. *Corrosion* **1999**, *55*, 915–918. [CrossRef]
39. Sanchez, J.M.; Cottis, R.A.; Botana, F.J. Shot noise and statistical parameters for the estimation of corrosion mechanisms. *Corros. Sci.* **2005**, *47*, 3280–3299. [CrossRef]
40. Kelly, R.G.; Inman, M.E.; Hudson, J.L. Analysis of electrochemical noise for type 410 stainless steel in chloride solutions. In *Electrochemical Noise Measurement for Corrosion Applications*; ASTM International: West Conshohocken, PA, USA, 1996.
41. *Standard Reference Test Method for Making Potentiostaticc and Potentiodynamic Anodic Polarization Measurements*; ASTM-G5-13E2; ASTM International: West Conshohocken, PA, USA, 2013.
42. *Standard Practice for Calculation of Corrosion Rates from Electrochemical Measurements*; ASTM-G102-89; ASTM International: West Conshohoken, PA, USA, 2010.
43. Ha, H.Y.; Kang, J.Y.; Yang, J.; Yim, C.D.; You, B.S. Limitations in the use of the potentiodynamic polarisation curves to investigate the effect of Zn on the corrosion behaviour of as-extruded Mg–Zn binary alloy. *Corros. Sci.* **2013**, *75*, 426–433. [CrossRef]

44. Treseder, R.S. *NACE Corrosion Engineers Reference Book*, 2nd ed.; NACE International: Houston, TX, USA, 1991.
45. Bertocci, U.; Yang-Xiang, Y. An examination of current fluctuations during pit initiation in Fe-Cr alloys. *J. Electrochem. Soc.* **1984**, *131*, 1011–1017. [CrossRef]
46. Bertocci, U.; Koike, M.; Leigh, S.; Qiu, F.; Yang, G. A statistical analysis of the fluctuations of the passive current. *J. Electrochem. Soc.* **1986**, *133*, 1782. [CrossRef]
47. Miyata, Y.; Handa, T.; Takazawa, H. An analysis of current fluctuations during passive film breakdown and repassivation in stainless alloys. *Corros. Sci.* **1990**, *31*, 465–470. [CrossRef]
48. Savas, T.P.; Wang, A.Y.L.; Earthman, J.C. The effect of heat treatment on the corrosion resistance of 440C stainless steel in 20% HNO_3 + 2.5% $Na_2Cr_2O_7$ solution. *J. Mater. Eng. Perform.* **2003**, *12*, 165–171. [CrossRef]
49. Heyn, A.; Goellner, J.; Bierwirth, M.; Klapper, H. Recent applications of electrochemical noise for corrosion testing-Benefits and restrictions. In *CORROSION 2007, Proceedings of the Corrosion NACE Expo2007, Nashville, TN, USA, 11–15 March 2007*; NACE International: Houston, TX, USA, 2007.
50. Eden, A.D.; John, G.D.; Dawson, J.L. Corrosion Monitoring. International Patent WO1987007022A1, 19 November 1987.
51. Brennenstuhl, A.M.; Palumbo, G.; Gonzalez, F.S.; Quirk, P.G. The Use of Electrochemical Noise to Investigate the Corrosion Resistance of UNS Alloy N04400 Nuclear Heat Exchanger Tubes. In *Electrochemical Noise Measurement for Corrosion Applications*; ASTM International: West Conshohocken, PA, USA, 1996.
52. Padilla-Viveros, A.; Garcia-Ochoa, E.; Alazard, D. Comparative electrochemical noise study of the corrosion process of carbon steel by the sulfate-reducing bacterium *Desulfovibrio alaskensis* under nutritionally rich and oligotrophic culture conditions. *Electrochim. Acta* **2006**, *51*, 3841–3847. [CrossRef]
53. Webster, S.; Nathanson, L.; Green, A.G.; Johnson, B.V. The Use of Electrochemical Noise to Assess Inhibitor Film Stability. In *UK Corrosion 1992*; Institute of Corrosion: Manchester, UK, 1992.
54. Rothwell, A.N.; Edgemon, G.L.; Bell, G.E.C. *Data Processing for Current and Potential Logging Field Monitoring Systems*; CORROSION/1999, paper no. 192; NACE: Houston, TX, USA, 1999.
55. Girija, U.; Mudali, K.; Khatak, H.S.; Raj, B. The application of electrochemical noise resistance to evaluate the corrosion resistance of AISI type 304 *SS* in nitric acid. *Corros. Sci.* **2007**, *49*, 4051–4068. [CrossRef]
56. Homborg, A.M.; Tinga, T.; Van Westing, P.M.; Zhang, X.; Ferrari, G.M.; de Wit, J.H.W.; Mol, J.M.C. A Critical Appraisal of the Interpretation of Electrochemical Noise for Corrosion Studies. *Corrosion* **2014**, *70*, 971–987. [CrossRef]
57. Cottis, R.A.; Al-Awadhi, M.A.A.; Al-Mazeedi, H.; Turgoose, S. Measures for the detection of localized corrosion with electrochemical noise. *Electrochim. Acta.* **2001**, *46*, 3665–3674. [CrossRef]
58. Jurak, T.; Jamali, S.S.; Yue Zhao, Y. Theoretical analysis of electrochemical noise measurement with single substrate electrode configuration and examination of the effect of reference electrodes. *Electrochim. Acta* **2019**, *3011*, 377–389. [CrossRef]
59. Cottis, R.A. Interpretation of Electrochemical Noise Data. *Corrosion* **2001**, *57*, 265–285. [CrossRef]
60. Betova, I.; Bojinov, M.; Laitinen, T.; Mäkelä, K.; Pohjanne, P.; Saario, T. The transpassive dissolution mechanism of highly alloyed stainless steels: I. Experimental results and modelling procedure. *Corros. Sci.* **2002**, *44*, 2675–2697. [CrossRef]
61. Ye, W.; Li, Y.; Wang, F. The improvement of the corrosion resistance of 309stainless steel in the transpassive region by nano-crystallization. *Electrochem. Acta* **2009**, *54*, 1339–1349. [CrossRef]
62. Man, C.; Dong, C.; Cui, Z.; Xiao, K.; Yu, Q.; Li, X. A comparative study of primary and secondary passive films formed on AM355 stainless steel in 0.1 M NaOH. *Appl. Surf. Sci.* **2018**, *427*, 763–773. [CrossRef]
63. Bojinov, M.; Betova, I.; Fabricius, G.; Laitinen, T.; Saario, T. The stability of the passive state of iron–chromium alloys in sulphuric acid solution. *Corros. Sci.* **1999**, *41*, 1557–1584. [CrossRef]
64. Bojinov, M.; Fabricius, G.; Kinnunen, P.; Laitinen, T.; Mäkelä, K.; Saario, T.; Sundholm, G. The mechanism of transpassive dissolution of Ni–Cr alloys in sulphate solutions. *Electrochem. Acta* **2000**, *45*, 2791–2802. [CrossRef]
65. Lara Banda, M.; Gaona-Tiburcio, C.; Zambrano-Robledo, P.; Cabral, M.J.A.; Estupinan, F.; Baltazar-Zamora, M.A.; Almeraya-Calderon, F. Corrosion Behaviour of 304 Austenitic, 15-5PH and 17-4PH Passive Stainless Steels in acid solutions. *Int. J. Electrochem. Sci.* **2018**, *13*, 10314–10324. [CrossRef]
66. Huang, J.; Wu, X.; Han, E.H. Electrochemical properties and growth mechanism of passive films on Alloy 690 in high-temperature alkaline environments. *Corros. Sci.* **2010**, *52*, 3444–3452. [CrossRef]

67. Calinski, C.; Strehblow, H.H. ISS depth profiles of the passive layer on Fe/Cr alloys. *J. Electrochem. Soc.* **1989**, *36*, 1328–1331. [CrossRef]
68. Radhakrishnamurty, P.; Adaikkalam, P. pH-potential diagrams at elevated temperatures for the chromium/water systems. *Corros. Sci.* **1982**, *22*, 753–773. [CrossRef]

Article

Preparation and Corrosion Behavior in Marine Environment of MAO Coatings on Magnesium Alloy

Yuhong Yao [1,*], Wei Yang [1,*], Dongjie Liu [2], Wei Gao [1] and Jian Chen [1]

[1] School of Materials Science and Chemical Engineering, Xi'an Technological University, Xi'an 710032, China; eifa@sina.com (W.G.); chenjian@xatu.edu.cn (J.C.)

[2] School of Materials Science and Engineering, Xi'an University of Technology, Xi'an 710048, China; liudongjie@xaut.edu.cn

* Correspondence: yyhong0612@yahoo.com (Y.Y.); yangwei_smx@163.com (W.Y.)

Received: 1 December 2019; Accepted: 10 January 2020; Published: 12 January 2020

Abstract: To improve the corrosion performance of magnesium alloys in the marine environment, the MAO, MAO–$Cu_2CO_3(OH)_2{\cdot}H_2O$ and MAO–$Cu_2P_2O_7$ ceramic coatings were deposited on AZ91D magnesium alloys in basic electrolyte and the discoloration mechanism of the Cu-doped MAO coatings and the corrosion behavior of the three MAO coatings in the artificial seawater solution were investigated by SEM, EDS and XPS. The results indicated that the formation and discoloration mechanism of the brown MAO ceramic coatings were attributable to the formation of Cu_2O in the coatings. Though the three MAO coatings had a certain protective effect against the corrosion of AZ91D substrate in the artificial seawater, the distinctive stratification phenomenon was found on the MAO–$Cu_2P_2O_7$ coated sample and the corrosion model of the MAO–$Cu_2P_2O_7$ coatings in the immersion experiment was established. Therefore, the brown Cu-doped MAO coatings were speculated to significantly reduce the risk of the magnesium parts in marine environments.

Keywords: magnesium alloy; MAO coating; corrosion behavior; stratification phenomena; marine environments

1. Introduction

The magnesium alloy is the lightest structure metal material, and is considered as the green engineering material in the 21st century [1]. Now it is widely applied to the electronic industry, aerospace industry, and auto industry [2,3]. However, for its poor corrosion resistance, magnesium alloy, particularly as the magnesium alloy parts for outdoor applications, is confronted with great challenges [4–7]. Many literatures have verified that the surface modification technique, such as chemical conversion coatings, anodic oxidation, micro arc oxidation (MAO), organic coatings, vapor phase processes, etc, is an effective way to change the surface composition and improve its corrosion resistance of magnesium alloy [8–12]. MAO is a newly developed technology for the preparation of ceramic coatings on aluminum, magnesium or titanium alloys to improve the corrosion resistance [10,13–16]. At present, the synthesis of white MAO coatings on magnesium alloy is a mature and universal technology [16], but it has the same disadvantage as the chemical conversion coating technology, which often causes the formation of the light spots on the surfaces of magnesium diecast components and hardly meets the market demand of 3C (computer, communication and consumer) electronic products. It has been reported that the colored MAO coatings can be formed by adding metal salts in the electrolytes [17–24]. The black MAO coating containing V_2O_3 can be obtained on an aluminum alloy surface by adding ammonium metavanadate into the commonly used $(NaPO_3)_6$ (sodium hexametaphosphate) and Na_2SiO_3 solution [21] and the black MAO coating with excellent properties can also be prepared in the electrolyte with dichromate addition [22]. Moreover, it has been reported that by adding potassium fluotitanate or sodium stannate into the base electrolyte, a yellow or grey MAO coating can be formed on the surface of Mg alloys, respectively [23,24].

Brown is a very important and common decorative color. Lee et al. [25] has reported that with the addition of 3% and 5% Cu in the base electrolyte, the color of the MAO coating changes from brown to dark brown and the corrosion resistance of the AZ91 alloy is significantly improved after being treated with the micro-arc oxidation process, but the corrosion process and coloring mechanism of this brown MAO coating are still not clear. Furthermore, the magnesium alloy parts with MAO coating for lightweight are sometimes exposed to marine environment and the marine corrosion behavior of the coatings has not been clarified. As a result, the application of magnesium alloy in a marine environment is seriously restricted. In this study, a convenient process to fabricate the MAO coatings with brown color on Mg alloys is introduced by adding alkaline copper carbonate and copper pyrophosphate in the electrolyte and the white and two brown MAO coatings are prepared. Then, the microstructure, formation mechanism and seawater corrosion behaviors of the coatings are investigated in detail. Finally, the seawater corrosion mechanism of the MAO coatings is revealed, which is helpful for the surface protection of magnesium alloy applied in marine environments.

2. Materials and Methods

AZ91D magnesium alloys were used as the substrate discs in the size of ϕ30 mm × 5 mm and its nominal chemical composition (in wt. %) was Al 8.5–9.5 %, Zn 0.5–0.9 %, Mn 0.17–0.27 %, Cu $\leq$0.01, Ni $\leq$0.01, Si $\leq$0.01, Fe $\leq$0.004 and Mg balance. Before the micro arc coatings, the specimens were prepared by means of standard metallographic procedure, such as coarse grinding, accurate grinding, polishing with alumina waterproof abrasive paper up to 1200 grit and then ultrasonically degreased in acetone for 10 min followed by rinsing with distilled water.

The MAO coatings were prepared on the specimen surface by using of micro arc oxidation equipment (JHMAO-60, China) with the constant voltage of 420 V, the frequency of 400 Hz, the duty cycle of 10% and the treatment time of 8 min. The base electrolyte solution was composed of 8.0 g/L sodium silicate ($Na_2SiO_3{\cdot}9H_2O$), 5.0 g/L potassium hydroxide (KOH), 5.0 g/L potassium fluoride (KF), 1.0 g/L EDTA ($C_{10}H_{16}N_2O_8$) and 3.5 g/L potassium sodium tartrate ($C_4H_4O_6KNa{\cdot}4H_2O$). The two Cu-doped brown coatings were prepared by respectively adding 2.5 g/L basic copper carbonate ($Cu_2CO_3(OH)_2{\cdot}H_2O$) and 2.5 g/L copper pyrophosphate (Cu2P2O7) into the base solution and the temperature of the electrotype with pH of about 13 was kept below 35 °C during the MAO process.

The thickness of the MAO coatings was measured by using TT240 eddy current thickness meter with an accuracy of 0.1 μm. Six measurements were carried out evenly on the whole sample surface. The surface morphologies and element distribution of the MAO coatings were analyzed by scanning electron microscope (SEM) with Oxford Inca X-Max energy dispersive spectrometry (EDS). X-ray photoelectron spectroscopy (XPS) with Al (mono) Kα irradiation at pass energy of 160 eV (AXIS UTLTRADLD) was used to characterize the chemical bonds of the coatings. The binding energies were referenced to the C 1 s line at 284.6 eV from adventitious carbon. The corrosion behavior of the coated AZ91D magnesium alloy was evaluated by the immersion tests in the artificial seawater, whose composition was shown in Table 1. Before the immersion test, the three MAO-coated specimens were treated with epoxy resin to avoid the effect of defects at the edge of the samples, then immersed in the artificial seawater solution for 14 days and the corrosion morphologies of the samples were observed by SEM.

Table 1. Composition of artificial seawater in immersion test.

Chemical Reagents	Concentration (g/L)
NaCl	24.53
$MgCl_2{\cdot}6H_2O$	11.11
Na_2SO_4	4.09
$CaCl_2$	1.16
KCl	0.70
$NaHCO_3$	0.20
KBr	0.10

3. Results and Discussions

The MAO–$Cu_2CO_3(OH)_2 \cdot H_2O$ and MAO–$Cu_2P_2O_7$ coatings on AZ91D magnesium alloys in Figure 1 are prepared with the thicknesses of 8.4 μm, 9.1 μm and 9.6μm, respectively. It reveals that for the more intense micro arc discharge many defects are observed at the edge of the sample in Figure 1b. The microstructures of the three MAO coatings are given in Figure 2. It can be seen from Figure 2 that the surfaces of the three MAO coatings are characterized by lots of micropores with the size range from submicron to several micro scale, and different from the MAO and MAO–$Cu_2P_2O_7$ coatings, there are some bumps on the MAO–$Cu_2CO_3(OH)_2 \cdot H_2O$ coating surface, which are formed by the spark discharge and gas bubbles throughout the discharge channels during MAO process [26,27]. So it reveals that the addition of $Cu_2CO_3(OH)_2 \cdot H_2O$ into the base electrolyte results in a strongly intense micro arc discharge, which promotes the formation of large molten deposits (Figure 2c,d). The elemental compositions of the three MAO coatings are examined by EDS in Table 2. Carbon (C) is considered to be an impurity from the atmosphere or the electrolyte, F and Na are also presumed to originate from the electrolyte or the AZ91D alloy substrate, P and Si are from the electrolyte and Zn comes from the substrate which indicates that the thickness of the three MAO coatings is very thin. Thus the three MAO coatings are mainly composed of Mg, O, Si and a little amount of C, F, Na species. The presence of Si and O reveals that the components of the electrolyte have intensively incorporated into the micro arc oxidation reactions to from the ceramic coatings [16]. Moreover, by addition of $Cu_2CO_3(OH)_2 \cdot H_2O$ or $Cu_2P_2O_7$ into the electrolyte, a very small amount of Cu or Cu, P has been respectively doped into the MAO–$Cu_2CO_3(OH)_2 \cdot H_2O$ and MAO–$Cu_2P_2O_7$ coatings to make the color of the coatings change from white to brown. The discoloration mechanism of the MAO coatings will be further discussed by using SEM + EDS and XPS.

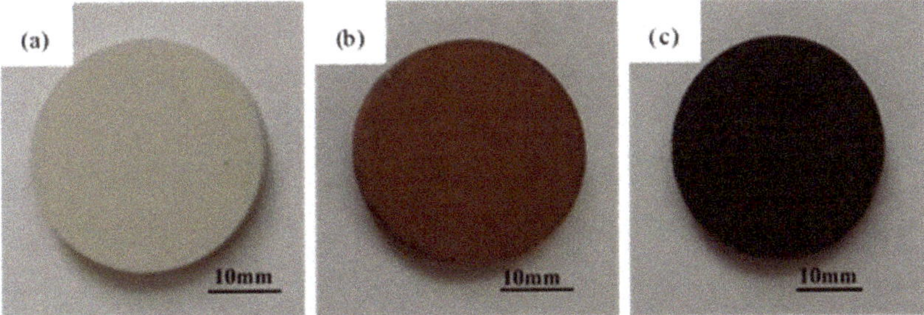

Figure 1. Macrograph of three MAO coatings on AZ91D magnesium alloy, (**a**) MAO, (**b**) MAO–$Cu_2CO_3(OH)_2 \cdot H_2O$ and (**c**) MAO–$Cu_2P_2O_7$.

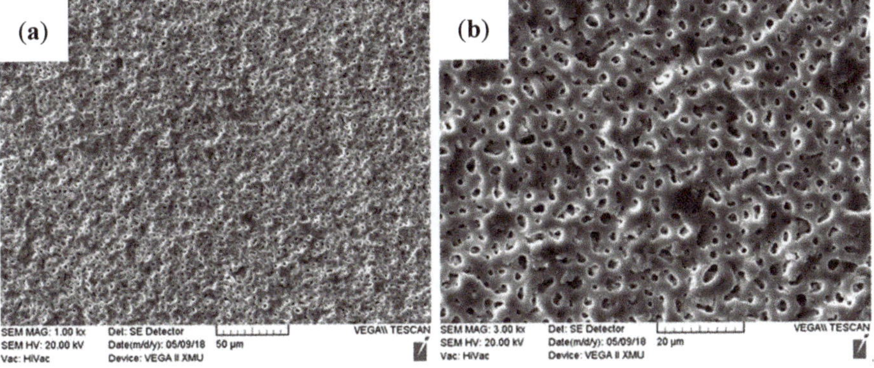

Figure 2. *Cont.*

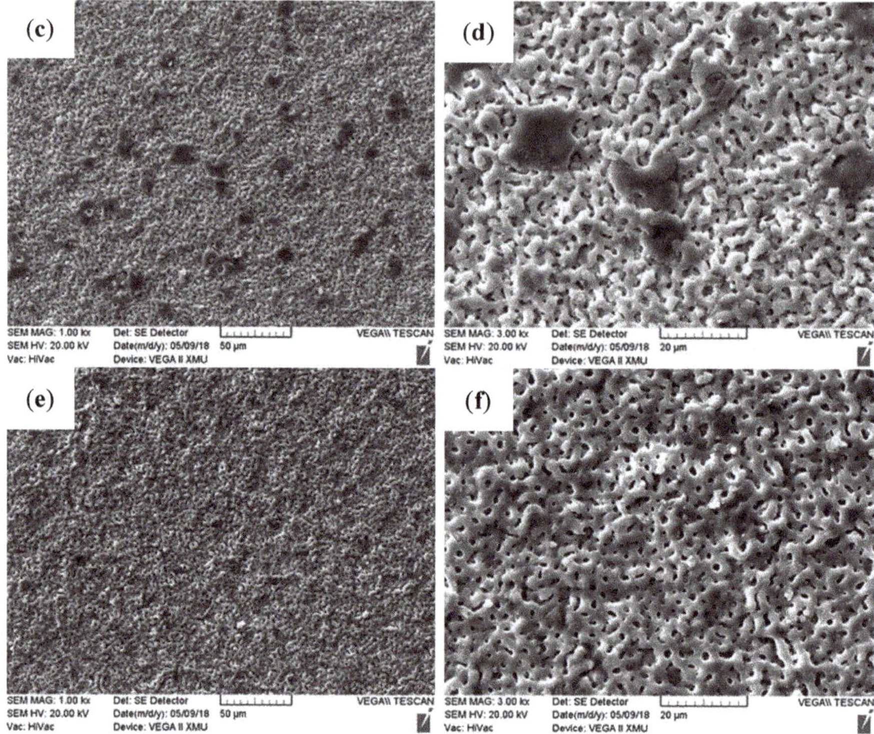

Figure 2. Surface morphologies of three MAO coatings on AZ91D magnesium alloy, (**a**) and (**b**) MAO, (**c**) and (**d**) MAO–$Cu_2CO_3(OH)_2{\cdot}H_2O$ and (**e**) and (**f**) MAO–$Cu_2P_2O_7$.

Table 2. EDS results of the MAO, MAO–$Cu_2CO_3(OH)_2{\cdot}H_2O$ and MAO–$Cu_2P_2O_7$ coatings on AZ91D magnesium alloy (at. %).

Coatings	C	O	F	Na	Mg	Si	Zn	Cu	P
MAO	1.84	49.64	3.63	0.46	33.98	9.78	0.67	-	-
MAO–$Cu_2CO_3(OH)_2{\cdot}H_2O$	2.95	48.18	3.76	0.86	35.28	8.05	0.39	0.53	-
MAO–$Cu_2P_2O_7$	2.35	49.42	4.45	0.21	32.42	7.27	1.17	0.74	2.81

The surface morphologies of the MAO–$Cu_2CO_3(OH)_2{\cdot}H_2O$ coatings formed at different oxidation times are shown in Figure 3 and the corresponding EDS analysis results are listed in Table 3. As shown in Figure 3a and Table 3, when the oxidation time is about 70 s, the surface morphology of the AZ91D substrate is inhomogeneous at the moment of starting arc with two distinct regions: region I with a damaged area caused by electrical break-down involving amount of O, F, Si and Cu elements from the electrolyte and Mg and Zn alloying species from the substrate, and region II with a smooth surface morphology and the scratches in the substrate including lower content of O, Si, Cu and much higher content of Mg than those in region I, which indicates that the electrical breakdown phenomenon does not occur in region II. With the prolongation of oxidation time, the surface of the samples presents a typical porous feature and the pore size of the MAO coatings increases with the oxidation time in Figure 3b–d; a similar micro arc process and the mechanism of the pore initiation and the pore development are reported by some literatures [10,21]. However, it is worth noting that the concentrations of O, F, Si, Cu, Mg and Zn elements are similar in the MAO–$Cu_2CO_3(OH)_2{\cdot}H_2O$ coatings, the color of the coatings gradually becomes deeper with oxidation time, so it is very meaningful to analyze the discoloration mechanism of the coatings.

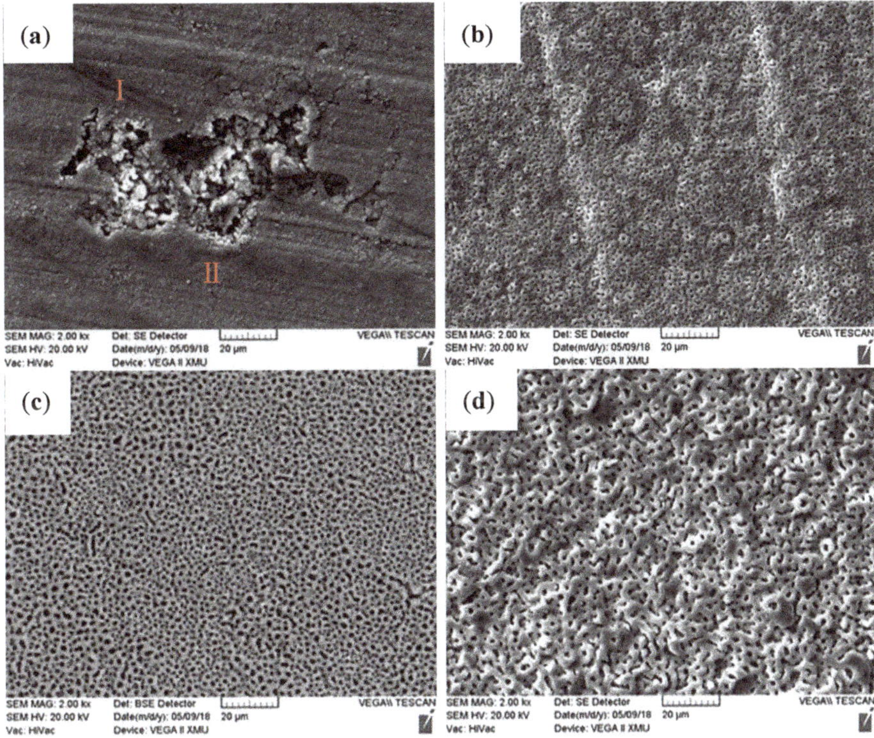

Figure 3. Surface morphologies of MAO–$Cu_2CO_3(OH)_2{\cdot}H_2O$ coatings formed with different oxidation time on AZ91D magnesium alloy, (**a**) 70 s, (**b**) 100 s, (**c**) 120 s and (**d**) 180 s.

Table 3. EDS results of MAO–$Cu_2CO_3(OH)_2{\cdot}H_2O$ coatings formed with different oxidation time on AZ91D magnesium alloy (at. %).

Oxidation Time	C	O	F	Mg	Si	Cu	Zn
70 s—I	2.92	45.97	4.45	36.30	7.73	0.26	2.37
70 s—II	2.05	29.67	2.77	60.81	2.45	-	1.76
100 s	1.04	49.26	5.97	34.99	7.71	0.28	0.75
120 s	1.00	52.41	5.80	32.41	7.65	0.28	0.45
180 s	1.80	50.68	4.55	34.28	8.16	0.12	0.40

The chemical states of Cu, Mg and Si are investigated by using XPS in Figure 4. From the wide spectra demonstrated in Figure 4a, Mg, O, Si and F elements are all found in the three MAO coatings and Cu is detected in the MAO coating with MAO–$Cu_2CO_3(OH)_2{\cdot}H_2O$ or $Cu_2P_2O_7$ addition by XPS. The Cu2p3/2 spectrum consists of two peaks at the binding energies of 932 eV associated with Cu_2O and 939.7 eV corresponding to CuF_2 in the brown MAO coating with $Cu_2CO_3(OH)_2{\cdot}H_2O$ addition in Figure 4b. Figure 4c,d illustrate that the Mg2p peak at the binding energy of 47.8 eV and the Si2p peak of 100.25 eV are individually assigned to MgO and SiO2. So, it can be concluded that the solute ions (such as Cu and Si) from the electrolyte are involved in the growth process of the MAO coatings, and the same results are found in the growth process of the MAO coatings on Ti substrate [28,29]. Therefore, it is the formation of the red Cu_2O in the MAO coatings with MAO–$Cu_2CO_3(OH)_2{\cdot}H_2O$ addition that results in the discoloration of the coatings.

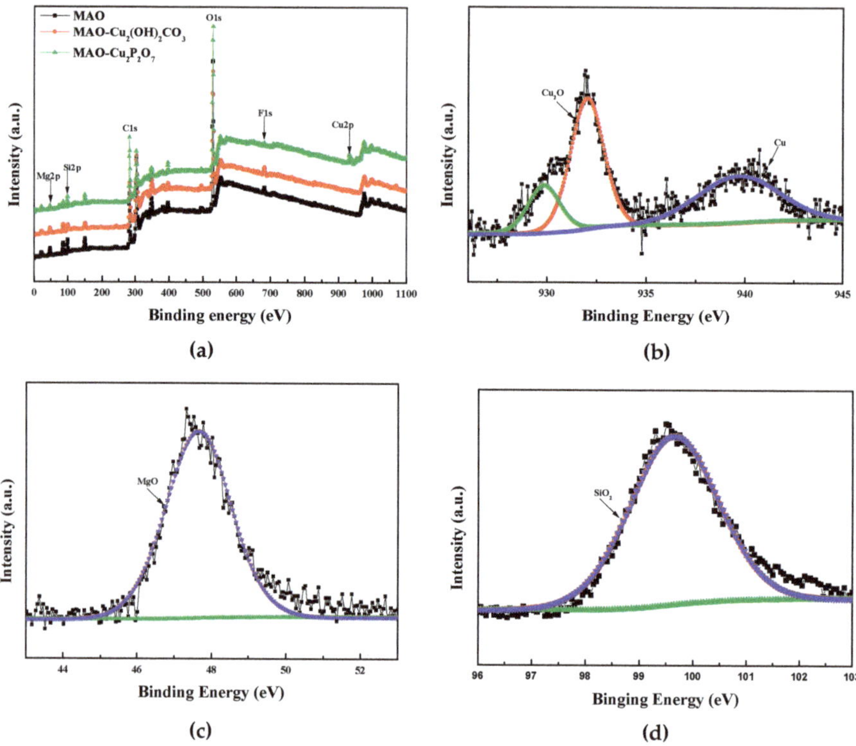

Figure 4. (**a**) XPS survey spectra of three MAO coatings, (**b**) Cu2p, (**c**) Mg2p and (**d**) Si2p typical high-resolution XPS spectrum of MAO–$Cu_2CO_3(OH)_2{\cdot}H_2O$ coatings.

Figures 5 and 6 are the macro and micro surface appearances of the three MAO coatings immersed in the artificial seawater solution for 14 days, respectively. Compared with Figure 1, it can be observed from Figure 5 that the color of the three MAO coatings obviously changes and some corrosion products are formed on the immersed coatings. The corrosion is more serious at the edge of the three MAO samples due to the defects caused by micro arc discharge (Figure 1b) or the damage during the embedding. It has been reported that once the corrosion reaction is initiated on the sample through a pit or minute pore, the corrosive medium can come into contact with the substrate to form some corrosion pits [30]. The elemental compositions of the three MAO coatings detected by EDS after the immersion test are listed in Table 4. The three corroded MAO coatings are mainly composed of Mg from the substrate and the artificial seawater and O from the electrolyte during the micro arc discharge, and Si, Cu and F from the electrolytes and Cl, K and Ca elements from the artificial seawater are also found in the corroded MAO coatings, which indicate that the MAO coatings are not completely destroyed. From Figure 6, it is quite clear that the microstructures of the MAO coatings have a significant change before and after the immersion test. The white MAO coating exhibits a relatively uniform surface appearance with a high degree of porosity, some cracks and corrosion products as shown in Figure 6a, which indicates that the AZ91D substrate is still protected by the coatings. The MAO–$Cu_2CO_3(OH)_2{\cdot}H_2O$ coatings are relatively rougher and exhibit a stacking structure with limited number of pores in Figure 6b. For the MAO–$Cu_2P_2O_7$ coatings, besides many micropores and some micro-cracks, there is a sedimentary layer with a lot of cracks in the coatings in Figure 6c and SEM morphology and EDS analysis results of this MAO coating after the immersion test are shown in detail in Figure 7 and Table 5. It can be seen from Figure 7 that the corroded coatings are divided into three regions: region I (the inner layer near the Mg substrate), region II (the middle layer attached to the surface coating) and region III (the top layer). From Table 5, Mg, O, F and Si elements are found in the region I and the atom percentage of these elements is similar to those in the MAO coating before the immersion test. Moreover, Cl and Ca elements from the artificial seawater are not observed, indicating that the MAO ceramic coatings in region I have not been destroyed. Region II is a dense layer attached to the MAO coating, where the contents of Mg, Si and F elements dramatically decrease whereas the contents of O element significantly increase compared with that in region I; both Cl and Ca elements have been detected. It indicates that there is the interaction between the MAO coating and the corrosive medium. In the case of region III, this layer is relatively thick and composed of some loose and porous structure sediments, mainly containing O and Ca elements. It is well known that the deposition of corrosion products can hinder the transfer of the charge and increase the inhibition of corrosion, so it is believed that this thick sediment layer is very helpful to improve the corrosion behavior of the MAO–$Cu_2P_2O_7$ coatings in artificial seawater.

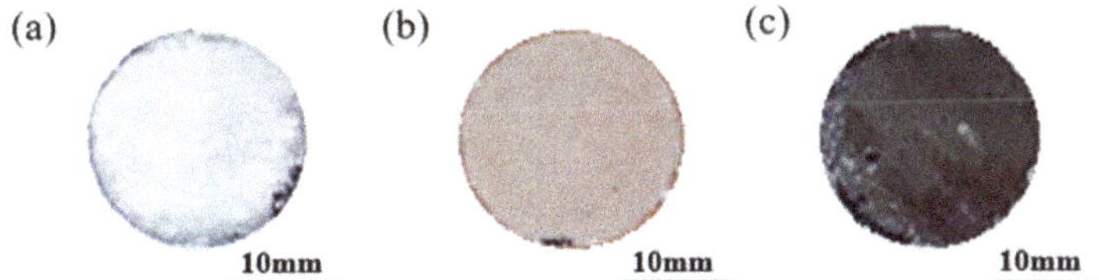

Figure 5. Macrograph of three MAO coatings, (**a**) MAO, (**b**) MAO–$Cu_2CO_3(OH)_2{\cdot}H_2O$ and (**c**) MAO–$Cu_2P_2O_7$, on AZ91D magnesium alloy after 14 days immersion test in the artificial seawater.

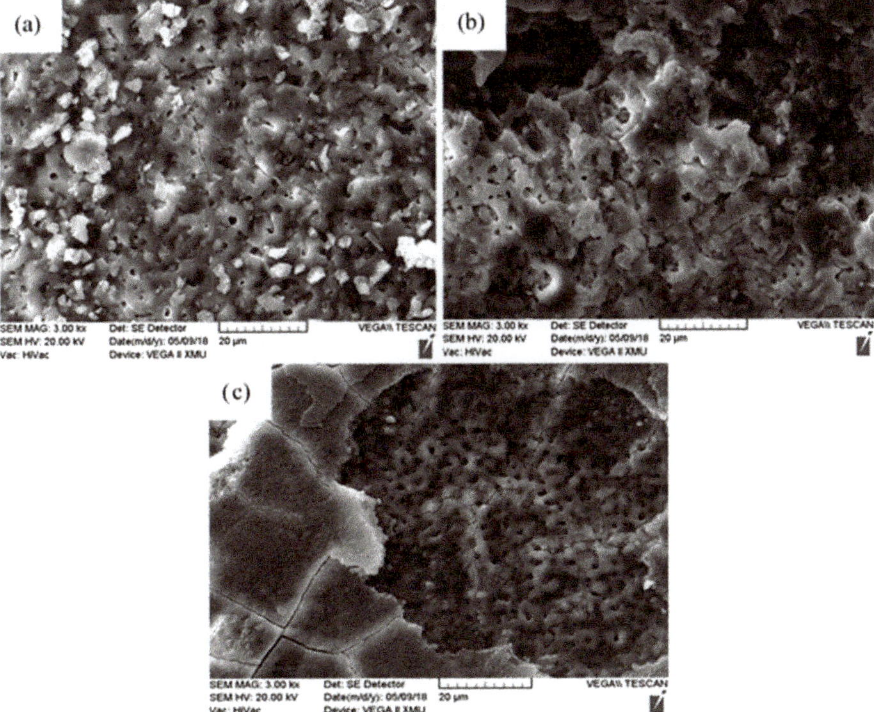

Figure 6. Micro morphologies of three MAO coatings by SEM after immersion test in the artificial seawater for 14 days, (**a**) MAO, (**b**) MAO–$Cu_2CO_3(OH)_2{\cdot}H_2O$ and (**c**) MAO–$Cu_2P_2O_7$.

Table 4. EDS results of the three MAO coatings immersed in artificial seawater for 14 days (at. %).

Coatings	O	Mg	Zn	Cu	Si	F	Cl	K	Ca
MAO	48.03	29.17	1.66	-	5.72	7.05	0.35	0.26	0.81
MAO–$Cu_2CO_3(OH)_2{\cdot}H_2O$	35.25	38.13	2.02	-	4.07	5.92	0.41	-	-
MAO–$Cu_2P_2O_7$	57.89	27.09	0.82	1.04	6.30	-	0.59	-	0.55

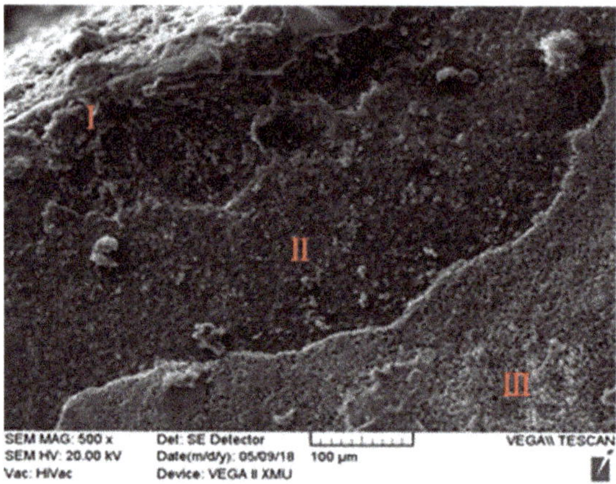

Figure 7. SEM surface morphologies and EDS results of MAO–$Cu_2P_2O_7$ coating after immersion test.

Table 5. EDS results of the three layers of the MAO–$Cu_2P_2O_7$ coatings immersed in artificial seawater for 14 days (at. %).

Regions	O	Mg	Si	F	Cl	Ca
Region I	47.08	30.15	5.72	7.05	-	-
Region II	60.47	25.42	2.37	-	0.73	2.57
Region III	71.34	0.61	-	-	-	22.09

From Figure 7 and Table 5, it has been learnt that the inner layer of the coatings is not destroyed near the substrate during the corrosion process, a middle dense layer is formed in the corrosion solution and the top thick deposition layer is of porous characteristics. Thus, a schematic diagram of the MAO–$Cu_2P_2O_7$ coatings during the immersion corrosion in the artificial seawater solution is shown in Figure 8. Due to the eruption and condensation of molten materials caused by micro arc discharge, the MAO coating normally has porous characteristics [31]. As a result, some interconnected micro closed-pores exist inside the MAO ceramic layer in Figure 8a. As shown in Figure 8b, the corrosion medium, especially Cl-, can seep into the interface between the electrolyte and the coatings through the micropores during the corrosion process; these micropores are exposed to the corrosive medium due to the oxygen concentration polarization between the interior holes and the interface. Then the corrosion products accumulate in the micropores and form a dense corrosion product layer on the MAO coating with the prolonging of the immersion corrosion time to prevent further corrosion of the coatings. Finally, the further interaction between the substrate and the artificial seawater has effectively been prevented by the corrosion product layer and the top calcium oxide-like thick porous layer forming on the middle dense layer in the following corrosion process in Figure 8c, which can further prevent the corrosion medium into the substrate to enhance its corrosion performance in the artificial seawater. Thus, it can be concluded that the stratification phenomena have been found on the MAO–$Cu_2P_2O_7$-coated sample and similar results have also been reported in the literatures [32,33]. However, although there is no Mg detected in the top sedimentary layer, for its loose and porous structure there is maybe a risk of corrosion for AZ91D substrate with further extension of the corrosion process.

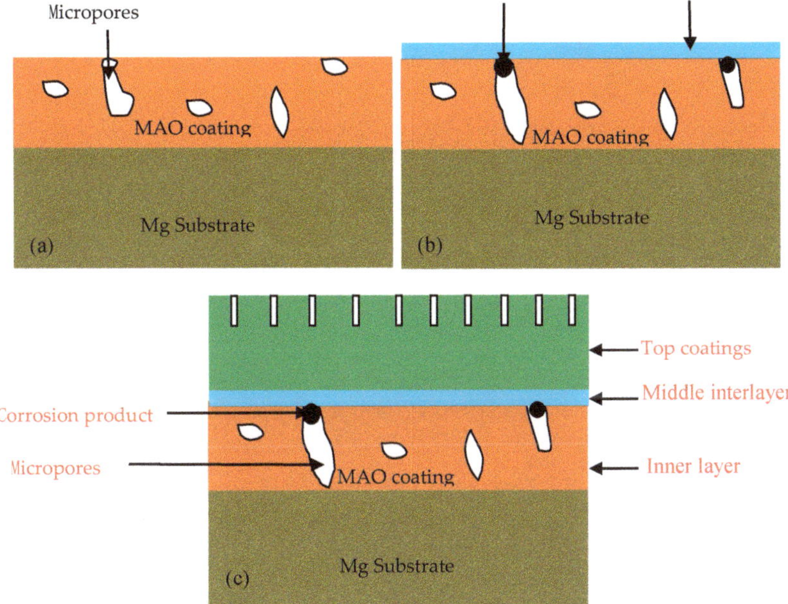

Figure 8. Schematic diagrams of the immersion corrosion of MAO–$Cu_2P_2O_7$ coating in the artificial seawater. (**a**) Morphology of MAO coating before corrosion, (**b**) Morphology of MAO coating at the initial stage of corrosion, (**c**) Morphology of MAO coating at the latter stage of corrosion.

4. Conclusions

(1) Brown MAO coatings on magnesium alloy can be prepared in the Na_2SiO_3 alkaline electrolyte with copper pyrophosphate or copper carbonate as the colorant, and for strongly intense micro arc discharge, some bumps appear on the Cu-doped MAO coatings.

(2) With the increase of reaction time, Cu in the colorant is fully involved in the formation of the MAO–$Cu_2CO_3(OH)_2{\cdot}H_2O$ and MAO–$Cu_2P_2O_7$ coatings on AZ91D magnesium alloys, and the discoloration mechanism of these coatings are attributed to the formation of Cu_2O in the coatings.

(3) After immersion in seawater for 14 days, the stratification corrosion microstructure with the three layers named the inner layer, the middle dense layer and the top calcium oxide-like layer are formed on the MAO–$Cu_2P_2O_7$ coated sample, which is helpful to prolong the service life of AZ91D magnesium alloys in artificial seawater.

Author Contributions: Designing the experiments, Y.Y., and W.Y.; performing the experiments, Y.Y., W.Y., D.L., and W.G.; contributing the reagents, materials and analysis tools, J.C.; analyzing the data, Y.Y., W.Y., D.L., and W.G.; writing—original draft preparation, Y.Y., and W.Y.; writing—review and editing, W.Y, and Y.Y. All authors have read and agreed to the published version of the manuscript.

Funding: This research was funded by the Key Research and Development Plan of Shaanxi Province—Industrial project (Grant No. 2018GY-127).

Conflicts of Interest: The authors declare no conflict of interest. The funders had no role in the design of the study; in the collection, analyses, or interpretation of data; in the writing of the manuscript, or in the decision to publish the results.

References

1. Ding, W.J. *Science and Technology of Magnesium Alloy*; Science Press: Beijing, China, 2007; pp. 1–10. (In Chinese)
2. Yi, A.H.; Du, J.; Wang, J.; Mu, S.L.; Zhang, G.G.; Li, W.F. Preparation and characterization of colored Ti/Zr conversion coating on AZ91D magnesium alloy. *Surf. Coat. Technol.* **2015**, *276*, 239–247. [CrossRef]

3. Li, O.; Tsunakawa, M.; Shimada, Y.; Nakamura, K.; Nishinaka, K.; Ishizaki, T. Corrosion resistance of composite oxide film prepared on Ca-added flame-resistant magnesium alloy AZCa612 by micro-arc oxidation. *Corros. Sci.* **2017**, *125*, 99–105. [CrossRef]
4. Zhang, C.Y.; Zeng, R.C.; Liu, C.L.; Gao, J.C. Comparison of calcium phosphatecoatings on Mg-Al and Mg-Ca alloys and their corrosion behavior in Hank's solution, *Surf. Coat. Technol.* **2010**, *204*, 3636–3640.
5. Cui, Z.Y.; Ge, F.; Lin, Y.; Wang, L.W.; Lei, L.; Tian, H.Y.; Yu, M.D.; Wang, X. Corrosion behavior of AZ31 magnesium alloy in the chloride solution containing ammonium nitrate. *Electrochim. Acta* **2018**, *278*, 421–437. [CrossRef]
6. Sadeghi, A.; Hasanpur, E.; Bahmani, A.; Shin, K.S. Corrosion behaviour of AZ31 magnesium alloy containing various levels of strontium. *Corros. Sci.* **2018**, *141*, 117–126. [CrossRef]
7. Qu, Q.; Li, S.L.; Li, L.; Zuo, L.M.; Ran, X.; Qu, Y.; Zhu, B.L. Adsorption and corrosion behaviour of Ttrichoderma harzianum for AZ31B magnesium alloy in artificial seawater. *Corros. Sci.* **2017**, *118*, 12–23. [CrossRef]
8. Arthanari, S.; Shin, K.S. A simple one step cerium conversion coating formation on to magnesium alloy and electrochemical corrosion performance. *Surf. Coat. Technol.* **2018**, *349*, 757–772. [CrossRef]
9. Wu, L.; Yang, D.N.; Zhang, G.; Zhang, Z.; Zhang, S.; Tang, A.T.; Pan, F.S. Fabrication and characterization of Mg-M layered double hydroxide films on anodized magnesium alloy AZ31. *Appl. Surf. Sci.* **2018**, *431*, 177–186. [CrossRef]
10. Zhang, R.F.; Zhang, S.F. Formation of micro-arc oxidation coatings on AZ91HP magnesium alloys. *Corros. Sci.* **2009**, *51*, 2820–2825. [CrossRef]
11. Yang, W.; Xu, D.P.; Wang, J.L.; Yao, X.F.; Chen, J. Microstructure and corrosion resistance of micro arc oxidation plus electrostatic powder spraying composite coating on magnesium alloy. *Corros. Sci.* **2018**, *136*, 174–179. [CrossRef]
12. Yamauchi, N.; Ueda, N.; Okamoto, A.; Sone, T.; Tsujikawa, M.; Oki, S. DLC coating on Mg-Li alloy. *Surf. Coat. Technol.* **2007**, *201*, 4913–4918. [CrossRef]
13. Yang, W.; Gao, Y.; Guo, P.; Xu, D.P.; Wang, A.Y. Adhesion, biological corrosion resistance and biotribological properties of carbon films deposited on MAO coated Ti substrates. *J. Mech. Behav. Biomed.* **2020**, *101*, 103448. [CrossRef] [PubMed]
14. Liu, D.J.; Jiang, B.L.; Liu, Z.; Ge, Y.F.; Wang, Y.M. Preparation and catalytic properties of Cu_2O-CoO/Al_2O_3 composite coating prepared on aluminum plate by microarc oxidation. *Ceram. Int.* **2014**, *40*, 9981–9987. [CrossRef]
15. Durdu, S.; Usta, M. Characterization and mechanical properties of coatings on magnesium by micro arc oxidation. *Appl. Surf. Sci.* **2012**, *261*, 774–782. [CrossRef]
16. Guo, H.F.; An, M.Z.; Huo, H.B.; Xu, S.; Wu, L.J. Microstructure characteristic of ceramic coatings fabricated on magnesium alloys by micro-arc oxidation in alkaline silicate solutions. *Appl. Surf. Sci.* **2006**, *252*, 7911–7916. [CrossRef]
17. Gao, Y.H.; Li, W.F.; Du, J.; Zhang, Q.L.; Jie, J. Preparation and micro-structures of yellow ceramic coating by micro-arc oxidation. *J. Mater. Sci. Eng.* **2005**, *23*, 542–545. (In Chinese)
18. Yan, F.Y.; Fan, S.Y.; Zhang, W.Q.; Zhang, Y.H. Preparation of green micro-arc oxidation ceramic coating on magnesium alloy. *Mater. Prot.* **2008**, *41*, 4–6. (In Chinese)
19. Han, J.X.; Cheng, Y.L.; Tu, W.B.; Zhan, T.Y.; Cheng, Y.L. The black and white coatings on Ti-6Al-4V alloy or pure titanium by plasma electrolytic oxidation in concentrated silicate electrolyte. *Appl. Surf. Sci.* **2018**, *428*, 684–697. [CrossRef]
20. Tu, W.B.; Cheng, Y.L.; Wang, X.Y.; Zhan, T.Y.; Han, J.X.; Cheng, Y.L. Plasma electrolytic oxidation of AZ31 magnesium alloy in aluminate-tungstate electrolytes and the coating formation mechanism. *J. Alloy. Compd.* **2017**, *25*, 199–216. [CrossRef]
21. Li, J.M.; Cai, H.; Jiang, B.L. Growth mechanism of black ceramic layers formed by micro arc oxidation. *Surf. Coat. Technol.* **2007**, *201*, 8702–8708. [CrossRef]
22. Yang, W.; Xu, D.P.; Chen, J.; Liu, J.N.; Jiang, B.L. Characterization of self-sealing MAO ceramic coatings with green or black color on an Al alloy. *RSC Adv.* **2017**, *7*, 1597–1605. [CrossRef]
23. Yang, W.; Wang, J.L.; Xu, D.P.; Li, J.H.; Chen, T. Characterization and formation mechanism of grey micro-arc oxidation coatings on magnesium alloy. *Surf. Coat. Technol.* **2015**, *283*, 281–285. [CrossRef]

24. Yang, W.; Xu, D.P.; Yao, X.F.; Wang, J.L.; Chen, J. Stable preparation and characterization of yellow micro arc oxidation coating on magnesium alloy. *J. Alloy. Compd.* **2018**, *745*, 609–616. [CrossRef]
25. Lee, S.J.; Do, L.H.T. Effects of copper additive on micro-arc oxidation coating of LZ91 magnesium-lithium alloy. *Surf. Coat. Technol.* **2016**, *307*, 781–789. [CrossRef]
26. Li, Q.B.; Yang, W.B.; Liu, C.C.; Wang, D.A.; Liang, J. Correlations between the growth mechanism and properties of micro-arc oxidation coatings on titanium alloy: Effects of electrolytes. *Surf. Coat. Technol.* **2017**, *316*, 162–170. [CrossRef]
27. Chen, W.W.; Wang, Z.X.; Sun, L.; Lu, S. Research of growth mechanism of ceramic coatings fabricated by micro-arc oxidation on magnesium alloys at high current mode. *J. Magn. Alloy.* **2015**, *3*, 253–257. [CrossRef]
28. Yang, W.; Xu, D.P.; Guo, Q.Q.; Chen, T.; Chen, J. Influence of electrolyte composition on microstructure and properties of coatings formed on pure Ti substrate by micro arc oxidation. *Surf. Coat. Technol.* **2018**, *349*, 522–528. [CrossRef]
29. Tao, X.J.; Li, S.J.; Zheng, C.Y.; Fu, J.; Guo, Z.; Hao, Y.L.; Yang, R.; Guo, Z.X. Synthesis of a porous oxide layer on a multifunctional biomedical titanium by micro-arc oxidation. *Mat. Sci. Eng. C Mater.* **2009**, *29*, 1923–1934. [CrossRef]
30. Veys-Renaux, D.; Barchiche, C.E.; Rocca, E. Corrosion behavior of AZ91 Mg alloy anodized by low-energy micro-arc oxidation: Effect of aluminates and silicates. *Surf. Coat. Technol.* **2014**, *251*, 232–238. [CrossRef]
31. Shokouhfar, M.; Allahkaram, S.R. Formation mechanism and surface characterization of ceramic composite coatings on pure titanium prepared by micro-arc oxidation in electrolytes containing nanoparticles. *Surf. Coat. Technol.* **2016**, *291*, 396–405. [CrossRef]
32. Yan, W.G.; Jiang, B.L.; Li, H.T.; Shi, W.Y. Exfoliation of ceramic layers formed by micro-arc oxidation under cathode environment. *Hot Working Technol.* **2017**, *46*, 158–161.
33. Shen, Y.; Wang, H.X.; Pan, Y.P. Effect of current density on the microstructure and corrosion properties of MAO coatings on aluminum alloy shock absorber. *Key Eng. Mater.* **2018**, *764*, 28–38. [CrossRef]

Article

Corrosion Resistance and Apatite-Forming Ability of Composite Coatings formed on Mg–Al–Zn–Ca Alloys

Anawati Anawati [1,*], Hidetaka Asoh [2] and Sachiko Ono [2]

[1] Department of Physics, Faculty of Mathematics and Natural Sciences, Universitas Indonesia, Depok 16424, Indonesia
[2] Department of Applied Chemistry, Kogakuin University, 2665-1 Nakano, Hachioji, Tokyo 192-0015, Japan
* Correspondence: anawati@sci.ui.ac.id; Tel./Fax: +62-21-3193-8136

Received: 15 June 2019; Accepted: 10 July 2019; Published: 14 July 2019

Abstract: The properties of composite coatings formed by plasma electrolytic oxidation (PEO) were affected by the alloy composition. The corrosion resistance and apatite-forming ability of PEO coatings formed on Mg–6Al–1Zn–xCa alloys with a variation of Ca content were investigated. Potentiodynamic polarization and electrochemical impedance spectroscopy (EIS) measurements showed an order magnitude improvement of corrosion resistance in the AZ61 alloy as a result of the coating. A higher enhancement in polarization resistance was obtained in the Mg–6Al–1Zn–1Ca and Mg–6Al–1Zn–2Ca alloys due to thicker coatings were formed as a result of the incorporation of calcium oxide/hydroxide. However, the underlying substrates were more prone to localized corrosion with increasing Ca content. The microstructure investigation revealed an enlargement in precipitates (Al_2Ca, Mg_2Ca) sizes with increasing Ca content in the alloys. The growth of larger size precipitates increased the danger to micro galvanic corrosion. Apatite layers were formed on all of the coatings indicating high apatite-forming ability, but the layers formed on the Mg–6Al–1Zn–1Ca and Mg–6Al–1Zn–2Ca alloys contained higher Mg, possibly due to the accumulation of corrosion product, than that on the Mg–6Al–1Zn alloy. The alloying element Ca should be limited to 1 wt.% as the excess tended to degrade the corrosion resistance and apatite-forming ability of the PEO coating.

Keywords: magnesium; microstructure; coating; corrosion; polarization; apatite

1. Introduction

The use of magnesium (Mg) and its alloys for biodegradable materials rely on the surface treatment such as plating, coating, and anodizing, due to the high corrosion rate of Mg in aqueous environments [1–3]. Among the available techniques, plasma electrolytic oxidation (PEO) becomes famous for corrosion protection of Mg alloys considering its flexibility to coat complex geometry and environmentally friendly process [4,5]. The PEO coating that forms as a result of high-voltage (hundreds volt) anodization in an alkaline electrolyte, with the incorporation of the electrolyte species, provides superior corrosion and wear resistance [6–8]. A ceramic-like composite layer consisting of crystalline and amorphous oxides is developed under the exposure of a high-temperature plasma discharge. The corrosion resistance of the PEO coating depends on the processing parameters [9], alloy composition [10], and the type of electrolyte used [11]. Among various alkaline electrolytes, the PEO coating formed in phosphate-based solution exhibited the highest corrosion resistance due to the significant formation of crystalline magnesium phosphate [12]. Magnesium phosphate is a biocompatible compound and has attracted much attention during the past few years as a material for bone replacement [13]. However, both the in vitro [7,14] and in vivo [15,16] tests have indicated that the deposition of bone mineral apatite ($Ca_{10}(PO_4)_6$) on the PEO coating was limited. The apatite-forming ability defines the bioactivity of biomaterials.

Various methods have been proposed to accelerate the bioactivity of the PEO coating in physiological solution such as by coating with apatite [17,18], incorporating Ca [19,20] or apatite [21,22] in the coating, and introducing Ca as an alloying element [23]. The unstable Ca compounds easily dissolves in the electrolyte during plasma discharge and therefore adding Ca in the electrolyte bath is not beneficial. Incorporation of apatite in the PEO coating improved the corrosion resistance, but long-term immersion in simulated body fluid (SBF) indicated no further growth of the apatite [19,20]. Apart from the coating modification during the PEO process, the effect of Ca as an alloying element in Mg in modifying the bioactivity of the coating has not been thoroughly investigated. Our previous result [23] showed an increase in the bioactivity of the PEO coating on Mg–6Al–0.26 Mn–xCa alloys with increasing Ca concentration in the alloys. An embodiment of Ca in the coating and its presence in the alloy accelerated the growth of apatite in simulated body fluid (SBF). However, short-term (8 min) grown coatings on Mg–6Al–1Zn–xCa alloys revealed no bioactivity in SBF and significantly high corrosion rate [24]. The reason for such behavior was still unclear. The alloying element Ca might have different effect on the microstructure, and hence the resulting PEO film of the Mg–Al–Zn alloys as compared to that of the Mg–Al–Mn alloys. The development of an Mg based-biodegradable implant included Zn as alloying element to improve its mechanical strength [25–27]. This work aims to throw light on explaining the corrosion behavior and apatite-forming ability of the coatings formed on the Mg–6Al–1Zn–xCa alloys as a result of alteration in the microstructure with a variation of Ca content.

2. Materials and Methods

2.1. Specimen Preparation

Rolled plates of Mg–6Al–1Zn–xCa alloys with the composition listed in Table 1 were used as substrates. For the sake of simplicity, the alloy was further designated as AZ61, AZX611, and AZX612 for the alloys containing 0, 1, and 2 wt.% Ca, respectively. The plate was cut into 5 cm^2 working area. For surface observation, the specimen was embedded in a resin and then ground to 1200-grit silicon carbide (SiC) paper. The specimen was then degreased in acetone in an ultrasonic bath for 3 min. The etching treatment was applied on the specimen using 4 vol.% HNO_3 in ethanol for 20 s to reveal the microstructure.

Table 1. Chemical composition of the alloys.

Element	AZ61	AZX611	AZX612
Mg	Bal.	Bal.	Bal.
Al	6.50	5.8–7.2	5.8–7.2
Zn	0.92	0.4–1.5	0.4–1.5
Mn	0.32	0.15–0.5	0.15–0.5
Cu	≤0.05	≤0.05	≤0.05
Ni	≤0.005	≤0.005	≤0.005
Fe	≤0.005	≤0.005	≤0.005
Si	≤0.1	≤0.1	≤0.1
Ca	-	1.0	2.0

2.2. Plasma Electrolytic Oxidation

To dissolve the natural oxide layer on the surface, the specimen was dipped in a mixed acid solution of 8 vol.% HNO_3-1 vol.% H_3PO_4 for 20 s, and then in 5 wt.% NaOH solution at 80 °C for 1 min. The PEO coating was developed under a galvanostatic mode at 200 Am^{-2} in 0.5 mol·dm^{-3} Na_3PO_4 solution at 25 °C for 20 min. A regulated DC power supply was used as the current source. The specimen was placed as an anode while a pair of carbon rod was used as cathode. The voltage output was measured by digital multimeter bench from Keithley series 2700. A coating thickness gauge of dual type (SME-1) from Sanko was used for determining the coating thickness.

2.3. Electrochemical Corrosion

The electrochemical corrosion measurement was performed by conducting potentiodynamic polarization and electrochemical impedance spectroscopy (EIS) tests in a physiological solution 0.9 wt.% NaCl solution at 37 °C based on ASTM G5 [28]. A defined surface with an area of 2.5 cm^2 was exposed to the solution. The electrochemical tests were performed by using a potentiostat instrument from IviumStat. Pt wire was used as a counter electrode, and silver chloride (Ag/AgCl) was used as a reference electrode. The electrochemical cell condition and arrangement were similar to as reported earlier [24]. Potentiodynamic polarization test was conducted from 100 mV below open circuit potential (OCP) and terminated when the current output reached 30 mA at a sweep rate of 1 $mV{\cdot}s^{-1}$. The corrosion potential and current density were estimated by using Tafel extrapolation. The EIS was performed on the uncoated substrates at an open circuit potential (OCP) over a frequency ranging from 10^{-2} Hz to 10^4 Hz. Prior to the tests, the specimen was left at an OCP for 40 min to stabilize the potential. The coated specimen did not give spectra at low frequency and therefore was analyzed at frequency range 10^2 to 10^7 Hz.

2.4. Apatite-Forming Ability Test

The apatite-forming ability test was performed by immersing the specimens individually in SBF at 37 °C with a surface-to-volume ratio of 20 mL/cm^2 for 14 days. An SBF10, with the ionic concentration shown in Table 2, was used. The solution preparation followed the Reference [29]. The fresh solution pH was adjusted to 7.4 at 37 °C. The solution was replaced after 1, 3, 5, 7, 10, and 12 days, and was collected each time after each replacement for further analysis. The Mg concentration dissolved in the solution was analyzed by using the titration method. A digital titration kit from Hach, which included a digital titrator, buffer solutions, and indicators, was used. The solution was titrated with 0.08 M of sodium ethylenediaminetetraacetic acid (EDTA) using a digital titrator. The method was able to detect Mg concentrations within a 1 ppm margin of error.

Table 2. Ionic composition of simulated body fluid (SBF).

Ion	Na^+	K^+	Mg^+	Ca^+	Cl^-	HCO_3^-	HPO_4^{2-}	SO_4^{2-}
Concentration (mM)	142	5	1	2.5	126	10	1	1

2.5. Surface Analyses

The surface microstructure and the elemental composition was studied by an energy dispersive X-ray spectroscopy (EDS, JEOL EX-54175JMU, Tokyo, Japan) attached to the SEM (JEOL JSM-6380LA, Tokyo, Japan). The crystalline phases in the specimens were analyzed using thin coating X-ray diffraction analysis (TF-XRD, Rint 2000 Rigaku type, Tokyo, Japan) at an incident angle of 1°. The phases existed in the specimens were analyzed by indexing the peaks in the XRD pattern by referring to the JCPDS cards. Depth profile analysis on the coatings was performed on a circular area with a diameter of 4 mm using glow-discharge optical emission spectroscopy (GDOES, Jobin-Yvon JY5000RF, Horiba, Ltd., Kyoto, Japan). Sputtering was done using Ar^+ ion at 40 W.

3. Results

3.1. Alloy Microstructure

Figure 1 shows the X-ray diffraction (XRD) pattern of AZ61, AZX611, and AZX612 alloys and the optical microscope images showing the microstructure of the alloys. All of the alloys consisted of primary α-Mg phase and secondary β-phase ($Al_{12}Mg_{17}$). Precipitation of Ca-containing intermetallic, Al_2Ca and Mg_2Ca, was detected in the AZX611 and AZX612 alloys. The peaks for Al_2Ca and Mg_2Ca phases were quite small and often overlapped with the peaks for other phases. As opposed to the microstructure of AM60 alloy which contained a low number of precipitates [23], the Mg matrix of the

AZ61 alloy exhibited an ultrafine equiaxed grain with the average grain size ~3 μm × 5 μm decorated by numerous spherical intermetallic $Al_{12}Mg_{17}$ phase with diameters approximately 0.5 μm as shown in Figure 1b. The presence of 1 wt.% Ca expanded the grain size about an order of magnitude, but no further enlargement with increasing Ca content to 2 wt.%. The XRD pattern of AZX611 and AZX612 alloys displayed a higher intensity in the Mg peaks confirming larger metallic grain sizes than that of the base alloy. The equiaxed grains size in both AZX611 and AZX612 alloy was approximately 50 μm × 250 μm. The grain boundaries became thicker with increasing Ca content in the alloys due to segregation of the intermetallic precipitates (Figure 1c,d). The XRD pattern of Ca-containing alloys revealed some additional β-phase peaks emerged at 23.7°, 40.0°, 42.0°, 43.7°, 61.6°, 63.3°, 64.9°, 66.4°, and 67.8° indicating a higher amount of β phase existed in the alloys.

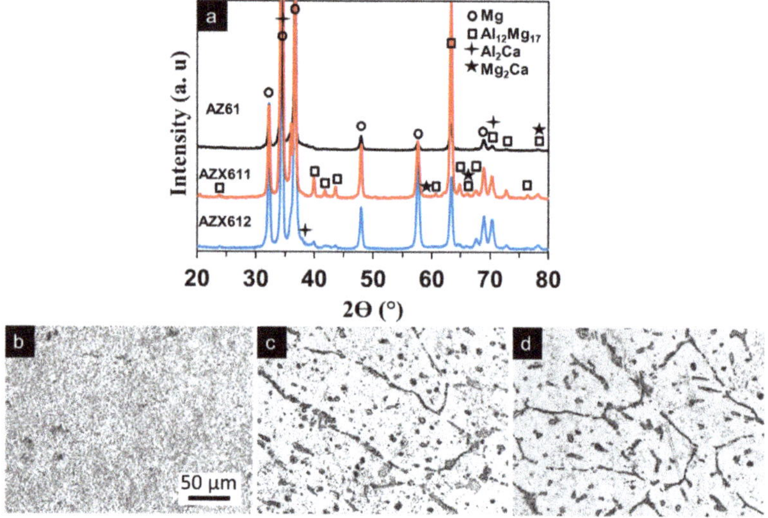

Figure 1. (**a**) X-ray diffraction pattern and surface microstructure of (**b**) AZ61, (**c**) AZX611, and (**d**) AZX612 alloys. The scale bar in image (b) applied to images (c) and (d).

Figure 2 shows the SEM images and the corresponding EDS maps for the elements Mg, Al, and Ca, showing the element composition of the three alloys. In the maps, blue represents Mg, red represents Al, and the green represents Ca. The AZ61 alloy shows a high density of precipitates on its surface. The maps displayed a nearly uniform blue and red maps corresponded to the β precipitate smeared in the matrix. The size of precipitates became larger with increasing Ca content in the alloys, and the matrix became clearer from precipitates. The maps for AZX611 and AZX612 showed that the intermetallic consisted of Mg, Al, and Ca. This analysis proved the unambiguous presence of Al_2Ca and Mg_2Ca precipitates besides the β phase in both AZX611 and AZX612 alloys. The amount of Al_2Ca phase dominated over Mg_2Ca phase.

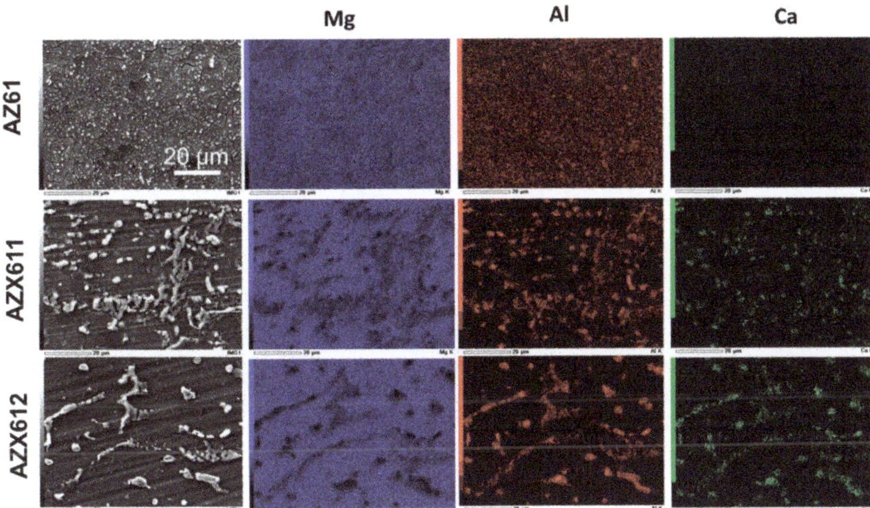

Figure 2. Plane-view SEM images and the corresponding EDS maps for Mg (blue), Al (red), and Ca (green), of AZ61, AZX611, and AZX612 alloys. The scale bar applied to all images.

3.2. PEO-Coating Structure and Composition

The voltage–time curves recorded during the PEO coating on the three alloys are depicted in Figure 3. Above the breakdown voltage (90 V), plasma discharge was developed at a few spots on the specimen surface, indicating the beginning of the electrolytic process. After the discrete plasma formed uniformly on the surface, the voltage stabilized at ~150 V. Strong plasma discharge began to occur at a critical voltage after about 9 min where large oscillation in the voltage (150–250 V) was attained. The presence of 1 wt.% Ca in the alloys shifted the critical voltage towards a longer time of approximately 2 min; however, the curve did not shift any further when the amount of Ca was increased to 2 wt.%.

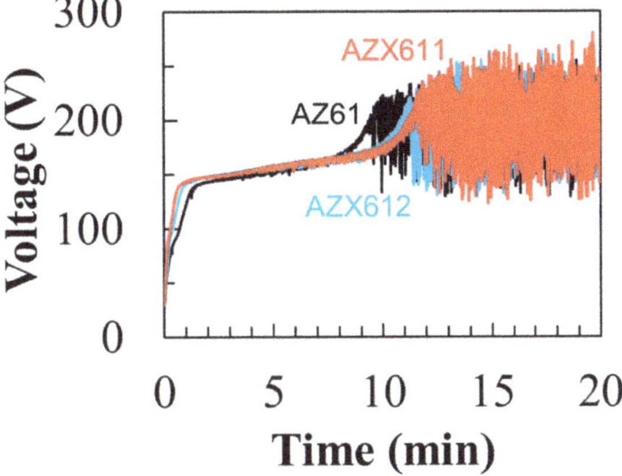

Figure 3. Voltage–time curves during 20 min anodization of AZ61, AZX611, and AZX612 alloys in 0.5 M Na_3PO_4 solution at 25 °C.

Figure 4 displays the XRD pattern of the resulting coatings formed on the three alloys and the cross-sectional FE-SEM images of the coatings. Most of the peaks for substrates no longer appeared in the XRD pattern of the coated specimens indicating an excellent blocking effect of the coatings from the X-ray. The composition of the coatings formed on the three alloys were similar, which were composed of both crystalline and amorphous phases as shown by the appearance of the broad peak between 20° and 40°, which attributed to the amorphous phase, and the small peaks inside the broad peak corresponded to the crystalline oxide phases of $Mg_3(PO_4)_2$ and $Mg(PO_3)_2$. The coatings that formed on all of the specimens exhibited a similar uneven structure, as shown in Figure 4b–d. Pores and cracks, which are the footprint of PEO coating, were observed on both the surface and inner part of the coating. Fine cracks were developed during the PEO process, while heavy cracks occurred unintentionally during specimen preparation. The coating formed on Ca-containing alloys was slightly thicker than that of the base alloy. In agreement with the cross-section images, the average coating thickness measured by coating thickness gauge were 25, 32, and 30 μm for the coating formed on AZ61, AZX611, and AZX612, respectively. The XRD analysis did not detect the presence of Ca in the coating, similar to the earlier result on AM60 [23]. The Ca compounds may have been present in an amorphous state.

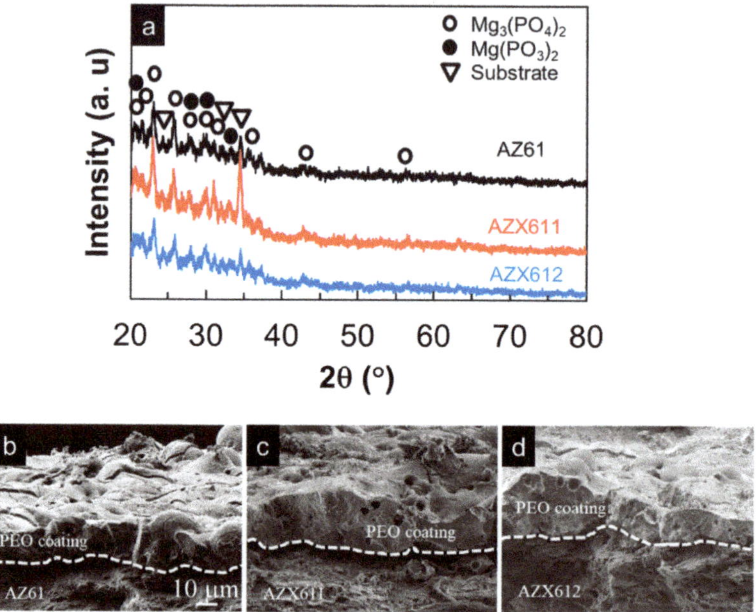

Figure 4. (**a**) X-ray diffraction pattern of plasma electrolytic oxidation (PEO) coatings and the corresponding cross-section FE-SEM images of the coating on (**b**) AZ61, (**c**) AZX611, and (**d**) AZX612 resulting from 20 min anodization in 0.5 M Na_3PO_4 solution at 25 °C. The scale bar applied to all images.

The coatings composition was further analyzed by depth profile GDOES, and the results are displayed in Figure 5. The dashed line marks the coating–metal interface at which the O and Mg profiles crossed each other. All of the coatings were composed of Mg, P, O, and Al. Confirming XRD results, the primary oxide phases in the coatings consisted of Mg–O compounds. The GDOES profile suggested the presence of other oxide phases, including MgO and Al_2O_3 in the amorphous state. The incorporation of Ca in the coating of AZX611 and AZX612 alloys was confirmed by the Ca profiles in Figure 5b,c. The intensities of Ca signal in the AZX611 and AZX612 coatings was about half of their bulk values, which verified the presence of calcium oxide/hydroxide phase in the coatings. The coating-metal interface shifted twice towards longer sputtering times from 400 s for AZ61 to 950 s for AZX611 and 900 s for AZX612 specimens. The shift indicated an increase in coating thickness as the thicker coating needs longer sputtering time to reach the bulk substrate.

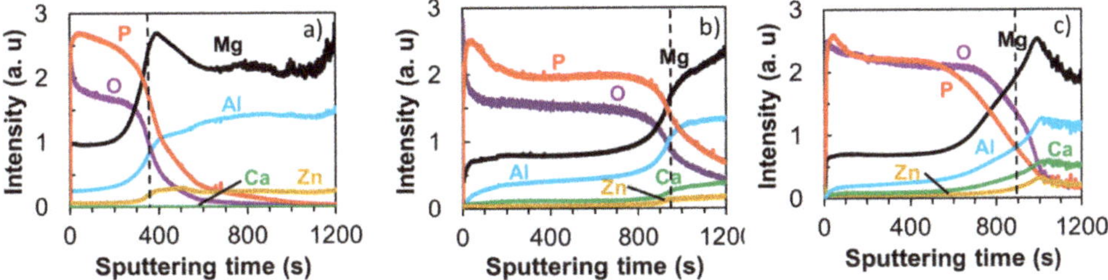

Figure 5. Glow-discharge optical emission spectroscopy (GDOES) elemental depth profiles of PEO coatings on (**a**) AZ61, (**b**) AZX611, and (**c**) AZX612 specimens after anodization in 0.5 mol·dm^{-3} Na_3PO_4 solution at 25 °C for 20 min. The dashed lines indicate the oxide–metal interface.

3.3. Electrochemical Corrosion

The corrosion behavior of the PEO coating was investigated by potentiodynamic polarization test in 0.9 wt.% NaCl solution at 37 °C. Figure 6 shows the corrosion potential and current densities data obtained from the polarization curves of the specimens before and after coating. The corrosion potentials of the substrates became slightly nobler from −1.49 to −1.45 $V_{Ag/AgCl}$ with increasing Ca content in the alloys. The behavior was preserved after coating that the corrosion potential increased with Ca content in the alloys from −1.66 to −1.64 $V_{Ag/AgCl}$ (Figure 6a). The corrosion current densities of the coated specimens were 10 times lower than that of the substrates. The substrates exhibited corrosion current densities in the range of 2.0 × 10^{-5} to 3.5 × 10^{-5} A·cm^{-2} while the coated specimens were in the range of 3.5 × 10^{-6} to 3.6 × 10^{-6} A·cm^{-2}. Depression of corrosion potential, which was accompanied by a reduction in corrosion current densities as a result of PEO coating, indicated an improvement in corrosion resistance of the alloys. The inhibition of cathodic reaction on the surface suppressed the corrosion potential of the coated specimens to the negative direction.

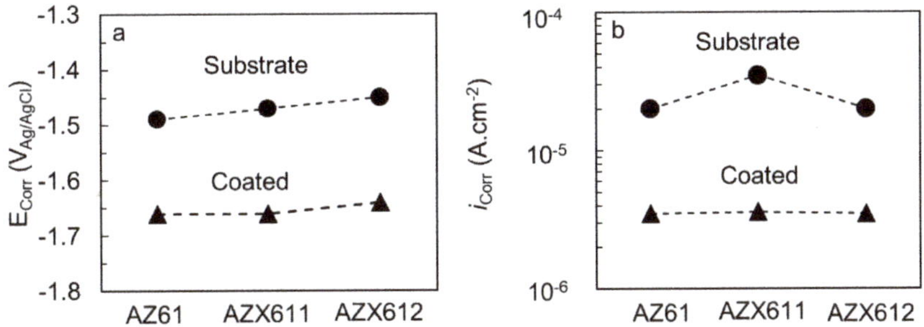

Figure 6. (**a**) Corrosion potentials and (**b**) corrosion current densities of the substrates and coated specimens of AZ61, AZX611, and AZX612 alloys.

The coating resistance was evaluated by EIS measurement in 0.9 wt.% NaCl solution at 37 °C. The results are displayed in Figure 7 for uncoated, and Figure 8 for the coated, specimens. The fitted parameters obtained from the EIS data are listed in Table 3. The n1 and n2 refers to constant phase for CPE1 and CPE2, respectively. The maximum n value is 1. The closer to 1 the n1 and n2 value is, indicates more capacitive behavior. The polarization resistance (R1) of the coated specimens was an order of magnitude higher than that of their substrates. The base alloy exhibited the highest polarization resistance in both uncoated and coated conditions relative to the Ca-containing alloys. The impedance spectrum of AZ61 substrate exhibited one apparent capacitive loop, while the Ca-containing alloys showed an additional inductive loop at low frequency. The corresponding inductive loop is an indication of localized corrosion [5,8]. The presence of inductive loop lowered the n2 values for AZX611 and AZX612 specimens compared to the AZ611 specimen. The presence of a constant phase element (CPE) in parallel with a resistance indicates the presence of a faradaic reaction (charge transfer) and a non-Faradaic reaction (charge accumulation at the interface) occurring at the interface. All of the three PEO coatings exhibited a similar trend of impedance spectrum. The L and R_L loops did not exist in the model for the PEO coating (Figure 8), indicating high resistance to metal dissolution. Moreover, a much higher impedance at low and medium frequencies of the Bode plot and an enlarged diameter of the semicircle in the Nyquist plot of the coated specimens are evidence of remarkable higher corrosion resistance than that of the substrates. Moreover, the capacitance values of the coated specimens were a hundred times lower than that of the substrates. The effect of the alloying Ca in the coating impedance was displayed by the Bode plot in Figure 8. The impedance at high frequencies, which represents the outer PEO coating of all the coated specimens, was similar. However, the impedance at low frequencies displayed slightly lower impedance than that of the AZ61 specimen. The results implied that the coating–metal interface of the Ca-containing specimens was more prone to localized corrosion than that of the base specimen.

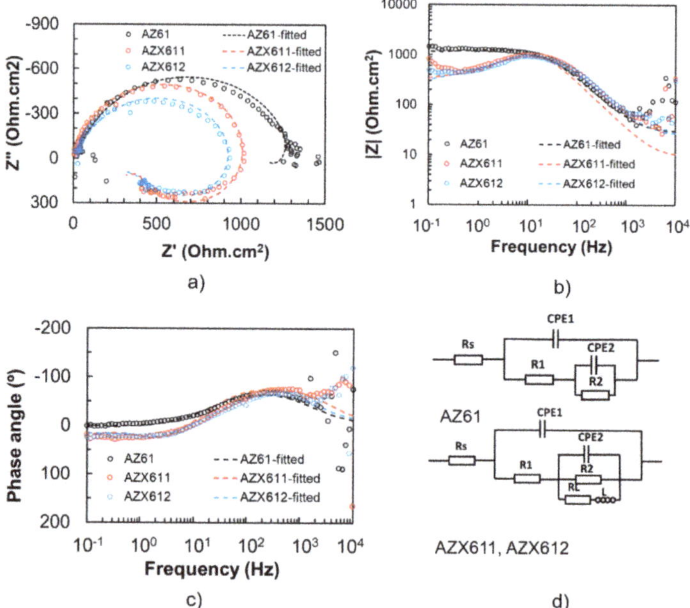

Figure 7. Electrochemical impedance spectra of uncoated specimens of AZ61, AZX611, and AZX612 specimens: (**a**) Nyquist plots, (**b**) bode plots, and (**c**) phase plots, and (**d**) the equivalent circuit.

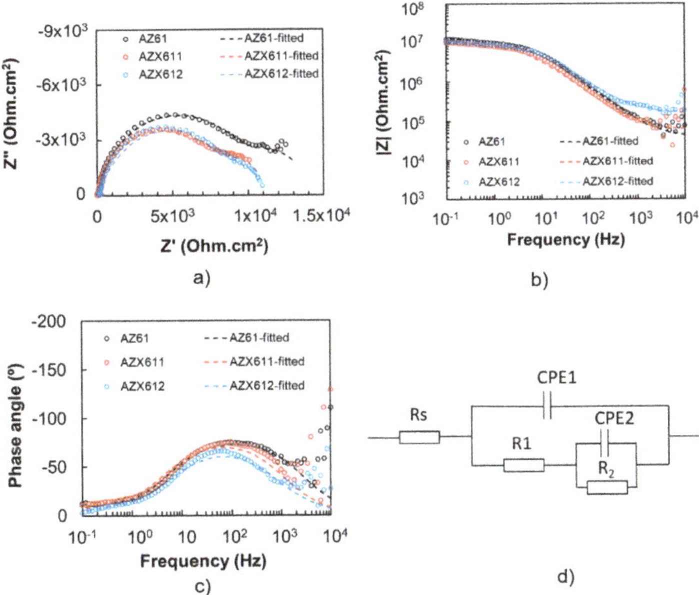

Figure 8. Electrochemical impedance spectra of coated specimens of AZ61, AZX611, and AZX612 specimens: (**a**) Nyquist plots, (**b**) bode plots, and (**c**) phase plots, and (**d**) the equivalent circuit.

Table 3. Parameter of the fitted plotting of EIS spectra.

Specimen	Rs (Ωcm^2)	CPE1 ($\Omega^{-1}s^n cm^{-2}$)	n_1	R1 (Ωcm^2)	CPE2 ($\Omega^{-1}s^n cm^{-2}$)	n_2	R2 (Ωcm^2)	L (H)	RL (Ωcm^2)
AZ61	28.25	9.78×10^{-6}	0.88	1.32×10^3	-1.00×10^{-3}	0.81	−200	-	-
AZX611	9.40	7.68×10^{-6}	0.94	1.10×10^3	-1.00×10^{-3}	0.75	−1000	15	700
AZX612	25.0	8.00×10^{-6}	0.88	1.00×10^3	-6.00×10^{-4}	0.75	−725	20	950
AZ61 (Coated)	40.5	1.09×10^{-9}	0.87	1.04×10^4	3.67×10^{-7}	0.91	3.61×10^3	-	-
AZX611 (Coated)	61.9	1.64×10^{-9}	0.87	8.4×10^3	2.48×10^{-7}	0.96	2.61×10^3	-	-
AZX612 (Coated)	163.8	1.97×10^{-9}	0.82	9.45×10^3	3.67×10^{-7}	0.99	1.41×10^3	-	-

3.4. Apatite-Forming Ability

The apatite-forming ability of the coated specimens was evaluated by immersion test in XRD. The concentration of Mg dissolved from the surface, and the pH solution during the immersion test in SBF was monitored. Figure 9 displays the concentration of Mg detected in the solution and the pH solution during 14 days of immersion. The AZ61 alloy released the lowest Mg concentration in the range of 100–150 mg/L indicating the lowest corrosion rate relative to the two Ca-containing alloys, which released 100–500 mg/L Mg into the solution (Figure 9a). The solution pH of the three alloys increased from 7.4 to 8 after one-day immersion indicating high corrosion activities (Figure 9c). The increase in pH occurred due to the release of hydrogen gas during hydrolysis of Mg ions following the corrosion attack. The pH change was not significant as the solution was buffered. The solution pH of AZ61 alloy decreased, approaching the fresh SBF while that of the AZX611 and AZX612 alloys fluctuated with increasing immersion time. On average, the lowest pH was attained in the AZ61 alloy. The PEO-coated specimens released much lower Mg into the solution relative to their substrates (Figure 9b). In agreement, the solution pH of the coated specimen was relatively lower compared to the substrates during the 14-day immersion. The dissolved Mg in the solution was attributed to both corrosion of the underlying substrate and the coating dissolution. The coated AZ61 specimen released 110 to 200 mg/L Mg at the initial immersion time and then stabilized at day 7 to 12. Dissolution no longer occurred at day 14. The coating on AZX611 specimen exhibited a relatively constant dissolution rate during

1–12 days followed by passivation at the end period of the test. The AZX612 specimen exhibited a dynamic dissolution resulting in a highly fluctuating concentration of the dissolved Mg in the range 300–450 mg/L. All of the coatings exhibited the tendency for passivation at day 14 with solution pH at 7.7, which was presumably due to surface coverage by the apatite layers on their surfaces.

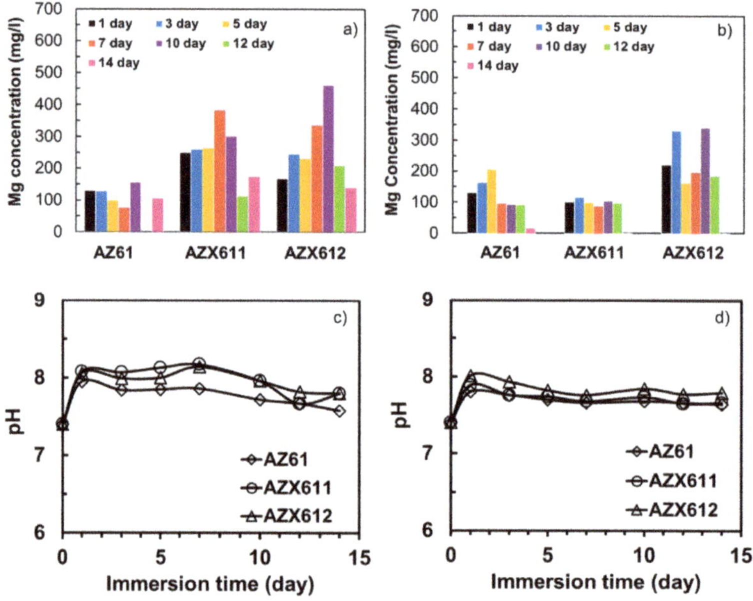

Figure 9. Concentration of Mg dissolved during immersion testing in SBF of the (**a**) substrates, (**b**) coated specimens, and the corresponding solution pH for (**c**) the substrates and (**d**) coated specimens.

The presence of apatite on the PEO coating was investigated by observing the surface morphology and analyzing the surface composition after the immersion test. Figure 10 shows the SEM images at different magnifications revealing the presence of uniform apatite layer on all of the three coatings. Most of the pores that existed in the coating of AZ61 specimen were no longer viewed, indicating surface coverage by thick apatite layer (Figure 10a). The layer formed uniformly following the coating structure, as shown clearly in Figure 10b. The spongy structure with scallop type of grain, which was a typical morphology of apatite formed in SBF was seen in a higher magnification image in the inset Figure 10b. The coating remained intact and protected by the apatite layer. Thinner apatite layers were formed on AZX611 and AZX612 specimens as inferred from the existence of open pores and cracks in the coating (Figure 10c,e). The scallop grain of apatite layer in the AZX611 and AZX612 specimens was smaller than that of formed in AZ61 specimen as shown in the inset image of Figure 9d,f. Localized coating damage was recognized at a few spots, pointed by arrows in the images. The AZX611 specimen showed relatively small coating deterioration while larger coating damage on an area approximately 150 μm × 300 μm was observed on AZX612 specimen.

Figure 11 shows the results of EDS analysis from the areas shown in Figure 10 as compared to that of fresh PEO coating. The observation areas for both cases were kept identical by using similar magnification scale. Prior to the immersion test, calcium was not detected in any of the coatings (Figure 11a). The signal attributed to the calcium oxide phase in the coating was weak relative to the other elements. After 14 days immersion in SBF, the Ca signal was detected as high as 8.0, 7.2, and 3.1 at.% in AZ61, AZX611, and AZX612 specimens, respectively, attributed to the apatite layer shown in Figure 10. The Mg concentration increased from 6.8 at.% in the AZ61 to 12.5 at.% in the AZX611 specimens. The concentration of P was relatively not altered at 12–13 at.%. The ratios of Ca/P of the apatite layers were 0.65, 0.60, and 0.24 for the layers formed on AZ61, AZX611, and AZX612 specimens, respectively. The ratio was half of the stoichiometry apatite, 1.67 [30]. The apatite layer that formed in the coating of AZX611 and AZX612 alloys contained more Mg than that of the AZ61 alloy. Moreover, the strong signal of P derived from the underlying PEO coating lowered the Ca/P ratio.

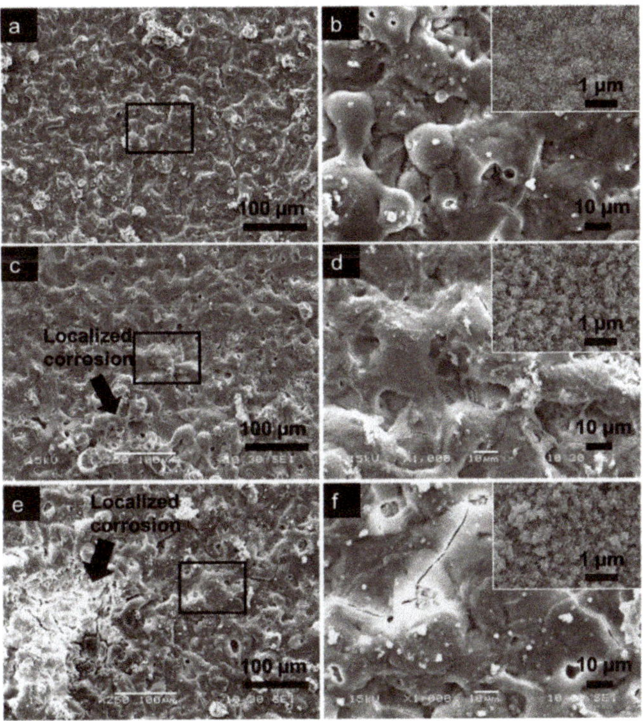

Figure 10. SEM images of PEO coatings on (**a**) and (**b**) AZ61, (**c**) and (**d**) AZX611, and (**e**) and (**f**) AZX612 specimens after 14 days of immersion in SBF showing the corrosion morphology. (b), (d), (f) are higher magnification images of the area inside the rectangle in images (a), (c), (e), respectively. The inset image shows the apatite layer structure.

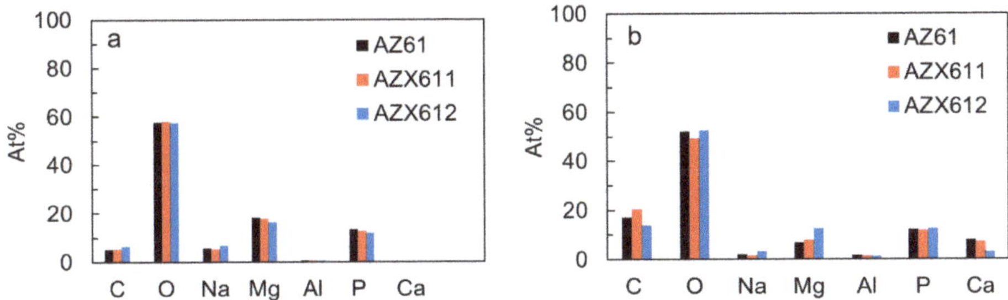

Figure 11. Atomic fraction of elements detected by EDS analysis on PEO coatings of AZ61, AZX611, and AZX612 specimens (**a**) before and (**b**) after 14 days immersion in SBF.

4. Discussion

Surface characterization showed that the coatings formed on all the three alloys exhibited similar uneven structure decorated with pores and cracks, but the composition and thickness were altered by the presence of Ca in the alloys. The coatings composed of a mixed crystalline and amorphous phase. The $Mg_3(PO_4)_2$ and $Mg(PO_3)_2$ existed in both crystalline and amorphous states as confirmed by XRD analysis, while MgO and Al_2O_3 were present in an amorphous state as detected by GD-OES and EDS analysis. The presence of Ca in the alloys induced the formation of amorphous calcium oxide/hydroxide phase in the coatings. The depth profile analysis by GD-OES (Figure 5) confirmed the presence of calcium oxide/hydroxide from the surface throughout the coating-metal interface. At identical anodization time, the resulting coating on the AZX611 alloy was 7 μm thicker than that of the base alloy. The coating did not thicken any further in the AZX612 alloy, as no further substantial change in the Ca content of the solid solution matrix.

The microstructure observation revealed that the alloying element Ca decreased the density of intermetallic precipitates in the AZ61 alloys by inducing larger precipitates size and enlarging the metallic grain. The grain size expanded about 30–50 times by the presence of 1 wt.% Ca in the alloy. The substitute Ca atom with atomic diameter larger than Mg atom [31] caused expansion in the matrix grain size. The high number of submicron-sized precipitates in the AZ61 alloy transformed into a low number of 10 micron-sized precipitates as a result of alloying with 1 wt.% Ca (Figure 2). The precipitates grew slightly larger, but the metallic grain size remained the same, with increasing Ca concentration to 2 wt.%. The enlargement of grain size was controlled by the amount of dissolved Ca in the solid solution matrix. The precipitates continued to grow as more excess of Ca existed in the AZX612 alloy. The limited solubility of Ca in Mg, which is only 0.8 wt.% [32], induced the formation of Al_2Ca precipitates. With increasing Ca content, the bigger precipitates grew larger at the expense of the smaller ones following Ostwald ripening [33]. Another intermetallic phase Mg_2Ca also existed in the AZX611 and AZX612 alloys, but only in a small number. The large misfit in atomic size between Ca and Mg in which the diameter of the Ca atom is 20% larger than Mg, is less favorable for the formation of Mg_2Ca phase [31]. The presence of heat-resistant Al_2Ca and Mg_2Ca phases delayed the formation of strong plasma discharge during electrolytic process resulting in a less crystalline oxide phases in the coatings. Moreover, the presence of Mg_2Ca phase is not beneficial from the corrosion viewpoints as it is more anodic than the Mg matrix [2].

During the PEO formation, Mg and Ca in the solid solution matrix was selectively oxidized. From thermodynamic viewpoints, Ca is easier to oxidize as it has lower Gibbs free energy than Mg [33]. However, since the concentration of Ca in the solid solution matrix was limited by its solubility, the resulting calcium oxide/hydroxide in the coating was relatively small compared to the magnesium phosphate/oxide phases. The GD-OES depth profile was proved to be a sensitive technique in detecting the calcium oxide/hydroxide. The calcium oxide/hydroxide phases presumably occurred during the 12 min exposure in fine plasma discharge. The strong discharge, which occurred in the latter stage, tended to dissolve the calcium oxide/hydroxide to the solution. The incorporation of calcium oxide/hydroxide contributed to the thickening of the coating formed on AZX611 and AZX612 alloys relative to the AZ61 alloy.

The higher resistivity offered by the thicker coating formed on the AZX611 and AZX612 alloys was counteracted by the vulnerability of their metal substrates to localized corrosion relative to the AZ61 alloy. The polarization coating resistance of the coated AZ61 specimen was higher than that of the AZX611 and AZX612 specimens. The enlargement of precipitates in the AZX611 and AZX612 alloys increased the contact area of galvanic coupling between the precipitates and the matrix, and therefore increased the danger to micro galvanic corrosion when exposed to the corrosive solution. Ennoblement of corrosion potentials in the polarization curves of the AZX611 and AZX612 alloys due to the increase number of cathodic intermetallic was balanced by the increased susceptibility to localized corrosion. The presence of inductive loop in the EIS spectra of AZX611 and AZX612 alloys (Figure 7a), which was not present in the spectra of AZ61 alloy, was attributed to the localized corrosion. A small amount of big precipitates accelerates corrosion, whereas finely distributed precipitates retarded corrosion [34]. The coating–metal interface in the vicinity of the precipitates was less protective. Once the corrosive solution penetrated the coatings, pockets of the solution were developed at the interface. Corrosion of the metal substrate occurred immediately as it was exposed to the solution inside the pockets. The increase in intermetallic precipitates size with Ca content in the alloys contributed to the enlargement of the attacked areas (Figure 10).

The higher corrosion rate of the underlying substrate limited the apatite-forming ability of the coatings on the AZX611 and AZX612 specimens. Even though a uniform apatite layer grew on all of the three coatings as a result of immersion in SBF, the layers that formed on the AZX611 and AZX612 coatings was thinner and contained higher Mg than that formed on the AZ61 coating. The incorporation of calcium oxide/hydroxide in the coatings did not necessarily enhance the bioactivity of the coatings, similar to that reported in Reference [19]. The corrosion rate of the underlying substrate is a more important factor in supporting the growth of apatite. An alternate corrosion rate of the underlying substrate, which released fluctuated hydrogen gas, tends to disturb the deposition of apatite on the coatings. In the case of AM60 alloys [23], the corrosion resistance of the PEO-coated specimens were relatively unaltered, and therefore the formation of apatite was not affected by the variation of Ca content in the alloys. During initial immersion time, thinning of the PEO coating occurred selectively at the amorphous part, leaving a rough crystalline oxide phase. The resulting sub-micron surface roughness became a preferential site for nucleation of apatite [14]. The apatite layer was likely to form on the coatings during 7–14 days immersion time where a constant dissolution rate of Mg was attained (Figure 9b). The released substances from the specimens contributed to increasing the degree of supersaturation of the SBF and the local pH, which triggered the deposition of apatite [30].

The following discussion led to the conclusion that the corrosion resistance of the PEO coatings formed on the AZ61 alloys was influenced by the concentration of alloying element Ca, in particular the concentration of the dissolved Ca in the solid solution matrix. Therefore, it is not necessary to add Ca above its solubility in Mg matrix. The Ca-containing precipitates that tended to shorten the lifetime of strong discharge during the PEO process resulted in the formation of lower crystalline oxide phases in the coating. A thinner PEO coating with a high-volume ratio of crystalline to amorphous phases gives much better corrosion resistance than a thicker coating with a lower ratio [12]. Addition of Ca above the solubility limit did not have a beneficial effect on the coating properties. On the contrary, the excess of Ca in the alloys resulted in the enlargement of Al_2Ca and Mg_2Ca precipitates, which tended to accelerate localized corrosion.

5. Conclusions

The PEO coating significantly enhanced the corrosion resistance of the Mg-6Al-1Zn-xCa (AZX) alloys relative to their substrates. The presence of 1 and 2 wt.% Ca in the alloys resulted in the formation of thicker coating, which contributed in higher enhancement of the coating resistance. However, the EIS spectra and the surface observation after the immersion test indicated an increased danger to localized corrosion with increasing Ca content in the alloys. The coating formed on AZX alloys exhibited lower polarization resistance than that which formed on the AZ61 alloy. After the 14-day immersion tests in SBF, the coating on AZ61 alloy remained intact while the coatings on AZX611 and AZX612 showed noticeable local damage. The microstructure investigation revealed that the alloying element Ca caused enlargement of metallic grain size and intermetallic precipitates relative to the AZ61 alloy. The increase in corrosion activities underneath the coatings lowered the apatite-forming ability of the AZX coatings, resulting in a thinner apatite layer formed during immersion in SBF.

Author Contributions: Conceptualization, A.A.; Methodology, A.A. Software, A.A. and S.O.; Validation, S.O.; Formal Analysis A.A., H.A. and S.O.; Investigation, A.A.; Resources, A.A. and S.O.; Data Curation: A.A.; Writing-Original Draft Preparation, A.A.; Writing-Review & Editing, A.A., H.A. and S.O.; Visualization, A.A.; Supervision, S.O.; Project Administration, A.A. and S.O.; Funding Acquisition, A.A. and S.O.

Funding: This research was supported by Direktorat Riset dan Pengabdian Masyarakat (DRPM) Universitas Indonesia through Q1Q2 Research grant 2019 (NKB-0266/UN2.R3.1/HKP.05.00/2019). Grant-in-Aid for Scientific Research from the Japan Society for the Promotion of Science and the Light Metal Education Foundation of Japan is acknowledged. We also thank a Strategic Research Foundation Grant-aided Project for Private Universities matching fund subsidy from the Ministry of Education, Culture, Sports, Science, and Technology of Japan.

Conflicts of Interest: The authors declare no conflict of interest.

References

1. Song, G. Control of biodegradation of biocompatable magnesium alloys. *Corros. Sci.* **2007**, *49*, 1696–1701. [CrossRef]
2. Kirkland, N.; Lespagnol, J.; Birbilis, N.; Staiger, M.; Kirkland, N. A survey of bio-corrosion rates of magnesium alloys. *Corros. Sci.* **2010**, *52*, 287–291. [CrossRef]
3. Witte, F.; Hort, N.; Vogt, C.; Cohen, S.; Ulrich, K.; Willumeit, R.; Feyerabend, F. Current Opinion in Solid State and Materials Science Degradable biomaterials based on magnesium corrosion. *Curr. Opin. Solid State Mater. Sci.* **2008**, *12*, 63–72. [CrossRef]
4. Patel, J.L.; Saka, N. Microplasmic ceramic coating. *Int. Ceram. Rev.* **2001**, *50*, 398–401.
5. White, L.; Koo, Y.; Neralla, S.; Sankar, J.; Yun, Y. Enhanced mechanical properties and increased corrosion resistance of a biodegradable magnesium alloy by plasma electrolytic oxidation (PEO). *Mater. Sci. Eng. B* **2016**, *208*, 39–46. [CrossRef] [PubMed]
6. Echeverry-Rendon, M.; Duque, V.; Quintero, D.; Robledo, S.M.; Harmsen, M.C.; Echeverria, F. Improved corrosion resistance of commercially pure magnesium after its modification by plasma electrolytic oxidation with organic additives. *J. Biomater. Appl.* **2018**, *33*, 725–740. [CrossRef] [PubMed]
7. Matykina, E.; Garcia, I.; Arrabal, R.; Mohedano, M.; Mingo, B.; Sancho, J.; Merino, M.; Pardo, A. Role of PEO coatings in long-term biodegradation of a Mg alloy. *Appl. Surf. Sci.* **2016**, *389*, 810–823. [CrossRef]
8. Lu, J.; He, X.; Li, H.; Song, R. Microstructure and Corrosion Resistance of PEO Coatings Formed on KBM10 Mg Alloy Pretreated with $Nd(NO_3)_3$. *Materials* **2018**, *11*, 1062. [CrossRef]
9. Li, L.; Gao, J.; Wang, Y.; Srinivasan, P.B.; Liang, J.; Blawert, C.; Störmer, M.; Dietzel, W.; Asoh, H.; Ono, S.; et al. Effect of current density on the microstructure and corrosion behaviour of plasma electrolytic oxidation treated AM50 magnesium alloy. *Surf. Coat Technol.* **2015**, *272*, 182–189. [CrossRef]
10. Hussein, R.; Northwood, D.; Nie, X. The effect of processing parameters and substrate composition on the corrosion resistance of plasma electrolytic oxidation (PEO) coated magnesium alloys. *Surf. Coat. Technol.* **2013**, *237*, 357–368. [CrossRef]
11. Ono, S.; Moronuki, S.; Mori, Y.; Koshi, A.; Liao, J.; Asoh, H. Effect of Electrolyte Concentration on the Structure and Corrosion Resistance of Anodic Films Formed on Magnesium through Plasma Electrolytic Oxidation. *Electrochim. Acta* **2017**, *240*, 415–423. [CrossRef]
12. Asoh, H.; Matsuoka, S.; Sayama, H.; Ono, S. Anodizing under sparking of AZ31B magnesium alloy in Na3PO4 solution. *J. Jpn. Inst. Light Met.* **2010**, *60*, 608–614. [CrossRef]
13. Kanter, B.; Vikman, A.; Brückner, T.; Schamel, M.; Gbureck, U.; Ignatius, A. Bone regeneration capacity of magnesium phosphate cements in a large animal model. *Acta Biomater.* **2018**, *69*, 352–361. [CrossRef] [PubMed]
14. Anawati; Asoh, H.; Ono, S. Enhanced uniformity of apatite coating on a PEO film formed on AZ31 Mg alloy by an alkali pretreatment. *Surf. Coatings Technol.* **2015**, *272*, 182–189. [CrossRef]
15. Han, Y.; Hong, S.H.; Xu, K. Structure and in vitro bioactivity of titania-based films by micro-arc oxidation. *Surf. Coatings Technol.* **2003**, *168*, 249–258. [CrossRef]
16. Gu, X.; Li, N.; Zhou, W.; Zheng, Y.; Zhao, X.; Cai, Q.; Ruan, L. Corrosion resistance and surface biocompatibility of a microarc oxidation coating on a Mg–Ca alloy. *Acta Biomater.* **2011**, *7*, 1880–1889. [CrossRef] [PubMed]
17. Kannan, M.B.; Mathan, B.K. Electrochemical deposition of calcium phosphates on magnesium and its alloys for improved biodegradation performance: A review. *Surf. Coat. Technol.* **2016**, *301*, 36–41. [CrossRef]
18. Gao, Y.; Yerokhin, A.; Matthews, A. Deposition and evaluation of duplex hydroxyapatite and plasma electrolytic oxidation coatings on magnesium. *Surf. Coat. Technol.* **2015**, *269*, 170–182. [CrossRef]
19. Srinivasan, P.B.; Liang, J.; Blawert, C.; Stormer, M.; Dietzel, W. Characterization of calcium containing plasma electrolytic oxidation coatings on AM50 magnesium alloy. *Appl. Surf. Sci.* **2010**, *256*, 4017–4022. [CrossRef]
20. Yang, J.; Lu, X.; Blawert, C.; Di, S.; Zheludkevich, M.L. Microstructure and corrosion behavior of Ca/P coatings prepared on magnesium by plasma electrolytic oxidation. *Surf. Coat. Technol.* **2017**, *319*, 359–369. [CrossRef]

21. Lederer, S.; Sankaran, S.; Smith, T.; Fürbeth, W. Formation of bioactive hydroxyapatite-containing titania coatings on CP-Ti 4+ alloy generated by plasma electrolytic oxidation. *Surf. Coat. Technol.* **2019**, *363*, 66–74. [CrossRef]
22. Adeleke, S.; Ramesh, S.; Bushroa, A.; Ching, Y.; Sopyan, I.; Maleque, M.; Krishnasamy, S.; Chandran, H.; Misran, H.; Sutharsini, U. The properties of hydroxyapatite ceramic coatings produced by plasma electrolytic oxidation. *Ceram. Int.* **2018**, *44*, 1802–1811. [CrossRef]
23. Anawati, A.; Asoh, H.; Ono, S. Effects of alloying element ca on the corrosion behavior and bioactivity of anodic films formed on AM60 mg alloys. *Materials* **2017**, *10*, 11. [CrossRef] [PubMed]
24. Anawati, A.; Asoh, H.; Ono, S. Degradation Behavior of Coatings Formed by the Plasma Electrolytic Oxidation Technique on AZ61 Magnesium Alloys Containing 0, 1 and 2 wt.% Ca. *Int. J. Technol.* **2018**, *9*, 622. [CrossRef]
25. Jun, J.H.; Park, B.K.; Kim, J.M.; Kim, K.T.; Jung, W.J. Effects of Ca Addition on Microstructure and Mechanical Properties of Mg-RE-Zn Casting Alloy. *Mater. Sci. Forum* **2009**, *488*, 107–110. [CrossRef]
26. Jang, Y.; Tan, Z.; Jurey, C.; Xu, Z.; Dong, Z.; Collins, B.; Yun, Y.; Sankar, J. Understanding corrosion behavior of Mg–Zn–Ca alloys from subcutaneous mouse model: Effect of Zn element concentration and plasma electrolytic oxidation. *Mater. Sci. Eng. C* **2015**, *48*, 28–40. [CrossRef] [PubMed]
27. Brady, M.P.; Fayek, M.; Leonard, D.N.; Meyer, H.M.; Thomson, J.K.; Anovitz, L.M.; Rother, G.; Song, G.L.; Davis, B. Tracer Film Growth Study of the Corrosion of Magnesium Alloys AZ31B and ZE10A in 0.01% NaCl Solution. *J. Electrochem. Soc.* **2017**, *164*, C367–C375. [CrossRef]
28. Channing, S. *Annual Book of ASTM Standards*; ASTM International: Conshohocken, PA, USA, 1991.
29. Müller, L.; Müller, F.A. Preparation of SBF with different HCO3-content and its influence on the composition of biomimetic apatites. *Acta Biomater.* **2006**, *2*, 181–189. [CrossRef]
30. Lu, X.; Leng, Y. Theoretical analysis of calcium phosphate precipitation in simulated body fluid. *Biomaterials* **2005**, *26*, 1097–1108. [CrossRef]
31. Pekguleryuz, M.; Kainer, K.; Kaya, A. *Fundamentals of Magnesium Alloy*; Woodhead Publishing Ltd.: Cambridge, UK, 2013.
32. Massalski, T.; Okamoto, H.; Subramanian, P.; Kacprzak, L. *Binary Alloy Phase Diagrams*, 2nd ed.; ASM: Materials Park, OH, USA, 1990.
33. Hosford, W.F. *Physical Metallurgy*, 2nd ed.; Taylor and Francis Group: Boca Raton, FL, UAS, 2010.
34. Lunder, O.; Nordien, J.; Nisancioglu, K. Corrosion Resistance of Cast Mg-Al Alloys. *Corros. Rev.* **1997**, *15*, 439–470. [CrossRef]

Article

Influence of Micro-Arc Oxidation Coatings on Stress Corrosion of AlMg6 Alloy

Lesław Kyzioł [1,*] and Aleksandr Komarov [2]

1 Faculty of Marine Engineering, Gdynia Maritime University, 81-225 Gdynia, Poland
2 Joint Institute of Mechanical Engineering of the National Academy of Sciences of Belarus, The State Scientific Institution, 220072 Minsk, Belarus; al_kom@tut.by
* Correspondence: leslawkyziol@gmail.com; Tel.: +48-694-476-390

Received: 16 December 2019; Accepted: 10 January 2020; Published: 12 January 2020

Abstract: This paper shows results of a study on the corrosion behavior of micro-arc oxidation (MAO) coatings sampled from the AlMg6 alloy. The alloy was simultaneously subjected to a corrosive environment and static tensile stress. For comparative purposes, the tests were run for both coated samples and samples without coatings. The research was conducted at a properly prepared stand; the samples were placed in a glass container filled with 3.5% NaCl aqueous solution and stretched. Two levels of tensile stress were accepted for the samples: $\sigma_1 = 0.8R_{0.2}$ $\sigma_2 = R_{0.2}$, and the tests were run for two time intervals: $t_1 = 480$ h and $t_2 = 1000$ h. Prolonged stress corrosion tests (lasting up to 1000 h) showed that the samples covered with ceramic coatings demonstrated significantly higher corrosion resistance than the samples without the coatings. Protective properties of the coating could be explained by its structure. Surface pores were insignificant, and their depth was very limited. The porosity level of the main coating layer was 1%. Such a structure of coating and its phase composition provided high protective properties.

Keywords: aluminum alloy AlMg6; Al_2O_3 coating; phase composition; corrosion; stress corrosion; micro-arc oxidation

---

1. Introduction

Al-Mg alloys combine good formability, rather high strength, corrosion resistance, and weldability. Therefore, such alloys are used in many structures exposed to weathering, and especially in shipbuilding and offshore structures. It should be noted that studies carried out on samples made of the AlMg6 alloy showed a good resistance to stress corrosion for this alloy, but one much lower than for alloy 5083 [1,2]. However, the strength of alloy 5083 containing 5% Mg is noticeably inferior to the strength of AlMg6 alloy.

Stress corrosion of materials manifests itself through the formation of cracks in the metal when exposed to corrosive environment and static tensile stress. The cracks appearing on the metal surface are perpendicular to the direction of tensile stress and can be of intercrystalline or mixed nature. Studies have shown that before the appearance of pronounced cracks, there is often an incubation period. The intensity of microcracks can be determined on the basis of changes in mechanical properties after time intervals of the stress corrosion test [1–6].

Studies have also shown that Al-Mg alloys with a content of Mg $\leq 3,5\%$ exhibit a low susceptibility to stress corrosion. This is due to the discontinuity of β-phase molecules at the grain boundaries, which in turn results from the low supersaturation of the solid solution [7–10].

Al-Mg alloys with a content of Mg $\geq 3,5\%$, especially above 5% (AlMg5, AlMg6), with a certain state of the structure and under specific external conditions may be resistant to intercrystalline, layer, and stress corrosion [3–5]. The study has shown that Al–Mg alloys with a content of more than 6 wt. % of Mg can be obtained, and they show good weldability and high mechanical properties [1,9–11]. An increase in the Mg content results in an increase in strength, but leads to higher susceptibility to local corrosion and reduced resistance to corrosion under stress [12]. It is assumed that the grain boundaries are a favorable place for β-phase formation due to their low diffusion barrier (presence of defects such as dislocations and vacancies) [13]. In many cases, overexposure of Al–Mg sheets or other products to a temperature range of 70 °C to 200 °C may impede production, causing precipitation of the β-phase rich in Mg at the grain boundaries. These alloys are susceptible to intercrystalline, stress, or pitting corrosion, because the β-phase is electrochemically more active than the aluminum matrix [14–18].

Structures with a high Mg content resistant to corrosion are obtained through the use of complex manufacturing methods [4]. The goal of the research was to determine the impact of the protective ceramic coating on the corrosion resistance of the tested alloy. Micro-arc oxidation (MAO) is a very promising process in making tight coatings and curing of metal elements. This is a coating technique capable of forming ceramic coatings on metals such as Al, Mg, Ti, and their alloys [17–22]. This environmental friendly technique allows ceramic layers to be grown, giving the pieces high level mechanical and tribological properties and good corrosion protection in a single step of processing [23–26]. A number of studies have shown that the coatings obtained by micro-arc oxidation are an effective means of protecting aluminum alloys from corrosion [27–33]. The corrosion properties of the MAO coating of the BS 6082 Al alloy were tested against different immersion periods in a 0.5M NaCl solution for up to 48 h [28]. The research has shown the importance of sealing the pores in MAO coatings with the use of the sol–gel technique, enhancing the short-term corrosion resistance in 0.6M NaCl solution. The above investigation also revealed that the MAO coatings improved the corrosion resistance of the Al alloy because of the lack of defects in coatings [29].

Despite numerous studies on the protective properties of coatings, their effectiveness in protecting aluminum–magnesium alloys from stress corrosion has not been scrutinized. The major objective of the present study is to evaluate the overall effectiveness of MAO coatings in terms of resistance to aqueous stress corrosion. Regarding the above objective, MAO coatings were deposited on a AlMg6 alloy and their corrosion behavior was evaluated in 3.5% NaCl solution. In the available literature, there are only a few sources referring to the importance of protective coatings for stress corrosion [34,35]. These tests are very important because often the material is subject simultaneously to corrosive environments and stresses. In this paper the stress corrosion resistance has been evaluated using a change of mechanical properties of samples with coatings and without coatings.

2. Materials and Methods

Samples of AlMg6 alloy without coating and coated with a ceramic coating by MAO were subjected to stress corrosion. The tests were carried out on identical samples without coatings and with coatings to determine the effectiveness of the ceramic coating that protects the material against stress corrosion.

Samples for the static tensile test and the stress corrosion were made from a sheet with a thickness of g = 3 mm. The samples were cut down in a direction transverse to the rolling direction. The chemical composition and parameters of heat treatment of the AlMg6 alloy plates are presented in Table 1. The shape and dimensions of the samples for the determination of mechanical and stress corrosion properties are shown in Figure 1.

Table 1. Chemical composition and heat treatment parameters of the AlMg6 alloy sheet.

Material	Sheet Thickness, mm	Parameters of Production Technology	Chemical Composition, %							
			Mg	Mn	Ti	Zn	Si	Fe	Cu	Al
AlMg6	6	annealing at temp. 319 °C/10 h	6.15	0.61	0.05	0.05	0.16	0.27	0.05	rest

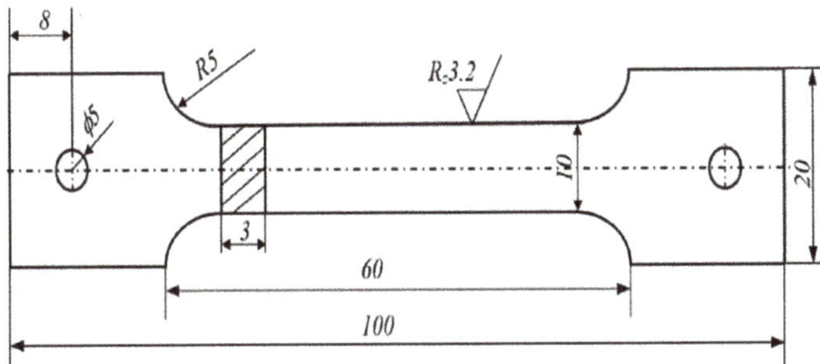

Figure 1. The shape and dimensions of samples for static and stress corrosion tests of the AlMg6 alloy.

The samples were polished with abrasive paper 600# and degreased with acetone followed by rinsing with distilled water before coating formation. The electrolyte was an aqueous solution of 2 g/L KOH and 5 g/L Na_2SiO_3. The electrolyte was agitated with compressed air. A forty litter bath from stainless steel was used. The specimen was served as the anode and the bath wall was served as the contrary electrode. The system was cooled by cold water pumped through double walls of the bath. The electrolyte temperature was controlled at 25–30 °C throughout the process. The MAO treatment was carried out using a pulsed AC power source. The current density, voltage, frequency, duty cycle, and duration time were 25A/dm^2, 280 V, 50 Hz, 50% and 60 min, respectively. After the treatment, the samples were rinsed in distilled water and dried in air.

Then, using abrasive paper 600#, the upper loose porous layer of about 20 µm thickness was removed from the surface of the samples, after which the samples were washed in distilled water with an ultrasonic bath SONOREX for 15 min and dried in air. The thickness of the coating on the samples prepared for testing was 150 ± 5 µm. The surface morphologies and cross-section morphologies of the coatings were examined with an optical microscope Axiovert 25 as well as with a scanning electron microscope (ZEISS, Jena, Germany). X–ray diffraction studies were carried out on a DRON–3M X–ray diffractometer (S. Petersburg, Russia) in scanning mode in 0.1 increments using Cu–Kα radiation in the Bragg–Brentano mode. The time of a set of pulses at a point was 15 s.

Stress corrosion tests of the AlMg6 alloy were carried out on samples for σ = const. The samples were placed in a glass container filled with 3.5% NaCl aqueous solution and stretched. Samples with and without coatings were subjected to stress corrosion tests for two levels of stress:

1—$\sigma_1 = 0.8R_{0.2}$, 2—$\sigma_2 = 1.0R_{0.2}$, where $R_{0.2}$ is the average value of the experimentally determined yield strength of samples without coating and with coating equal to 212 MPa and 217 MPa, respectively.

The corrosion tests were conducted for two time intervals: 1—480 h, 2—1000 h. Figure 2 presents a stand for testing stress corrosion resistance.

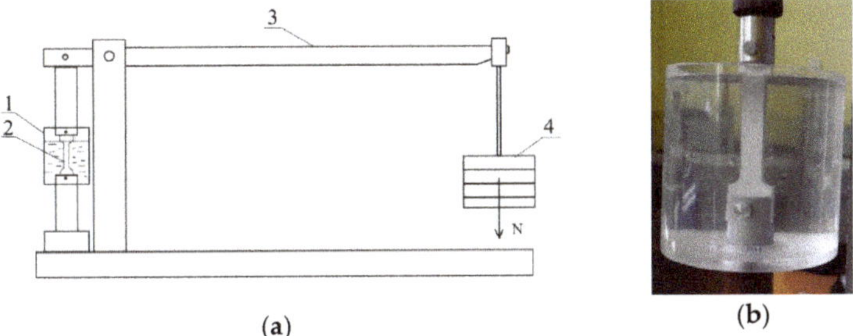

Figure 2. Stand for testing the stress corrosion resistance of AlMg6 alloy, (**a**) test bench scheme for σ = const., 1—container filled with 3.5% of NaCl, 2—sample, 3—lever arm, 4—load, (**b**) container filled with 3.5% of NaCl in which the sample is located.

After stress corrosion tests the samples were subjected to a static tensile test to determine the mechanical changes of the AlMg6 test material. A static tensile test on samples without coatings and with coatings was carried out on a ZwickRoell testing machine (Ulm, Germany). The research was carried out at the Faculty of Mechanical Engineering of the Gdańsk University of Technology (Gdańsk, Poland).

3. Research Results and Their Analysis

Figure 3 presents images of samples for mechanical and stress corrosion tests without coating and coated. The formed coating has a light gray color and a uniform surface. The study of the surface morphology of the samples showed that the surface layer of the coating is characterized by the presence of pores up to 10 microns in size (Figure 4). Such a structure is typical for coatings obtained with the use of the micro-arc oxidation method, in which there is an upper loose porous layer and a dense, low porosity basic coating layer [5,6].

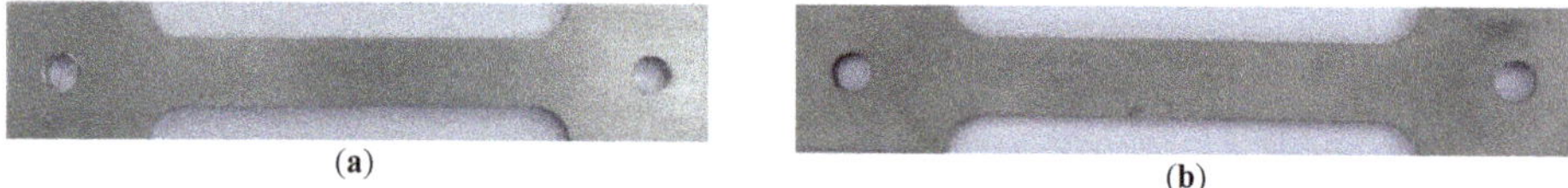

(**a**) (**b**)

Figure 3. Images of samples for stress corrosion tests of AlMg6 alloy with sheet thickness g = 6 mm, (**a**) without coating, (**b**) with coating.

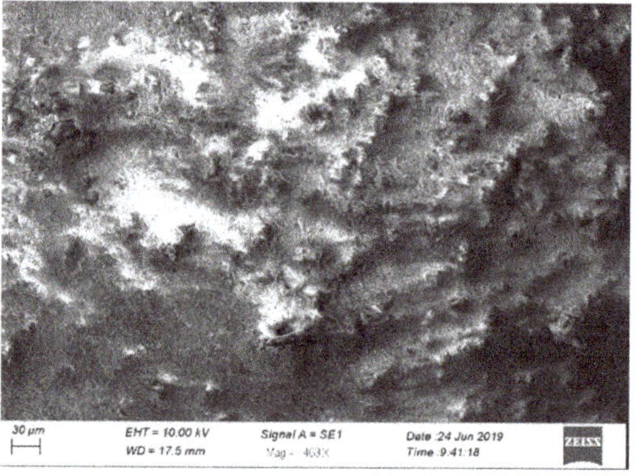

Figure 4. SEM image of the coating surface prior to testing.

Figure 5 presents images of samples without coating and coated after mechanical and stress corrosion tests. On samples without coatings, there are visible deposits and small corrosion centers (Figure 5a). There is no evidence of corrosion on samples with coating (Figure 5b). Studies performed on the cross-section of the coating after corrosion tests also showed no signs of corrosion (Figure 6).

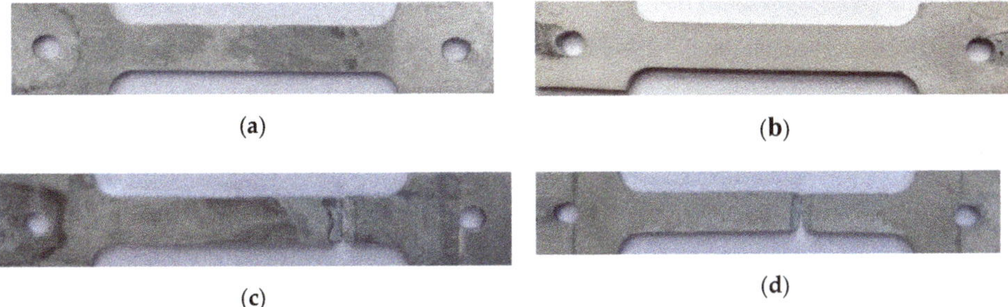

Figure 5. Images of samples after stress corrosion tests of AlMg6 alloy with sheet thickness g = 6 mm, (**a**) without coating, (**b**) with coating, (**c**) without cover after breaking, (**d**) with a coating after breaking.

High protective properties of the coating can be explained by its structure. As follows from the Figure 6, the pores on the surface of the coating have an insignificant size and depth and are limited to a loose surface layer. The main coating layer has a very low porosity (less than 1%), while the permeable porosity is absent. Such coating structures, as well as its phase composition, represented by chemically resistant aluminum oxide in the $\gamma-Al_2O_3$ and $\alpha-Al_2O_3$ modifications (Figure 7), provide high protective properties.

On the contrary, the surface of uncoated specimens of AlMg6 alloy was subjected to intense corrosion, as evidenced by the sources of material etching and corrosion products (Figure 5a,c).

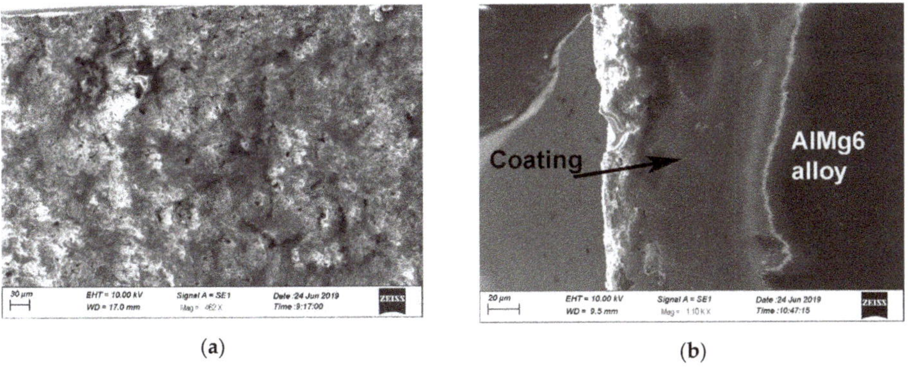

Figure 6. SEM image of the surface (**a**) and cross-section (**b**) of the sample with coating after stress corrosion tests.

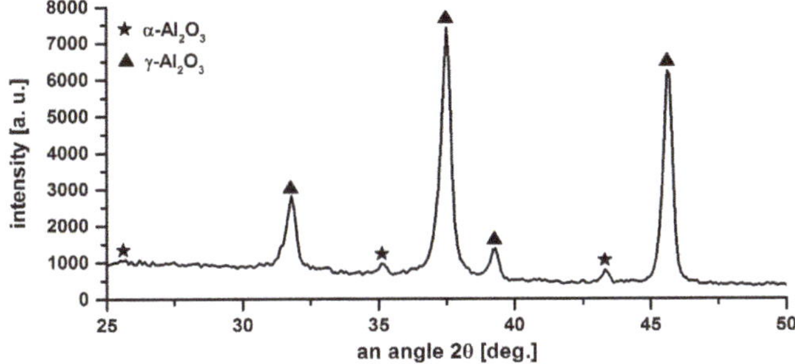

Figure 7. A fragment of the X−ray diffraction pattern of coating on a sample for the stress corrosion test.

Table 2 shows results of the test for static tensile stress of the AlMg6 alloy without coatings and with coatings. The obtained results indicate identical mechanical properties of the tested materials. A very thin layer of ceramics (about 150 μm), in relation to the thickness of the sample (g_s = 3 mm) has no influence on the mechanical properties of the tested materials.

Table 2. Mechanical properties of AlMg6 alloy samples without coatings and with coatings.

State of the Samples.	Yield Strength $R_{0.2}$	Tensile Strength R_m	Relative Elongation A_5
	MPa	MPa	%
without coatings	211	373	13.25
	215	372	13.45
	210	371	13.20
average	212	372	13.30
with coatings	215	369	13.65
	218	371	13.45
	218	370	13.40
average	217	370	13.50

The stress corrosion test is very rigorous because the material is simultaneously subject to a corrosive environment and stress. None of the 24 samples tested broke. The results of the tensile test of samples after exposure to stress corrosion for two time intervals and two stress levels are presented in Tables 3 and 4.

Table 3 shows results of the test for static tensile stress in samples with and without coatings at t = 480 h exposed to stress corrosion at the level of $\sigma = 0.8R_{0.2}$ and $\sigma = R_{0.2}$.

The results of the tests showed that for the level of stress $\sigma = 0.8R_{0.2}$:

- The samples without coating exhibited a decline of $R_{0.2}$ by 15%, R_m by 6%, and A_5 by over 20%;
- The samples with coating did not exhibit any deterioration of mechanical properties. The protective coating served its purpose well, protecting the material against corrosion.

For the level of stress $\sigma = R_{0.2}$:

- The samples without coating exhibited a decline of $R_{0.2}$ by 20%, R_m by 8%, and A_5 by over 36%;
- The samples with coating exhibited an insignificant decline of $R_{0.2}$ by 5%, no change, and A_5 by 7%.

For such a high level of stress the samples without coating exhibited a significant decline in mechanical properties caused by simultaneous exposure to stress and a corrosive environment.

In the case of the coated samples, there was an insignificant decline in mechanical properties caused by the high level of stress. The coating very tightly protected the sample surface, and thus the material of the sample was not subjected to corrosive environment.

Table 3. Mechanical properties of samples with and without coatings of AlMg6 alloy after exposure to stress corrosion at time t = 480 h.

State of the Samples	Stress Level	$R_{0.2}$	R_m	A_5
		MPa	MPa	%
Without coatings	$\sigma = 0.8R_{0.2}$	185	352	10.25
		182	351	10.32
		178	347	10.33
Average		180	350	10.30
With coatings		212	365	12.85
		209	374	13.18
		209	371	12.97
Average		210	370	13.00
Without coatings	$\sigma = 1.0R_{0.2}$	167	338	8.45
		171	343	8.65
		172	339	8.40
Average		170	340	8.50
With coatings		208	368	12.40
		210	361	12.65
		206	366	12.45
Average		208	365	12.50

Table 4 shows the results of the samples exposed to stress corrosion at similar levels of stress, i.e., $\sigma = 0.8R_{0.2}$ and $\sigma = R_{0.2}$ after the time of t = 1000 h.

For the level of stress $\sigma = 0.8R_{0.2}$:

- the samples without coating exhibited a decline of $R_{0.2}$ by 25%, R_m by 18%, and A_5 by over 50%;
- the samples with coatings exhibited a decline of $R_{0.2}$ by 3%, R_m by 6%, and A_5 by 14%.

With the amount of stress corrosion time, a further decline in the mechanical properties of uncoated AlMg6 alloy was observed.

The coated samples showed an insignificant decline in mechanical properties. The decline was not caused by the corrosive environment, but rather by crawling of the material, which was subjected to a high level of stress.

For the level of stress $\sigma = R_{0.2}$:

- The samples without coating exhibited a decline of $R_{0.2}$ by over 30%, R_m by 23%, and A_5 by nearly 60%;
- The samples with coatings exhibited a decline of $R_{0.2}$ by 4%, R_m by 8%, and A_5 by 18%.

For such a high level of stress and a long exposure to a corrosive environment, a further decline in mechanical properties was observed in the samples with no protective coatings. The decline occurred due to a simultaneous concurrence of high stress and corrosion caused by the 3.5% NaCl solution.

Having been exposed to a corrosive environment for t = 1000 h and highly stressed for up to $\sigma = R_{0.2}$, the coated samples sustained an insignificant decrease in mechanical properties. The limited decline is caused by the high level of stress.

Table 4. Mechanical properties of samples with and without coatings of AlMg6 alloy after exposure to stress corrosion at time t = 1000 h.

State of the Samples	Stress Level	$R_{0.2}$ MPa	R_m MPa	A_5 %
Without coatings	$\sigma = 0.8R_{0.2}$	155	303	6.48
		162	307	6.54
		163	305	6.48
Average		160	305	6.50
With coatings		213	354	11.62
		209	342	11.48
		208	348	11.40
Average		210	348	11.50
Without coatings	$\sigma = 1.0R_{0.2}$	140	288	5.48
		148	280	5.53
		147	287	5.55
average		145	285	5.52
With coatings		208	341	10.98
		205	336	11.12
		202	341	10.90
Average		205	340	11.00

Figures 8–10 shows the influence of corrosive environment and stress level on the reduction of mechanical properties of AlMg6 alloy samples with and without a ceramic coating. The research shows that there was almost a 30% reduction in yield strength and tensile strength, and nearly a 60% reduction in the plasticity of AlMg6 alloy samples without a ceramic coating for stress level $\sigma = R_{0.2}$ and corrosion exposure time t = 1000 h. For identical corrosion conditions and stress levels, there was a 4% decrease in yield strength and tensile strength as well as a 8% decrease in plasticity of AlMg6 alloy samples coated with ceramic coating.

The samples of AlMg6 alloy without coatings showed a concurrence of static tensile stress and corrosive environment. This kind of corrosion–stress synergy is the very reason for the accelerated degradation of the material. However, the samples with ceramic coatings showed a high resistance to simultaneous corrosion and stress factors as the coatings protected the material against corrosion. An insignificant decrease in mechanical properties occurred due to the high level of stress. Figures 8–10 shows a very narrow area of the decline in mechanical samples of coated AlMg6 alloys.

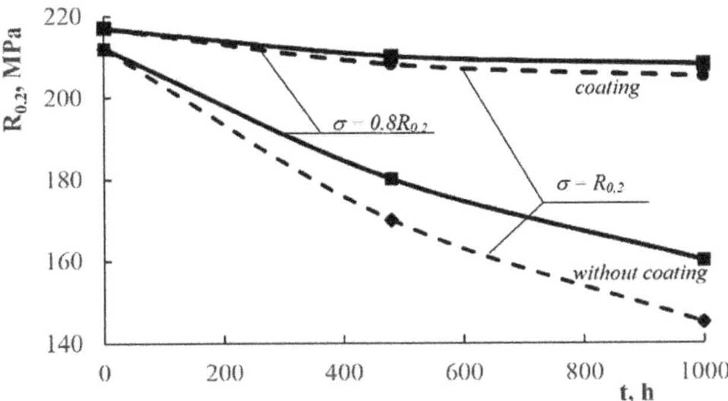

Figure 8. Influence of the corrosive environment and stress level on the yield strength reduction of AlMg6 alloy samples without coatings and with coatings.

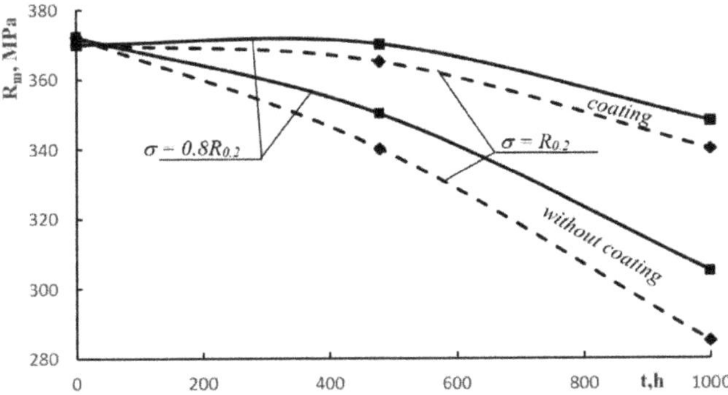

Figure 9. Influence of the corrosive environment and stress level on the tensile strength reduction of AlMg6 alloy samples without coatings and with coatings.

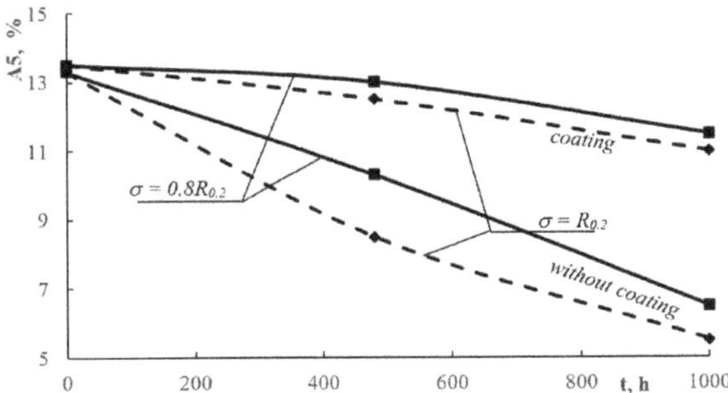

Figure 10. Influence of the corrosive environment and stress level on the relative elongation reduction of AlMg6 alloy samples without coatings and with coatings.

4. Summary

Studies have been carried out to evaluate the protective properties of the coating obtained by micro-arc oxidation on aluminum alloy AlMg6 from stress corrosion. The protective properties of the coating were evaluated by changing the mechanical characteristics (yield strength $R_{0.2}$, tensile strength R_m and relative elongation A_5) after holding the samples without coating and coated in a 3.5% NaCl solution under a stress of $\sigma_1 = 0.8R_{0.2}$ and $\sigma_2 = R_{0.2}$ for 480 h and 1000 h. The results have showed that the MAO coating provides a fairly effective protection of the aluminum alloy from stress corrosion if the selected test conditions are met. No corrosive centers were observed on samples with a ceramic coating, but small corrosion centers may be observed on samples without a coating. The effect of corrosion led to a decrease in the mechanical characteristics of the uncoated samples. So, after the 480-hour experiment at both stress levels, there was no noticeable decrease in the mechanical characteristics of coated samples, while the mechanical characteristics of uncoated samples decreased 1.1 (R_m) to 1.5 (A_5) times. After holding the samples for 1000 h, a more noticeable difference in mechanical characteristics was observed. In this case, the decrease in the characteristics of coated samples did not exceed 1.07 times (R_m) to 1.12 times (A_5), while the decrease in the properties of uncoated samples reached 1.3 (R_m) to 2.4 times (A_5). The high protective properties of the coating can be explained by the chemical inertness of its composition ($\gamma-Al_2O_3$ and $\alpha-Al_2O_3$) and low porosity (less than 1%) in the absence of through porosity.

Author Contributions: Conceptualization, L.K. and A.K.; methodology, L.K. and A.K.; formal analysis, L.K. and A.K.; investigation, L.K. and A.K.; resources, L.K and A.K.; data curation, L.K. and A.K.; writing—original draft preparation, L.K. and A.K.; writing—review and editing, L.K. and A.K.; visualization, L.K and A.K.; supervision, L.K. and A.K.; funding acquisition, L.K. All authors have read and agreed to the published version of the manuscript.

Funding: This research received no external funding.

Conflicts of Interest: The authors declare no conflict of interest.

References

1. Kyzioł, L. Wpływ obróbki cieplnej na odporność na korozje naprężeniową i wytrzymałość zmęczeniowo–korozyjną stopów AlMg6 i AlZn5Mg2CrZr przeznaczonych na kadłuby okrętów. Ph.D. Thesis, Akademia Marynarki Wojennej, Gdynia, Poland, August 1990.
2. Kyzioł, L.; Zganiacz, F. Wstępne badania korozji naprężeniowej stopów AlMg6 i AlMg4.5Mn. *Zeszyty Naukowe AMW, Gdynia* **1988**, *2*, 85–104.
3. Cudny, K.; Jasiński, R.; Kyzioł, L. Badanie stopów AlMg6 i AlZn5Mg2CrZr w celu ich zastosowania w budownictwie okrętowym. *Budownictwo Okrętowe i Gospodarka Morska, Gdańsk* **1991**, *9*, 10–14.
4. Sinavskij, V.S.; Volov, V.D.; Kalinin, V.D. *Korrozja i zaŝĉita aluminievych splavov*; Izdatielstvo Metallurgija: Moskva, Russia, 1986.
5. Jones, R.H.; Vetrano, J.S.; Windisch, C., Jr. Stress corrosion cracking of Al-Mg and Mg-Al alloys. *Corrosion* **2004**, *60*, 1144–1154. [CrossRef]
6. Torkan, A.; Rabiei, B.A.; Khakpour, I. Corrosion Behavior of AA5038 Nanostructured Aluminum Alloy Produced by Accumulative Roll-Bonding. *Nanosci. Nanometr.* **2018**, *4*, 34–40. [CrossRef]
7. Gao, J.; David, J.Q. Enhancement of the Stress Corrosion Sensitivity of AA5083 by Heat Treatment. *Metall. Mater. Trans. A* **2011**, *42A*, 356–364. [CrossRef]
8. Goswami, G.; Spanos, P.S.; Paoa, R.; Holtz, L. Precipitation behavior of the phase in Al-5083. *Mater. Sci. Eng.* **2010**, *527*, 1089–1095. [CrossRef]
9. Choi, D.-H.; Ahn, B.-W.; Quesnel, D.J.; Seung, B.J. Behavior of β-phase (Al_3Mg_2) in AA 5083 during friction stir welding. *Intermetallics* **2013**, *35*, 120–127. [CrossRef]
10. Radović, L.; Bučko, M.; Miladinov, M. Corrosion Behavior of TIG Welded AlMg6Mn Alloy. *Sci. Tech. Rev.* **2016**, *66*, 10–17. [CrossRef]
11. Romhanji, E.; Popović, M.; Radmilović, V. Room temperature deformation behaviour of AlMg6.5 alloy sheet. *Z. Metallkunde* **1999**, *90*, 305–310.
12. Timoshenko, Y.B. On the relation between the Luders deformation and grain boundary structure in aluminium alloy. *Revue Phys. Appl.* **1990**, *25*, 1001–1004. [CrossRef]
13. Radović, L.J.; Nikačević, M.; Jordović, B. Deformation behaviour and microstructure evolution of AlMg6Mn alloy during shear spinning. *Trans. Nonferrous Met. Soc. China* **2012**, *22*, 991–1000. [CrossRef]
14. Radović, L.J.; Nikačević, M. Microstructure and properties of cold rolled and annealed Al-Mg alloys. *Sci. Tech. Rev.* **2008**, *58*, 14–20.
15. Zhu, Y. Characterization of Beta Phase Growth and Experimental Validation of Long Term Thermal Exposure Sensitization of AA5xxx Alloys. Master's Thesis, The University of Utah, Salt Lake City, UT, USA, 2013.

16. Jones, R.H.; Baer, D.R.; Danielson, M.J.; Vetrano, J.S. Role of Mg in the Stress Corrosion Cracking of an Al-Mg Alloy. *Metall. Mater. Trans. A* **2001**, *32A*, 1699–1711. [CrossRef]
17. Yerokhin, A.L.; Voevodin, A.A.; Lyubimov, V.V.; Zabinski, J.; Donley, M. Plasma Electrolytic Fabrication of Oxide Ceramic Surface Layers for Tribotechnical Purposes on Aluminium Alloys. *Surf. Coat. Technol.* **1998**, *110*, 140–146. [CrossRef]
18. Vityaz', P.A.; Komarov, A.I.; Komarova, V.I. Triboengineering properties of ceramic oxide coatings under boundary friction against steel. *J Frict. Wear.* **2008**, *29*, 325–329. [CrossRef]
19. Nie, X.; Leyland, A.; Song, H.W.; Yerokhin, A.L.; Dowey, S.J.; Matthews, A. Thickness Effects on the Mechanical Properties of Micro-arc Discharge Oxide Coatings on Aluminium alloys. *Surf. Coat. Technol.* **1999**, *116–119*, 1055–1060. [CrossRef]
20. Gnedenkov, S.V.; Khrisanfova, O.A.; Zavidnaya, A.G.; Sinebrukhov, S.L.; Kovryanov, A.N.; Scorobogatova, T.M.; Gordienko, P.S. Production of Hard and Heat-Resistant Coatings on Aluminium Using a Plasma Micro-discharge. *Surf. Coat. Technol.* **2000**, *123*, 24–28. [CrossRef]
21. Wang, Y.K.; Sheng, L.; Xiong, R.Z.; Li, B.S. Study of Ceramic Coatings Formed by Microarc Oxidation on Al Matrix Composite Surface. *Surf. Eng.* **1999**, *15*, 112–114. [CrossRef]
22. Butyagin, P.I.; Khokhryakov, Y.V.; Mamaev, A.I. Microplasma Systems for Creating Coatings on Aluminium Alloys. *Mater. Lett.* **2003**, *57*, 1748–1751. [CrossRef]
23. Zhang, Y.; Yan, C.; Wang, F.; Lou, H.; Cao, C. Study on the Environmentally Friendly Anodizing of AZ91D Magnesium Alloy. *Surf. Coat. Technol.* **2002**, *161*, 36–43. [CrossRef]
24. Rama Krishna, L.; Somaraju, K.R.C.; Sundararajan, G. The Tribological Performance of Ultra-hard Ceramic Composite Coatings Obtained Through Microarc Oxidation. *Surf. Coat. Technol.* **2003**, *163–164*, 484–490. [CrossRef]
25. Voevodin, A.A.; Yerokhin, A.L.; Lyubimov, V.V.; Donley, M.S.; Zabinski, J.S. Characterization of Wear Protective Al-Si-O Coatings Formed on Al-based Alloys by Micro-arc Discharge Treatment. *Surf. Coat. Technol.* **1996**, *86–87*, 516–521. [CrossRef]
26. Tian, J.; Luo, Z.; Qi, S.; Sun, X. Structure and Antiwear Behavior of Micro-arc Oxidized Coatings on Aluminum Alloy. *Surf. Coat. Technol.* **2002**, *154*, 1–7. [CrossRef]
27. Kuhn, A.T. Plasma anodized aluminum-A 2000/2000 Ceramic Coating. *Met. Fnish.* **2002**, *100*, 44. [CrossRef]
28. Nie, X.; Meletis, E.I.; Jiang, J.C.; Leyland, A.; Yerokhin, A.L.; Matthews, A. Abrasive Wear/Corrosion Properties and TEM Analysis of Al_2O_3 Coatings Fabricated Using Plasma Electrolysis. *Surf. Coat. Technol.* **2002**, *149*, 245–251. [CrossRef]
29. Barik, R.C.; Wharton, J.A.; Wood, R.J.K.; Stokes, K.R.; Jones, R.L. Corrosion, Erosion and Erosion-Corrosion Performance of Plasma Electrolytic Oxidation (PEO) Deposited Al_2O_3 Coatings. *Surf. Coat. Technol.* **2005**, *199*, 158–167. [CrossRef]
30. Nitin Wasekar, P.; Jyothirmayi, A.; Rama Krishna, L.; Sundararajan, G. Effect of Micro Arc Oxidation Coatings on Corrosion Resistance of 6061-Al Alloy 2008. *J. Mater. Eng. Perform.* **2008**, *17*, 708–713. [CrossRef]
31. Sundararajan, G.; Rama Krishna, L. Mechanisms Underlying the Formation of Thick Alumina Coatings Through the MAO Coating Technology. *Surf. Coat. Technol.* **2003**, *167*, 269–277. [CrossRef]
32. Guangliang, Y.; Xianyi, L.; Yizhen, B.; Haifeng, C.; Zengsun, J. The Effects of Current Density on the Phase Composition and Microstructure Properties of Micro-arc Oxidation Coating. *J. Alloy Compd.* **2002**, *345*, 196–200. [CrossRef]
33. Sobolev, A.; Kossenko, A.; Zinigrad, M.; Borodianskiy, K. Comparison of plasma electrolytic oxidation coatings on Al alloy created in aqueous solution and molten salt electrolytes. *Surf. Coat. Technol.* **2018**, *344*, 590–595. [CrossRef]
34. Hua, S.; Song, R.G.; Zong, Y.; Cai, S.W.; Wang, C. Effect of solution pH on stress corrosion and electrochemical behavior of aluminum alloy with micro-arc oxidation coating. *Mater. Res. Express* **2019**, *6*, 096441. [CrossRef]
35. Lianxi, C.; Yinying, S.; Hanyu, Z.; Zhibin, L.; Xiaojian, W.; Wei, L. Influence of a MAO + PLGA coating on biocorrosion and stress corrosion cracking behavior of a magnesium alloy in a physiological environment. *Corros. Sci.* **2019**, *148*, 134–143.

Article

Enhanced Erosion–Corrosion Resistance of Tungsten by Carburizing Using Spark Plasma Sintering Technique

Yan Jiang [1], Junfeng Yang [2,*], Zhuoming Xie [2] and Qianfeng Fang [2,*]

[1] International Institute of Vanadium and Titanium, Panzhihua University, Panzhihua 617000, China; jiangyanzky@163.com

[2] Key Laboratory of Materials Physics, Institute of Solid State Physics, Chinese Academy of Sciences, Hefei 230031, China; zmxie@issp.ac.cn

* Correspondence: jfyang@issp.ac.cn (J.Y.); qffang@issp.ac.cn (Q.F.)

Received: 11 May 2020; Accepted: 12 June 2020; Published: 15 June 2020

Abstract: The biggest obstacle for the application of tungsten as the target materials in the spallation neutron source is its serious corrosion in the coolant of flowing water. For this reason, W–Cr–C clad tungsten was developed by tungsten carburizing in a spark plasma sintering device, with superior corrosion resistance in the static immersion and electrochemical corrosion test. This work focused on its erosion and corrosion performance in a flowing water system, based upon test parameters simulated under the service conditions. W–Cr–C clad tungsten showed superior corrosion resistance to that of bare tungsten due to the corrosion form changing from the intergranular corrosion of bare tungsten to pitting corrosion of W–Cr–C coating. The corrosion rate of tungsten was as high as tenfold that of the coated sample at 20 °C, and at most fourfold at 60 °C after testing for 360 h. Effects of water velocity and temperature on pitting and intergranular corrosion were investigated in detail and their corresponding corrosion mechanisms were analyzed and discussed.

Keywords: tungsten; W–Cr–C coating; carburization; intergranular corrosion; pitting corrosion

---

1. Introduction

Tungsten is being chosen as the most favorable spallation target material for the spallation neutron source (SNS) due to its obvious advantages including high melting point, high neutron yield, and superior thermal conductivity [1,2]. However, on the issue of compatibility with flowing water coolant, tungsten has long been challenged because of its serious corrosion in water, especially in the case of flowing water where both erosion and corrosion occur simultaneously [3–6].

To improve the corrosion resistance, tungsten cladding with high temperature and corrosion resistance coatings have been developed as a target material such as HIPed tantalum-clad tungsten [7,8]. However, it suffers from issues of interface defects [7] and the high decay heat of tantalum [8,9]. Alternatively, coatings such as tungsten carbide and chromium carbide have been generally adopted to protect the substrate from corrosion under harsh environments in engineering and industrial fields [10–15]. In the conventional preparation of cemented carbides, cobalt as binder phase is commonly introduced to densify the particles, which is known as a poorly resistant element to corrosion [16]. However, the development of binder-free tungsten carbide coatings needed to solve two problems: density and brittleness. To this end, carburizing by spark plasma sintering (SPS) technology was employed to produce tungsten carbide coatings on tungsten [17]. SPS is a high-temperature and fast-process sintering technique that provides a fast heating/cooling rate, short consolidation time, and controllable pressure, having been widely used in densification fabrication of nanomaterials, gradient functional materials, and ceramic materials. By SPS, the carburized layer over 20 μm thick was quickly achieved at 1600 °C, holding for 10 min with a pressure of 45 MPa [17], and a double thick layer was available when 20 min and 30 min were tried. On this basis, W–xCr–C (x = 0.5, 1, 2, 3, and 6 wt.%) composite layers with higher compactness and uniform grain size distribution were fabricated in the same way [18]. Electrochemical corrosion measurements on a WC clad tungsten system and W–Cr–C clad tungsten system indicated that the W–1%Cr–C clad tungsten sample performed well and exhibited the lowest corrosion current density [19].

As a result of these findings, W–1%Cr–C (W–Cr–C for short in the following text) clad tungsten by the SPS method was chosen as the potential material to undergo the following dynamic corrosion test in flowing water; flowing water (pH: 6~8) was adopted as the coolant in service condition to remove the heat in the target chamber. According to the simulation for the 100 kW neutron spallation source, a flow velocity of about 2 ms^{-1} is enough for the heat to be removed from the target, under which the highest target temperature is around 60 °C [2,20]. Therefore, this work investigated the corrosion performance of W–Cr–C clad tungsten at such conditions with the main purpose to explore and understand the behavior and mechanism in the erosion–corrosion process. Bare tungsten was taken into comparison.

2. Materials and Methods

A commercial grade tungsten disc (density: 18.48 g/cm^3, Zhuzhou Cemented Carbide Group Co. Ltd., Zhuzhou, China; the main elements included are listed in Table 1) with cylindrical form (Ø16 mm × 3 mm) was embedded in the mixed powder of graphite (99.95%, 8000 mesh, Aladdin) and 1 wt.% chromium (99.5%, ≥325 mesh, Aladdin, Shanghai, China), and then placed in a graphite mold and sintered in a SPS furnace (HPD 5, FCT Systeme GmbH, Rauenstein, Germany), as shown in Figure 1a. A target temperature of 1600 °C holding for 10 min with a pressure of 45 MPa was exploited on the graphite die (Ø20 mm). The diameter of the die could change from 10 mm to 100 mm according to the sample dimensions. Pressure was used to bring the powder and disc into full contact. The heating and cooling rates of SPS were both 100 °C/min. After sintering, the disc was taken out and cleaned ultrasonically in acetone. A scanning electron microscopy (SEM)(Sirion 200, FEI, Hillsboro, OR, USA) and electron backscatter diffraction (EBSD) detector (Aztec Nordlys Max3, Oxford Instruments, Oxford, UK, incorporated in the SEM device) was applied to characterize the surface morphology, phase composition, and distribution. The samples for the EBSD measurements were mechanically polished with W0.5 diamond paste and then electropolished in 2% NaOH solution. Phase identification was derived from the crystal structure database (HKL, ICSD, and NIST) installed on the EBSD system.

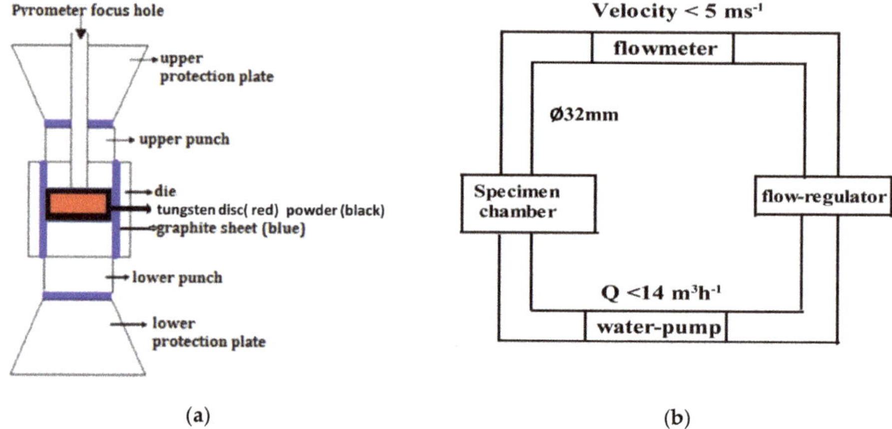

Figure 1. The schematic diagram of the sample placement in the spark plasma sintering setup (**a**) and the self-designed water flowing device (**b**).

The erosion–corrosion experiments were conducted in a home-made pipe flow circulating system as illustrated in Figure 1b. Diameter of all water passages was 32 mm. The water-pump was placed into an improved thermostatic water-container (0.6 × 0.6 × 1 m) where the water circulated. PH of the water was about 6~7. The oxygen content in the passages can be regarded as a saturation value since the water-container is a semi-open vessel. The tested samples were side by side fixed on the inner upper wall of the specimen chamber, leaving only one exposed surface (Ø16 mm) parallel to the flow direction.

Table 2 lists the test parameters with reference to the simulated service conditions. The flowing velocity and temperature was controlled by the flow regulator and the thermostatic bath, respectively. The error is also marked out at Table 2. The temperature value 20 °C was the entrance temperature of the flowing water. At the relative low temperature of 20 °C, a test at the largest velocity of 2 ms^{-1} was conducted, in light of the small difference in the corrosion behavior between 1 and 2 ms^{-1}. After the experiments, the corrosion performance was evaluated by analyzing the corrosion rate and the corroded morphology.

The weight was recorded every 12 h in the erosion–corrosion test, using a balance with an accuracy of 0.01 mg. The average corrosion rate v (g/cm^2·h) was calculated from the following equation:

$$v = (m_0 - m_t)/s \cdot t \tag{1}$$

where m_0 is the original mass of specimen, m_t the mass of specimen after corrosion for t hours, s (cm^2) the exposure area before corrosion and t (h) the corrosion time. As the change in s after corrosion is very small, it is considered as a constant. It is worth pointing out in the gravimetric measurements that it was not easy to accurately clean up the corrosion deposits of the corroded specimens without damaging the matrix. Therefore, after the tests, the specimens were directly dried in a vacuum oven at 100 °C for 12 h without any surface treatment, and then their mass m_t was measured. If $v > 0$, it means a mass loss and removal of the corrosion products from the surface. According to the variation of the v value, one can judge what happened to each specimen in the corrosion process. The corroded surface morphology was studied by means of SEM observations.

Table 1. The impurities with concentration over 0.001 wt.% in tungsten bulk.

Mo	Fe	Al	Si	P	C	N	O	W
0.009	0.001	0.001	0.001	0.001	0.001	0.001	0.0014	Bal

Table 2. The parameters set in the flowing water test.

Sample	Temperature (±3 °C)	Velocity (±0.1 ms^{-1})	Characterization
Tungsten and W–Cr–C coated tungsten	20 60	2 1, 1.5, 2	Corrosion rate and Morphology

3. Results

3.1. Morphology and Phases

Before corrosion measurement, the morphology, compositions, and phase distribution of the coated sample were characterized by the EBSD system. The fractured surface of the coatings showed good bonding with the substrate and high compact structures with a uniform thickness of around 25 μm (Figure 2a). The coating consisted of external W–Cr–C (hexagonal WC (ICSD [15406]) and $WC_{0.98}$ (ICSD [77738]) plus dispersed Cr_7C_3 (ICSD [52289]) (Figure 2b,c) and intermediate hexagonal W_2C (ICSD [77567]) (Figure 2e). Close to the carbon and chromium-rich zone, tungsten carbide reached stoichiometry WC (nonstoichiometric $WC_{0.98}$ is also commonly designated as WC) with an average grain size of 800 nm (Figure 2c,d), and chromium carbide is indexed as Cr_7C_3. Along the diffusion path, the composition becomes the carbon-depleted W_2C and nearly no chromium carbides are detected in the intermediate layer due to the preferential diffusion of carbon. The W_2C phase accounts for more than 95 vol.% of the coating due to its much lower Gibbs free energy than WC at high temperature [21]. Furthermore, a small fraction of $WC_{0.98}$ precipitates were found in the W_2C grains (Figure 2e), which were derived from the partial solid-state decomposition of W_2C below 1250 °C [22]. The vast majority of the W_2C phase remained stable due to the fast cooling rate (100 °C /min). Figure 2f shows the pole figures and inverse pole figures with respect to the growth direction of the W_2C layer. Interestingly, the W_2C layer presented obvious crystallographic preferred orientations: the <001>//Y axis (the sintering pressure direction). Strong fiber texture <001> suggests the formation of a columnar structure of W_2C coating via spark plasma sintering.

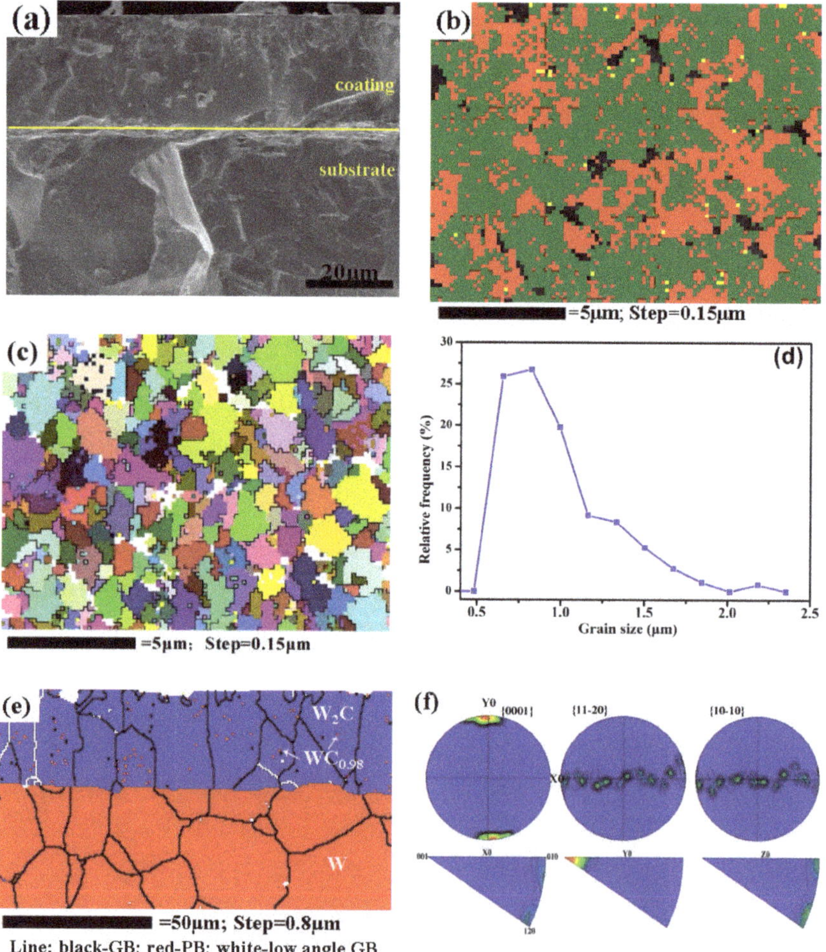

Figure 2. (**a**) Scanning electron microscopy (SEM) image of the cross-section of coating; (**b**) phase distribution of the coating surface (an indexing rate of about 91%, the black refers to the unindexed sites, the red to the WC phase, the green to $WC_{0.98}$, the yellow to Cr_7C_3); (**c**) Euler image and (**d**) grain size distribution of the WC and $WC_{0.98}$ phases; (**e**) phase distribution of the inner coating (the blue refers to the W_2C phase, the red to the W substrate, the pink dots are the $WC_{0.98}$ phases; GB: grain boundary; PB: phase boundary); (**f**) pore figure and inverse pore figure of the inner coating.

Such a gradient structure from base to bottom by carburizing through SPS technology could release thermal stress and avoid cracks in the coating. Other technology like HVOF (high velocity oxygen fuel) spray can also be applied in creating nano WC-based coatings made of powders [23]. The spray process acts like the high temperature sintering of the raw powders, and micro-cracks are caused by thermal stress in the spray process. More importantly, like other conventional preparations of a WC coating, the bonding phase of cobalt is commonly introduced to densify the WC, which is known as in-resistant element to corrosion.

3.2. Corrosion Rate

Under a water temperature of 20 °C, the corrosion rate as a function of time is shown in Figure 3. In general, the corrosion rate for tungsten clad with W–Cr–C decreased with increasing time and gradually went to a steady state after many fluctuations. The fluctuations imply that the accumulation and breakaway of the corrosion products occurred alternatively, namely coexistence of chemical and mechanically driven corrosion. The remarkable maximum corrosion rate belongs to bare tungsten, which was located uppermost over the entire time range and finally reached as high as tenfold that of the coated sample, which was relatively stable around the zero line.

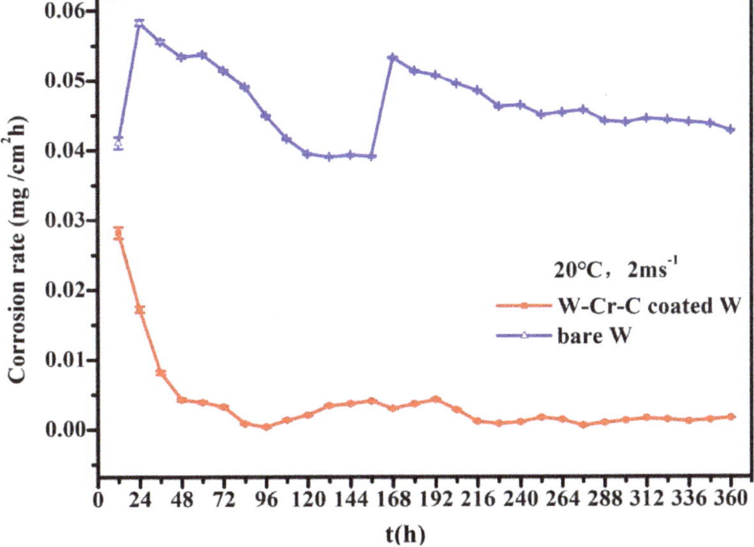

Figure 3. Variation of the corrosion rate with time for samples at the velocity of 2 ms^{-1} under the water temperature of 20 °C.

The corrosion rates under the water temperature of 60 °C are shown in Figure 4. The change in corrosion rate was positively related to the flow rate. When the flow velocity was 1 ms^{-1} and 1.5 ms^{-1}, the corrosion rate of the W–Cr–C coated W sample was less than 0.01 $mg{\cdot}cm^{-2}{\cdot}h^{-1}$. The corrosion rate of bare tungsten sample almost doubled and its fluctuation with time was larger. When the flow velocity increased to 2 ms^{-1}, the corrosion rates of all samples increased to higher values, varying in the range of 0.25–0.05 $mg{\cdot}cm^{-2}{\cdot}h^{-1}$ during the experiment time of 360 h.

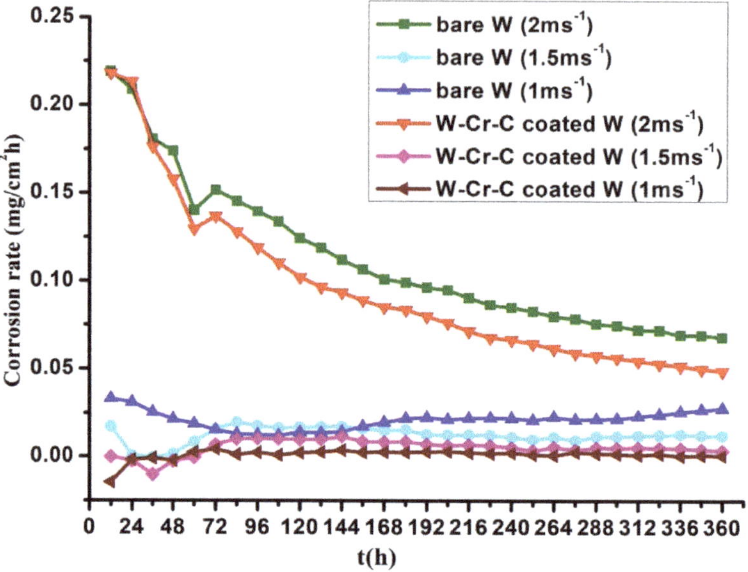

Figure 4. Variation of corrosion rate with time for samples at various flow velocities under the water temperature of 60 °C.

3.3. Corrosion Mechanism

The corroded appearances of tungsten clad with W–Cr–C are shown in Figure 5. The initial surface of the W–Cr–C coating (Figure 5a) presented deficiencies of small scattered pores left by SPS fabrication. After the 360 h test, the morphology changed slightly, but increased pores with increasing velocity or temperature, displaying a dominant feature of pitting corrosion (Figure 5b–e). Depth and the number of pores are difficult to quantify, so Figure 5f counts the area fraction of pores using image software (based upon the difference in primary contrast of images) by collecting six SEM graphs with the same magnification to describe the changes of pitting. Under the water temperature of 60 °C, the area fraction of the pores increased from the initial 0.4 ± 0.1% up to 4.5 ± 1% when the flow velocity varied from 0 to 2 ms^{-1}. However, there was no single linear or parabolic law between the pore area fraction and flow velocity for a remarkable rise at the flow velocity of 2 ms^{-1}. When the water temperature dropped to 20 °C, the reduction in pore area fraction reached 60%, almost close to the rate of temperature decline. Pitting area dependent upon velocity and temperature agreed well with the corrosion rate curves.

The corrosion behavior of bare tungsten was totally different, as shown in Figure 6. A rough metallographic structure with bimodal grain-size distribution appeared as a consequence of remarkable intergranular corrosion (Figure 6b–e) as well as slight dissolution of the tungsten grains (Figure 6f). The intergranular corrosion at high flow velocity (2 ms^{-1}) was serious, accompanied by the formation of deep groove and micro corrosion pits caused by grain exfoliation, indicating that a mechanical effect would promote intergranular corrosion, weakening the GBs cohesion and loosening the GBs, especially at the triple junction.

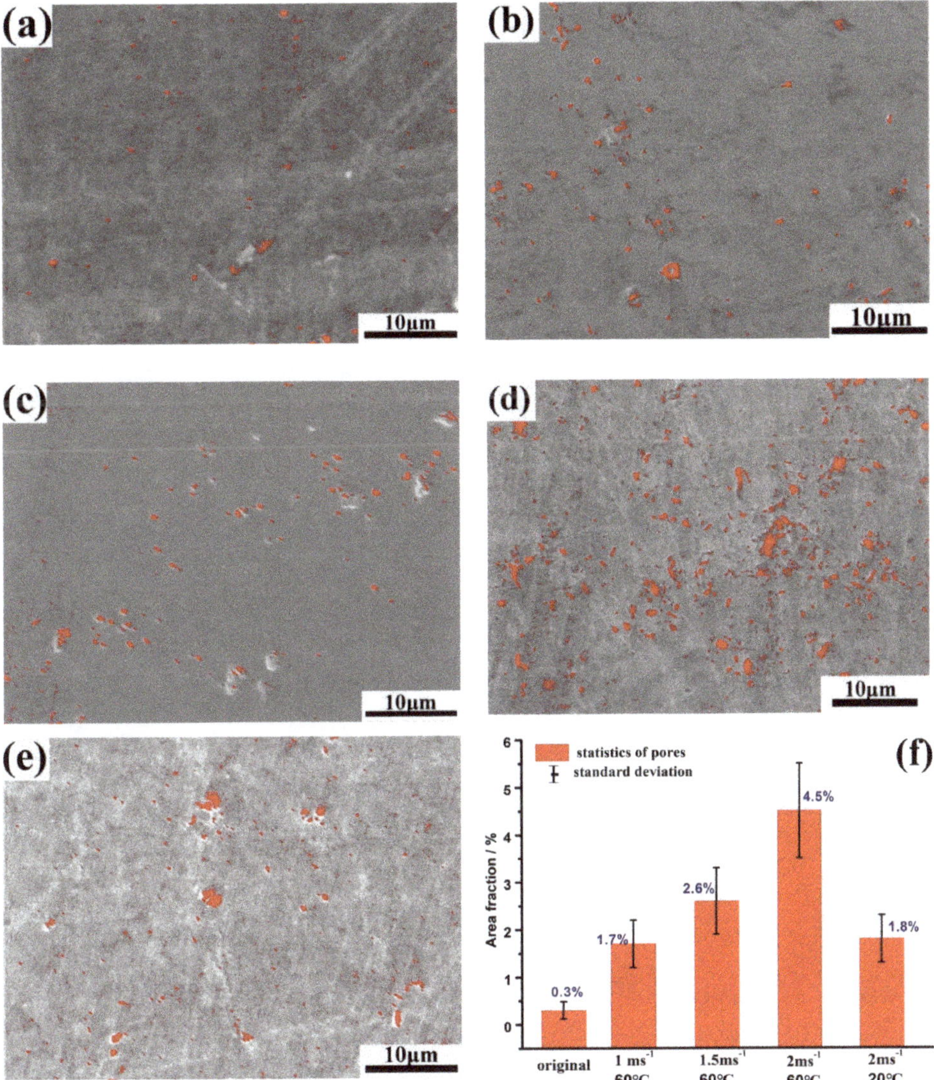

Figure 5. SEM image of the W–Cr–C coated tungsten samples (the red refers to the pores): (**a**) before corrosion; (**b**) corrosion at 1 ms^{-1}, 60 °C; (**c**) corrosion at 1.5 ms^{-1}, 60 °C; (**d**) corrosion at 2 ms^{-1}, 60 °C; (**e**) corrosion at 2 ms^{-1}, 20 °C; and (**f**) the bar chart of pitting fraction with velocity and temperature.

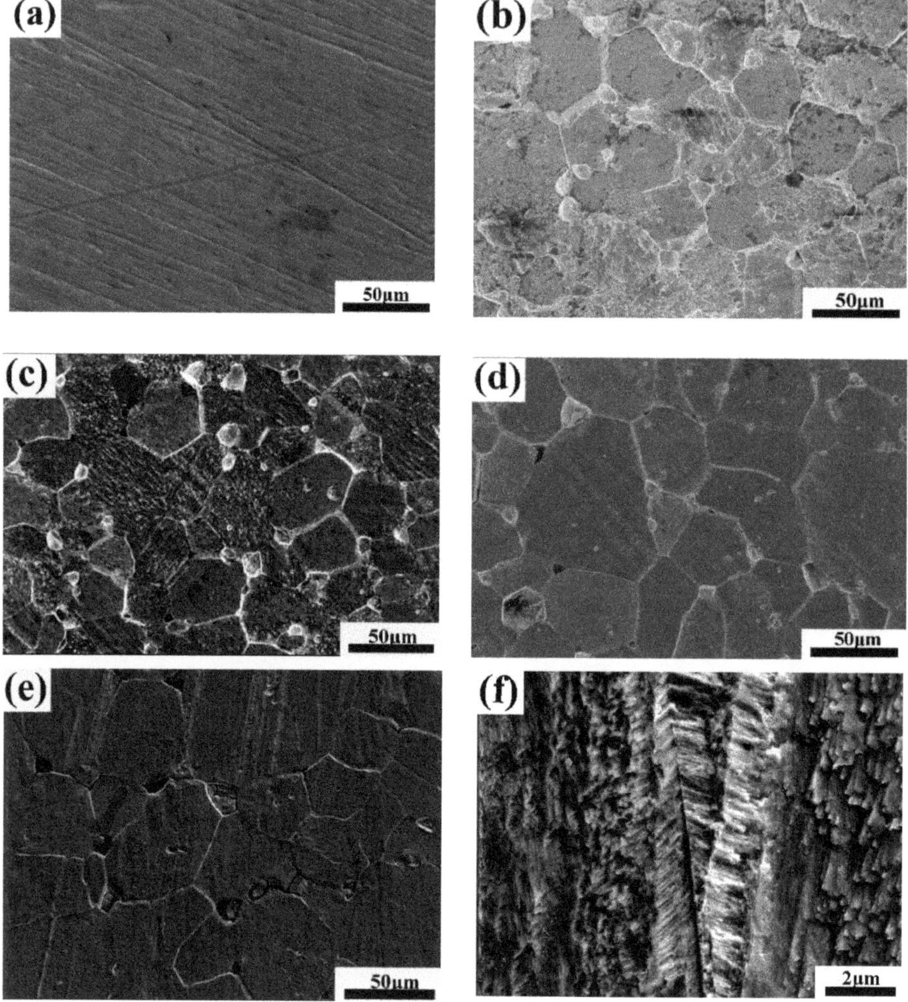

Figure 6. SEM images of bare tungsten: (**a**) before corrosion; (**b**) corrosion at 1 ms^{-1}, 60 °C; (**c**) corrosion at 1.5 ms^{-1}, 60 °C; (**d**) corrosion at 2 ms^{-1}, 60 °C; (**e**) corrosion at 2 ms^{-1}, 20 °C; and (**f**) corrosion on grains.

4. Discussion

Based upon the above results, the corrosion rate of W–Cr–C clad tungsten is much lower than that of bare tungsten due to the corrosion form changing from the intergranular corrosion of bare tungsten to the pitting corrosion of W–Cr–C coating. Pitting corrosion is a unique form of anodic reaction in the electrochemical reaction as well as a local corrosion. It causes less weight loss, so the change in corrosion rate with time gradually moves to a relatively steady state, as shown in Figures 3 and 4. The occurrence of pitting corrosion is known to be closely related to the environment and surface morphology and composition of materials. The medium used in this test containing special ions of oxygen and chlorine become the necessary trigger for pitting corrosion. In the service condition, chlorine and oxygen concentrations need to be controlled to reduce the pitting corrosion. Furthermore, reducing the porosity of W–Cr–C coating by fine polishing will help decrease the nucleation sites of pitting.

Occurrence of the intergranular corrosion on tungsten is essentially induced by the impurities, as listed in Table 1. It was already understood that the impurities were prone to segregate at grain boundaries (GBs), which resulted in a large composition difference between the GBs and tungsten grains. The nonmetal impurities such as phosphorus, silicon, oxygen, nitrogen, and their compounds acted as intergranular embrittlement elements and mild acidic electrolyte. The impurities iron, aluminum, and silicon, due to lower potential than tungsten, became the anode, which constituted a corrosion cell with tungsten around and induced pitting corrosion at the GBs. With the exacerbation and development of pitting corrosion, it evolved into macroscopic corrosion along the GBs, namely electrochemically-driven intergranular corrosion. In the regions away from GBs, chemically driven dissolution of tungsten proceeds through the interaction with oxygen and water. Understandably, the dissolution rate at the grain boundary is much higher than that on the tungsten grains. Temperature has an accelerated effect on both types of dissolution.

Intergranular corrosion at the triple junction was much more remarkable. From the perspective of defects, the volume percentage of the GBs and defect density at a triple or multiple-junction are much larger than that at conventional GBs, so corrosion would occur more easily at such regions. From the aspect of interface energy, the corrosion behavior of fine grains at the trigeminal boundaries is analogous to the pinning behavior of the second particles occupying similar sites of matrix. The pinning interaction of the second particles with the matrix grains depends on the contact area of the particles with boundaries. The smaller the second particle, the less the boundary area is occupied. For particles with a smaller size, there is a stronger unpinning tendency of the boundaries because of the lower decreased interface energy [24]. Once the pits at the triple junction of GBs are produced, it leads to a noticeable change in weight. This accounts for the large fluctuations of the corrosion rate for tungsten.

Regardless of whether it is intergranular corrosion or pitting corrosion, the effect of flow velocity on corrosion is complex. When flow velocity is elevated, on one hand, corrosion might be further exacerbated on account of sufficient oxygen supply and dissolution of the underlying metal caused by accelerated mechanical damage; on the other hand, corrosion may also be retarded due to the replenishment of inhibitors (corrosion products) to some dead spaces (grooves or pores), which hinder further corrosion [25]. Thus, in the low flow velocity range, the competition between the above-mentioned two effects approaches a balance and the corrosion rate increases slowly with increasing flow velocity. However, when flow velocity exceeds a critical value, the protruding mechanical action overshadows the inhibitors replenishing, and the materials are allowed to deteriorate at a higher corrosion rate, for instance, the corrosion at 2 ms^{-1}.

Evidently, temperature imposes a greater and nearly liner impact in promoting intergranular and pitting corrosion than velocity within 60 °C. Increase in temperature caused the acceleration of ion mobility and a corresponding rise in electrochemical reaction activity. Despite this, there is a critical corrosion temperature for materials. For instance, the corrosion rate of iron begins to decrease above 80 °C [26], while the corrosion rate of P110 steel reaches the maximum at 60 °C and then decreases [27]. For tungsten and W–Cr–C coatings in this study, the critical temperatures should be higher than 60 °C. Therefore, in the service temperature range, temperature promoted the corrosion rate of both samples.

In the test duration, W–Cr–C endures less weight loss and corrosion rate than intergranular corrosion of tungsten, showing an expected protective role. However, it is worth pointing out that pitting corrosion occurred on the W–Cr–C layer. With an increase in time, the pores will grow gradually, and when the pores stretch to the W_2C layer, the columnar crystal structure of W_2C will lead the corrosion mechanism to change into intergranular corrosion, which is not conducive to protecting the substrate. Furthermore, the W_2C phase is less stable than WC in neutral solution [28]. Therefore, changing the microstructure of the W_2C layer and increasing the proportion of the W–Cr–C layer by adjusting the carburizing parameters is the topic of ongoing work.

5. Conclusions

This work studied the corrosion behavior of tungsten, with and without W–Cr–C cladding prepared by the SPS technique, in a flowing water system, and the main results can be concluded as the follows:

(1) The commercial bare tungsten showed an electrochemically-driven type of intergranular corrosion, with a larger corrosion rate due to impurities segregated at the grain boundaries. After W–Cr–C cladding, the corrosion rate of tungsten dropped significantly due to the change of corrosion type from intergranular corrosion to pitting corrosion.

(2) The simultaneous increase of water temperature and flow velocity accelerated corrosion of the samples with and without the W–Cr–C cladding. The corrosion rate of the W–Cr–C coating was slower than that of tungsten under the same parameters, demonstrating its better corrosion-erosion resistance than that of pure tungsten.

Author Contributions: Data Curation and Formal Analysis, Z.X.; Methodology, J.Y.; Writing—original draft, J.Y.; Writing—review & editing, Y.J. and Q.F. All authors have read and agreed to the published version of the manuscript.

Funding: This research was funded by the National Key Research and Development Program of China (Grant No. 2017YFA0402800), the National Natural Science Foundation of China (Grant Nos. 51601189, 11674319, 11475216), the Central Government Guides Local Science and Technology Development Projects (2019ZYD026), and the Sichuan Key Laboratory of Comprehensive Utilization of Vanadium and Titanium Resources (2018FTSZ44).

Acknowledgments: The authors sincerely thank the National Natural Science Foundation of China for their support of this work.

Conflicts of Interest: The authors declare no conflicts of interest.

References

1. Zhang, J.J.; Yan, Q.W.; Zhang, C. Recent Progress of the Project of the Chinese Spallation Neutron Source. *J. Neutron Res.* **2005**, *13*, 11–14. [CrossRef]
2. Yan, Q.W.; Yin, W.; Yu, B.L. Optimized concept design of the target station of Chinese spallation neutron source. *J. Nucl. Mater.* **2005**, *343*, 45–52. [CrossRef]
3. Kawai, M.; Furusaka, M.; Kikuchi, K.; Kurishita, H.; Watanabe, R.; Li, J.F.; Sugimoto, K.; Yamamura, T.; Hiraoka, Y.; Abe, K. R&D of A MW-class solid-target for a spallation neutron source. *J. Nucl. Mater.* **2003**, *318*, 38–55.
4. Kawai, M.; Kikuchi, K.; Kurishita, H.; Li, J.F.; Furusaka, M. Fabrication of a tantalum-clad tungsten target for KENS. *J. Nucl. Mater.* **2001**, *296*, 312–320. [CrossRef]
5. Noji, N.; Kashiwagura, K.; Akao, N. Corrosion resistance of tungsten and tungsten alloys for spallation target in stagnant and flowing water. *J. Jpn. Inst. Met.* **2002**, *66*, 1107–1115. [CrossRef]
6. Maloy, S.A.; Lillard, R.S.; Sommer, W.F.; Butt, D.P.; Gac, F.D.; Willcutt, G.J.; Louthan, M.R., Jr. Water corrosion measurements on tungsten irradiated with high energy protons and spallation neutrons. *J. Nucl. Mater.* **2012**, *431*, 140–146. [CrossRef]
7. Nelson, A.T.; O'Toole, J.A.; Valicenti, R.A.; Maloy, S.A. Fabrication of a tantalum-clad tungsten target for LANSCE. *J. Nucl. Mater.* **2012**, *431*, 172–184. [CrossRef]
8. Yu, Q.Z.; Lu, Y.L.; Hu, Z.L.; Zhou, B.; Yin, W.; Liang, T.J. Decay heat calculations for a 500 kW W-Ta spallation target. *Nucl. Instrum. Methods B* **2015**, *351*, 41–45. [CrossRef]
9. Nio, D.; Ooi, M.; Takenaka, N.; Furusaka, M.; Kawai, M.; Mishima, K.; Kiyanagi, Y. Neutronics performance and decay heat calculation of a solid target for a spallation neutron source. *J. Nucl. Mater.* **2005**, *343*, 163–168. [CrossRef]
10. Espallargas, N.; Berget, J.; Guilemany, J.; Benedetti, A.V.; Suegama, P. Cr_3C_2–NiCr and WC–Ni thermal spray coatings as alternatives to hard chromium for erosion–corrosion resistance. *Surf. Coat. Technol.* **2008**, *202*, 1405–1417. [CrossRef]

11. Alzouma, O.M.; Azman, M.-A.; Yung, D.-L.; Fridrici, V.; Kapsa, P. Influence of different reinforcing particles on the scratch resistance and microstructure of different WC–Ni composites. *Wear* **2016**, *352*, 130–135. [CrossRef]
12. Chen, J.; Liu, W.; Deng, X.; Wu, S. Tool life and wear mechanism of WC–5TiC–0.5VC–8Co cemented carbides inserts when machining HT250 gray cast iron. *Ceram. Int.* **2016**, *42*, 10037–10044. [CrossRef]
13. Zhang, D.Q.; Liu, T.; Joo, H.G.; Gao, L.X.; Lee, K.Y. Microstructure and corrosion resistance of the brazed WC composite coatings in aerated acidic chloride media. *Int. J. Refract. Met. Hard Mater.* **2012**, *35*, 246–250. [CrossRef]
14. He, Y.Z.; Si, S.H.; Xu, K.; Yuan, X.M. Effect of Cr_3C_2 Particles on Microstructure and Corrosion-Wear Resistance of Laser Cladding Co-based Alloy Coating. *Chin. J. Lasers* **2004**, *31*, 1143–1148.
15. Durst, O.; Ellermeier, J.; Trossmann, T.; Berger, C. Erosion corrosion of graded chromium carbide coatings in multi layer structure. *Materialwiss. Werkstofftech.* **2009**, *40*, 756–768. [CrossRef]
16. Gao, J.X.; Fan, J.L. Research developments on the binderless WC-based cemented carbide. *China Tungsten Ind.* **2011**, *26*, 22–26.
17. Jiang, Y.; Yang, J.F.; Zhuang, Z.; Liu, R.; Zhou, Y.; Wang, X.P.; Fang, Q.F. Characterization and properties of tungsten carbide coatings fabricated by SPS technique. *J. Nucl. Mater.* **2013**, *433*, 449–454. [CrossRef]
18. Jiang, Y.; Yang, J.F.; Xie, Z.M.; Gao, R.; Fang, Q.F. Corrosion resistance of W–Cr–C coatings fabricated by spark plasma sintering method. *Surf. Coat. Tech.* **2014**, *254*, 202–206. [CrossRef]
19. Jiang, Y.; Yang, J.F.; Fang, Q.F. Effect of chromium content on microstructure and corrosion behavior of W-Cr-C coatings prepared on tungsten substrate. *Front. Mater. Sci.* **2015**, *9*, 77–84. [CrossRef]
20. Du, J.H. The thermal design of tungsten target for 100 kW neutron spallation source. *Mach. Des. Manuf.* **2008**, *10*, 86.
21. Tan, J.; Zhou, Z.J.; Liu, Y.Q.; Qu, D.D.; Zhong, M.; Ge, C.C. Effect of carbon nanotubes on the microstructure and mechanical properties of W. *Acta Metall. Sin.* **2011**, *47*, 1555–1560.
22. Kurlov, A.S.; Gusev, A.I. Tungsten carbides and W-C phase diagram. *Inorg. Mater.* **2006**, *42*, 121–127. [CrossRef]
23. Guzanová, A.; Brezinová, J.; Draganovská, D.; Maruschak, P.O. Properties of coatings created by HVOF technology using micro-and nano-sized powder. *Koroze A Ochr. Mater.* **2019**, *63*, 86–93. [CrossRef]
24. Grácio, J.J.; Picu, C.R.; Vincze, G.; Mathew, N.; Schubert, T.; Lopes, A.; Buchheim, C. Mechanical behavior of Al-SiC nanocomposites produced by ball milling and spark plasma sintering. *Metall. Mater. Trans. A* **2013**, *44*, 5259–5269. [CrossRef]
25. Copson, H.R. Effect of velocity on corrosion by water. *Ind. Eng. Chem.* **1952**, *44*, 1745–1752. [CrossRef]
26. Revie, R.W. *Corrosion and Corrosion Control*, 4th ed.; John Wiley & Sons: Hoboken, NJ, USA, 2008.
27. Zhu, S.D.; Yin, Z.F.; Bai, Z.Q.; Wei, J.F.; Zhou, G.S.; Tian, W. Influences of temperature on corrosion behavior of P110 steel. *J. Chin. Soc. Corros. Prot.* **2009**, *29*, 493–497.
28. Weidman, M.C.; Esposito, D.V.; Hsu, Y.C.; Chen, J.G. Comparison of electrochemical stability of transition metal carbides (WC, W2C, Mo2C) over a wide pH range. *J. Power Sources* **2012**, *202*, 11–17. [CrossRef]

Article

Investigations of the Deuterium Permeability of As-Deposited and Oxidized Ti_2AlN Coatings

Lukas Gröner [1,*], Lukas Mengis [2], Mathias Galetz [2], Lutz Kirste [3], Philipp Daum [1], Marco Wirth [1], Frank Meyer [1], Alexander Fromm [1], Bernhard Blug [1] and Frank Burmeister [1]

1 Department of Tribology, Fraunhofer-Institut für Werkstoffmechanik IWM, Woehlerstrasse 11, 79108 Freiburg, Germany; philipp.daum@iwm.fraunhofer.de (P.D.); marco.wirth@iwm.fraunhofer.de (M.W.); frank.meyer@iwm.fraunhofer.de (F.M.); alexander.fromm@iwm.fraunhofer.de (A.F.); bernhard.blug@iwm.fraunhofer.de (B.B.); frank.burmeister@iwm.fraunhofer.de (F.B.)

2 Department of High Temperature Materials, DECHEMA-Forschungsinstitut, Theodor-Heuss-Allee 25, 60486 Frankfurt am Main, Germany; lukas.mengis@dechema.de (L.M.); mathias.galetz@dechema.de (M.G.)

3 Department of Epitaxy, Fraunhofer-Institut für Angewandte Festkörperphysik IAF, Tullastraße 72, 79108 Freiburg, Germany; lutz.kirste@iaf.fraunhofer.de

* Correspondence: lukas.groener@iwm.fraunhofer.de; Tel.: +49-761-5142-488

Received: 31 March 2020; Accepted: 23 April 2020; Published: 1 May 2020

Abstract: Aluminum containing $M_{n+1}AX_n$ (MAX) phase materials have attracted increasing attention due to their corrosion resistance, a pronounced self-healing effect and promising diffusion barrier properties for hydrogen. We synthesized Ti_2AlN coatings on ferritic steel substrates by physical vapor deposition of alternating Ti- and AlN-layers followed by thermal annealing. The microstructure developed a {0001}-texture with platelet-like shaped grains. To investigate the oxidation behavior, the samples were exposed to a temperature of 700 °C in a muffle furnace. Raman spectroscopy and X-ray photoelectron spectroscopy (XPS) depth profiles revealed the formation of oxide scales, which consisted mainly of dense and stable α-Al_2O_3. The oxide layer thickness increased with a time dependency of ~$t^{1/4}$. Electron probe micro analysis (EPMA) scans revealed a diffusion of Al from the coating into the substrate. Steel membranes with as-deposited Ti_2AlN and partially oxidized Ti_2AlN coatings were used for permeation tests. The permeation of deuterium from the gas phase was measured in an ultra-high vacuum (UHV) permeation cell by mass spectrometry at temperatures of 30–400 °C. We obtained a permeation reduction factor (PRF) of 45 for a pure Ti_2AlN coating and a PRF of ~3700 for the oxidized sample. Thus, protective coatings, which prevent hydrogen-induced corrosion, can be achieved by the proper design of Ti_2AlN coatings with suitable oxide scale thicknesses.

Keywords: MAX phase; Ti_2AlN; PVD coating; oxidation; hydrogen permeation

---

1. Introduction

The increasing number of applications in which hydrogen is being used as a storage medium in energy conversion technologies demands the consideration of new construction materials, or at least a profound surface conditioning of established materials to prevent, e.g., hydrogen diffusion induced embrittlement or other forms of corrosion, especially the development of so-called "white etching cracks" [1]. One route for corrosion protection is the development and application of temperature-resistant coatings with excellent barrier properties for hydrogen. Recently performed studies indicate that MAX phase materials might fulfill these requirements [2–6]. The general formula, $M_{n+1}AX_n$, (short MAX) describes a family of materials consisting of an early transition metal (M), mostly a group 13 or 14 element (A) and nitrogen and/or carbon (X) with the stoichiometry of n = 1,2,3 [7].

The MAX phases crystallize in a hexagonal lattice within the space group D46h (P63/mmc) in which the octahedral $M_{n+1}X_n$ layers are separated by atomic monolayers of pure A-atoms. MAX phase materials are known to have a good oxidation resistance [3,8,9], a high damage tolerance as well as a high electrical and thermal conductivity [10].

The good oxidation resistance of Al containing MAX phases usually stems from the formation of dense and thermodynamically stable thermal grown oxides (TGO) consisting of α-Al_2O_3 on the coatings surface at relatively low temperatures of 600–700 °C. For comparison, the direct physical vapor deposition (PVD) of an α-Al_2O_3 in an industrial scale deposition process usually requires temperatures above 1000 °C [11]. Lower deposition temperatures of 500–600 °C have also been observed but at the expense of a brittle fracture behavior [12,13]. A further advantage of α-Al_2O_3 oxide scales thermally grown on MAX phase coatings is the well-known self-healing effect whereby small defects or cracks in the coating, which might serve as diffusion pathways, are blocked by oxide growth [2]. For this purpose, the oxidation kinetics of the TGO has to allow for a quick healing and oxidation of the surface, but has to prevent fast oxygen diffusion to the coating–substrate interface. The oxidation kinetics of Ti_2AlC at 1200 °C were modelled by G.M. Song et al. in [3]. This model contains the growth of oxide grains and assumes that the diffusion paths along the grain boundaries increase with time. This results in a time dependency for the increase in the thickness of the oxide scale $d_{Ox}(t)$ of:

$$d_{Ox}(t) = 2\sqrt{k_n} \times t^{1/4} \tag{1}$$

The growth factor $k_n = \Omega D_{GB} \Delta C \frac{3\delta}{d_0}$ contains a constant prefactor Ω, the diffusion coefficient for oxygen along the grain boundaries D_{GB}, the size of the grain boundaries δ, the initial lateral grain size d_0 and the gradient in the oxygen concentration ΔC.

This model of α-Al_2O_3-formation, as well as the structural properties of MAX phases, i.e., the sequence of dense MX-layers, motivated the present investigation on their barrier properties against hydrogen diffusion.

Although little information about the diffusion of hydrogen in MAX phases exists, similarly composed carbides or nitrides of early transition metals are already used as diffusion barriers for hydrogen [4,14]. It is expected that the anisotropic structure of MAX phases will induce a directional anisotropy of the hydrogen diffusion. In [5], F. Colonna and C. Elsässer presented the findings of an atomistic simulation of diffusion processes in Ti_2AlN using density-functional theory (DFT). Therein, interstitial diffusion paths of hydrogen and oxygen were examined. It was found that, for hydrogen, the migration perpendicular to the basal planes has a maximum barrier of ~3 eV, whereas the migration barrier parallel to the basal plane is one order of magnitude lower. The high migration barrier parallel to the c axis was explained by the presence of the Ti_2N double layer, where the interstitial octahedral sites of Al_3Ti_3 are already occupied by nitrogen atoms.

An experimental study on the hydrogen barrier properties of MAX phase coatings was presented by C. Tang et al. in [6]. Therein, ZrY-4 alloy cylinders were coated with Ti_2AlC and Cr_2AlC by a multilayer deposition followed by a subsequent annealing step. This process led to a {0001}-textured polycrystalline growth which could also be detected in [15] for Ti_2AlN. After loading the specimens in an Ar+H_2 atmosphere the cylinders were investigated by neutron radiography. It could be shown that a 5 µm thick Ti_2AlC and Cr_2AlC reduced the penetration of hydrogen under the detection limit.

To evaluate coatings in terms of their capability to reduce the hydrogen permeation a permeation reduction factor (PRF) can be calculated using the mass specific ion current j:

$$\mathrm{PRF} = \frac{j_{uncoated}}{j_{coated}}. \tag{2}$$

D. Levchuk et al. investigated Al-Cr-O coatings [16] and Er_2O_3 coatings [17] as hydrogen permeation barrier. Both coatings tend to form a dense crystalline structure, which is capable of effectively reducing the hydrogen permeation up to a PRF(Al-Cr-O) = 3500 and PRF(Er_2O_3) = 800.

2. Experimental Details

2.1. Deposition of Ti_2AlN

The Ti_2AlN MAX coatings were deposited on AISI 430 ferritic stainless-steel substrates (Fe81/Cr17/Mn/Si/C/S/P), which were polished (1400 grit) and cleaned in acetone and isopropanol using an ultrasonic bath prior to deposition. A custom build industrial sized magnetron sputter chamber SV400/S3 (FHR Anlagenbau GmbH, Ottendorf-Okrilla, Germany) equipped with rectangular titanium (purity 99.8%) and aluminum targets (99.999%) was utilized. To obtain a pronounced {0001}-texture with a parallel orientation of the basal planes and the substrate surface, we alternately deposited 150 single layers of Ti and AlN on the substrate, beginning with Ti. During the radio frequency-sputtering of the aluminum target, nitrogen (purity 99.9999%) was introduced in the chamber. A final subsequent annealing at 700 °C for 1 h in vacuum led to the formation of textured Ti_2AlN MAX phase coatings. Details of the deposition process are described elsewhere [15]. The coatings thickness was in the 2 μm–3 μm range.

2.2. Oxidation Procedure and Analysis

To investigate the oxidation kinetics of the Ti_2AlN coatings, comparable samples originating from the same batch were oxidized at 700 °C for 5 h, 10 h, 20 h and 100 h in a muffle furnace (Nabertherm GmbH, Lilienthal, Germany) in air. The samples were afterwards removed from the furnace and cooled in air. The crystallographic orientation and phase composition of oxidized and non-oxidized coatings were investigated by X-ray diffractometry (XRD) using a PANalytical Empyrean in parallel beam geometry (Empyrean, PANalytical, Almelo, The Netherlands) and Cu $K\alpha_1$ radiation with a 2-bounce Ge 220 monochromator. The samples were irradiated with primary X-rays using a line focus. The diffracted X-rays were detected using a PIXel-3D detector with a 1 mm slit for the phase analysis.

A surface sensitive phase analysis was performed using a confocal Raman spectrometer (Model inVia, Renishaw plc., Gloucestershire, United Kingdom) in backscatter geometry. The excitation wavelength $\lambda_{Nd:YAG}$ = 532 nm was used to determine possible changes in the Ti_2AlN phase upon thermal treatment whereas the wavelength λ_{HeNe} = 633 nm proofed to be suitable for exciting fluorescence bands in the thermally grown AlO_x phase. In all measurements, a 100-fold objective focused the laser on the surface to a spot diameter of about 2 μm.

XPS depth profiles were recorded with a PHI 5000 VersaProbe II (Ulvac-PHI, Inc., Chigasaki, Japan) equipped with an argon sputter option using Al Kα-rays. For analyzing the coarse elemental distribution close to the interface of coating and substrate a metallographic cross-section was prepared after an electrochemical deposition of Ni for protective purposes. The electron probe micro analysis (EPMA) was performed utilizing a JXA-8100 (Jeol, Akishima, Japan). The measurements were performed with an acceleration voltage of 15 kV and a dwell time of 30 ms.

2.3. Deuterium Permeation Setup

To investigate the diffusion of deuterium from the gas phase through coated and oxidized membranes, a permeation setup was developed following the works of C. Frank et al. [18], J. Gorman et al. [19] and D. Levchuk et al. [20]. The test rig consisted of two chambers separated by a thin steel membrane (see Figure 1). The high pressure side is filled with the diffusional species or the purging gas, the low pressure side is evacuated by a turbomolecular pump and an ion getter pump down to pressures of ~10^{-6} Pa. The latter is also equipped with a quadrupole mass spectrometer (Model PrismaPlus™ QMG 220, Pfeiffer Vacuum Technology AG, Aßlar, Germany) to determine the gas composition as well as the mass and time dependent ion current, which is detected by a secondary electron multiplier. With infrared transmissible windows on both sides, the membrane was heated by a focused halogen radiation heater and the temperature as well as the temperature distribution was recorded by a heat sensitive camera. The membranes, illustrated in Figure 1b, were water jet cut (Ø = 30 mm) from a 0.2 mm thick steel foil (AISI 430) and coated as described before. The uncoated back sides were corundum blasted to increase the absorption of infrared radiation. The membrane was mounted with conical copper gaskets with the coating facing the high pressure side. After a minimum base pressure of 1×10^{-5} Pa was reached, the measurement was started.

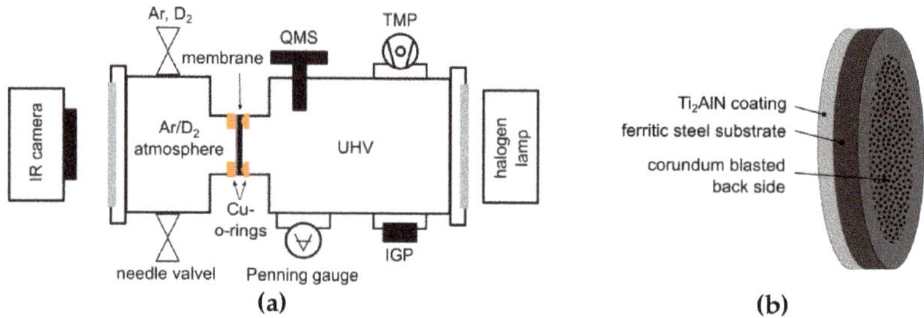

Figure 1. (**a**) Schematic illustration of the hydrogen permeation test rig with quadrupole mass spectrometer (QMS), turbomolecular pump (TMP) and ion getter pump (IGP). (**b**) Schematic illustration of the coated ferritic steel membrane.

To investigate the hydrogen barrier properties, the isotope deuterium was employed in order to avoid interpretation ambiguities due to contaminations with residual gases or water molecules. The permeation measurements were performed close to thermodynamic equilibrium. First, deuterium was injected on the atmospheric pressure side. Then the membrane temperature was set to a maximum and was reduced stepwise when a constant ion current was reached. The ion current of the atomic mass $m(D_2) = 4$ was recorded. The permeation reduction factor was calculated by (2) using the steady state values of the ion currents of a non-coated sample as a reference.

3. Results and Discussion

3.1. Oxidation

To investigate the influence of the oxygen exposure at high temperatures on the phase composition, XRD diffractograms were recorded for different exposure times and compared to the pristine sample. In Figure 2a, the diffractogram of the as-synthesized coating reveals an almost phase pure Ti_2AlN coating having a strong {0001}-texture. The peak at 42.5° is attributed to the (200) lattice plane of TiN. The Fe-bcc peaks at 44.6° and 65.0° are assigned to the steel substrate. Diffractograms of the oxidized samples are depicted in Figure 2 with an offset for better visibility.

These phase compositions appear almost unaltered upon thermal exposure. Only in the 2Θ-region between 42°–43° a slight change in the peak position is visible. This region is depicted in detail in Figure 2b. Due to broad peak widths and weak angle dependent interferences, the signals from TiN and α-Al_2O_3 cannot be clearly distinguished. Further ambiguities arise due to the small TGO layer thickness and its possibly amorphous structure. Hence, further surface sensitive Raman analysis was performed (see Figure 3). The Raman spectra of the coatings in Figure 3a still feature the characteristic Raman peaks for Ti_2AlN despite the oxidized surface, though an increase in the background is detected. The broad background and the peak between 500 cm^{-1}– 600 cm^{-1} might stem from the formation of surface oxides and/or oxycarbides [21] as well as from the formation of TiN close to the surface due to Al depletion [22]. Titanium oxides like anatase and rutile, which were reported in [23] after the oxidation of Ti_2AlN coatings at 750 °C, are not detected.

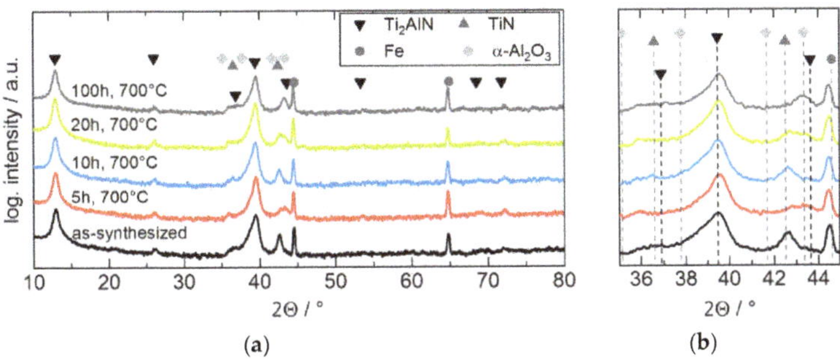

Figure 2. XRD diffractograms of as-synthesized and oxidized Ti_2AlN coatings on ferritic steel samples: (**a**) overview and (**b**) enlarged region around $2\theta° \approx 40°$.

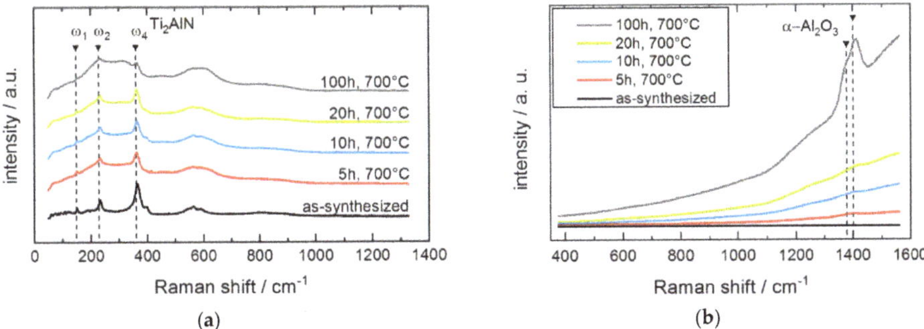

Figure 3. (**a**) Raman spectra ($\lambda_{Nd:YAG}$ = 532 nm) of the Ti_2AlN coatings before and after oxidation. (**b**) Raman fluorescence spectra (λ_{HeNec} = 633 nm) of the Ti_2AlN coatings before and after oxidation.

The fluorescence spectra in Figure 3b exhibit an increase in the background with an increasing oxidation time. Particularly after 100 h of oxidation, the formation of distinct peaks close to 1379 cm^{-1} and 1402 cm^{-1} can be detected. According to literature, these peaks are attributed to the fluorescence of Cr^{3+} and Fe^{3+} impurities in an α-Al_2O_3 environment [24–26].

After oxidation, XPS depth profiles of all coatings were created by convoluting the distribution of the binding energies. The underlying XPS spectra are not shown herein. In Figure 4 profiles of a sample oxidized for 100 h (a) and of an uncoated substrate (b) are presented for comparison. In the case of the MAX phase coating Figure 4a, oxidic Al2p bonds with a maximum in the energy of 74.3 eV were detected and ascribed to the formation of Al_2O_3 at the sample surface. To a smaller extend of about 8 at%, oxidic Ti2p bonds with a maximum in the energy of 458.2 eV were detected, which were distributed over two regional maxima located in a depth of ~45 nm in the Al_2O_3 scale and at the interface of Ti_2AlN/TGO. At the interface, the shift of the Ti2p-signal to nitridic binding energies of 454.3 eV and the shift of the Al2p-signal to metallic binding energies of 72.3 eV revealed the transition to the Ti_2AlN phase. With only ~16 at% of Al2p bonds and sustaining increasing signals at a depth of 200 nm, a transition regime close to the interface is observed where an Al depletion exists. The thickness d_{Ox} of the TGO scale was determined by the decrease of the O1s signal, and set to the sputter depth where the O-signal fell below 50% of the original ratio.

The numerical fit of the values for d_{Ox} according to (1) is plotted in Figure 5. The errors of the oxide thicknesses were estimated to 10 nm resulting from a temporal variation in the sputter rate during XPS measurements. The growth factor of the TGO was calculated to $k_n = 402 \pm 17 \frac{\text{mol}}{\text{m} \times \text{s}}$ with a quality factor of $R^2 = 0.9794$. The quality of the fit argues for the suitability of the mathematical description by (1 for the oxidation kinetics. However, no conclusion can be drawn so far as to whether the O or the Al diffuses along the grain boundaries to the oxidizing interface according to the above described model of G.M. Song et al.

The spectra of an uncoated ferritic steel substrate after 100h at 700 °C in Figure 4b revealed the formation of a TGO consisting of of Cr-, Mn- and Fe-oxides. The thickness of the TGO was calculated to 540 nm, which compares to approximately 100 nm in the case of a coated substrate.

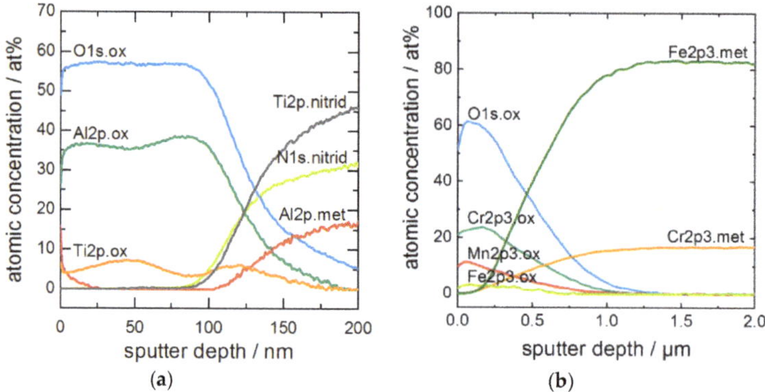

Figure 4. XPS depth profiles after oxidation for 100 h at 700 °C of (**a**) Ti_2AlN coating, and (**b**) uncoated ferritic steel substrate.

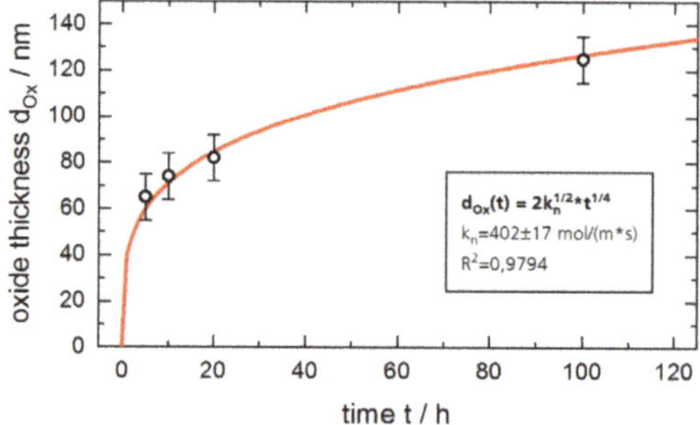

Figure 5. Measured oxide thicknesses by XPS depth profiles of the Ti_2AlN coatings after oxidation. Underlying fit was performed using (1).

The EPMA images of the sample oxidized for 100 h at 700 °C depicted in Figure 6 represent the elemental distributions of Ti (b), Al (c), N (d), O (e), Ni (f) and Fe (g), where the colors indicate the normalized elemental concentration. The measured distribution of Ti, Al and N across the coating thickness features a depletion in Ti and Al at the interface of Ti_2AlN/TGO and in the subsurface region. Accordingly, the substrate is locally enriched by Al and N and the formation of precipitates perpendicular to the surface is visible. In such areas, only a minor Fe-concentration is measured, as the microprobe signal is always to normalized 100% for all elements. The inward diffusion of Al and N is accompanying the outward diffusion of Fe into the coating according to the Fe elemental distribution map. Besides the thin oxide scale, which formed on top of the MAX phase coating, oxygen can be detected within the Ni-plating. This is caused by the formation of a longitudinal crack within the Ni-plating during the preparation of the cross-section.

The strong interdiffusion of the weakly bound A-element of MAX phases with the substrate is known to be a crucial aspect, when it comes to the chemical stability in high temperature applications [23]. Therefore, the interdiffusion should be suppressed by applying additional barrier films against Al diffusion between the substrate and coating. Further loss of the A-element also occurs during oxidation and annealing in vacuum due to Al_2O_3 formation and evaporation. In [27], Zhang et al. calculated that the Ti_2AlN MAX phase lattice structure is capable of accommodating Al vacancies down to a $Ti_2Al_{0.75}N$ stoichiometry.

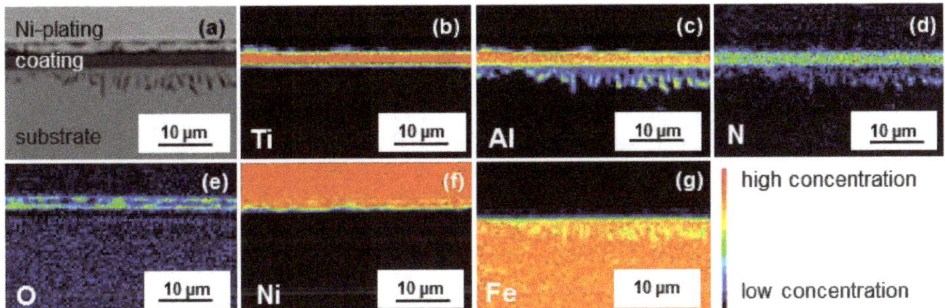

Figure 6. Scanning electron microscope image (**a**) and electron probe micro analysis (EPMA) scans of the cross-section after oxidation for 100 h at 700 °C for Ti (**b**), Al (**c**), N (**d**), O (**e**), Ni (**f**) and Fe (**g**).

3.2. Hydrogen Permeation

The D_2 ion current was measured by secondary ion mass spectrometry using the setup illustrated in Figure 1. The influence of the coatings on the permeation was investigated in a state close to the thermodynamic equilibrium. Three different membranes were investigated: the uncoated substrate material (substrate), the substrate coated with 2.7 µm of Ti_2AlN (substrate+Ti_2AlN) and the substrate coated with 2.7 µm of Ti_2AlN with a subsequent oxidation for 20 h at 700 °C (substrate+Ti_2AlN+TGO). In Figure 7, the D_2 ion currents are plotted in the temperature range from about 50 °C to 400 °C. Three measuring cycles were performed for each membrane. The results of the different cycles are denoted by the different symbols in the graph (square, circle and triangle). The small variance in the data points confirms the reproducibility of the permeation induced ion current.

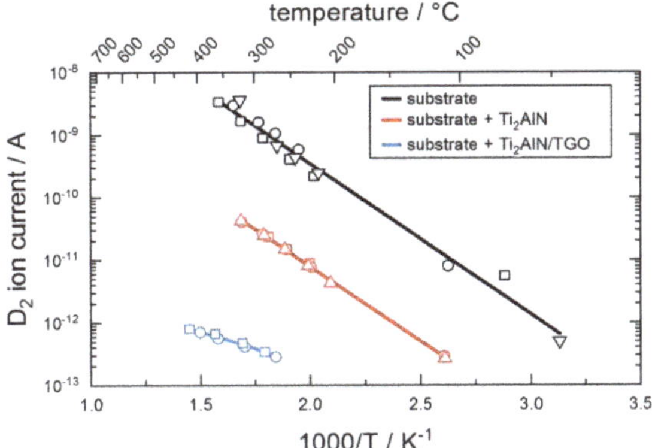

Figure 7. Arrhenius Plot of D_2 ion currents of ferritic membranes with and without Ti_2AlN or Ti_2AlN+thermal grown oxide (TGO) coating. The different symbols on each line refer to results of consecutive measuring cycles. The quasi-linear fit is performed to illustrate the Arrhenius type behavior.

The diffusion through all three membranes follows an Arrhenius type behavior, which confirms the assumption of a diffusion-controlled permeation. The deposition of a 2.7 µm thick Ti_2AlN coating already reduces the permeation of deuterium. The PRF at 300 °C is calculated to a factor of 45. As optical investigations on the coating after the measurements still revealed some minor cracks in the coating, the PRF might still be lower for a defect-free MAX-phase coating. However, the formation of the oxide scale leads to a further, significant reduction of the permeation. With an oxidation of the coated steel membrane for 20 h at 700 °C, the formation of TGO of 80 nm thickness is expected, compare Figure 5. With respect to the uncoated steel membrane, a PRF of about 3700 was achieved. This reduction can be explained by the low solubility of hydrogen in the α-Al_2O_3 phase as well as by the potential healing of small defects in Ti_2AlN, which blocks alternative migration paths with low energy barriers. The reduction of three orders of magnitude strongly supports the initially assumed suitability of Ti_2AlN coatings as high temperature hydrogen diffusion barriers.

4. Conclusions and Outlook

Ti_2AlN coatings were synthesized on ferritic steel samples by a repeated deposition of Ti/AlN double layers and a subsequent annealing in vacuum. The oxidation experiments at 700 °C in air revealed the formation of a thin TGO at the sample surface consisting mainly of α-Al_2O_3. By analyzing the thickness of the TGO, the kinetic confirms the findings of G.M. Song et al., which are described by a growth in thickness with a time dependency of ~$t^{1/4}$. Thereby, a thin protective oxide is quickly formed after exposure to air but further growth is strongly hindered by a slow diffusion of migrating particles through the dense oxide. It was shown that the TGO not only serves as an effective protective layer against further oxidation, but also serves as a diffusion barrier against hydrogen. Whereas a 2.7 μm thin Ti_2AlN coating reduces the permeation of deuterium to a factor of 45, the formation of an α-Al_2O_3 scale further reduces the permeation within three orders of magnitude. The healing of coating defects like pores and cracks at elevated temperatures upon oxidation is seen as an additional advantage of thermally grown diffusion barriers in comparison to directly deposited barrier coatings.

Further investigations need to focus on the interdiffusion process at the interface of coating and substrate in order to reduce the loss of the Al, which is required for the formation of α-Al_2O_3. Finding an optimum thickness of the TGO, which significantly reduces the hydrogen permeation and at the same time exhibits a sufficient thermal and mechanical stability is the crucial task for the utilization of Ti_2AlN as protective coatings in industrial applications.

In summary, Al containing MAX phase coatings, which tend to form a dense α-Al_2O_3 on the surface upon oxidation, seems to be effective protective coatings in high temperature applications, where oxygen and hydrogen corrode the substrate material.

Author Contributions: Conceptualization, L.G.; investigation, L.K., L.M., P.D. and L.G.; technical support, M.W.; writing—original draft preparation, L.G.; writing—review and editing, L.M., M.G., F.M. and B.B.; supervision, F.B.; funding acquisition, A.F. and F.B. All authors have read and agreed to the published version of the manuscript.

Funding: Financial support by the Baden-Württemberg-Stiftung gGmbH in the context of "CleanTech" (project CT-6 "LamiMat") is gratefully acknowledged.

Conflicts of Interest: The authors declare no conflict of interest.

References

1. Evans, M.-H. An updated review: White etching cracks (WECs) and axial cracks in wind turbine gearbox bearings. *Mater. Sci. Technol.* **2016**, *32*, 1133–1169. [CrossRef]
2. Farle, A.-S.; Kwakernaak, C.; van der Zwaag, S.; Sloof, W.G. A conceptual study into the potential of $M_{n+1}AX_n$-phase ceramics for self-healing of crack damage. *J. Eur. Ceram. Soc.* **2015**, *35*, 37–45. [CrossRef]
3. Song, G.M.; Schnabel, V.; Kwakernaak, C.; van der Zwaag, S.; Schneider, J.M.; Sloof, W.G. High temperature oxidation behaviour of Ti_2AlC ceramic at 1200C. *Mater. High Temp.* **2012**, *29*, 205–209. [CrossRef]
4. Hultman, L. Thermal stability of nitride thin films. *Vacuum* **2000**, *57*, 1–30. [CrossRef]
5. Colonna, F.; Elsässer, C. First principles DFT study of interstitial hydrogen and oxygen atoms in the MAX phase Ti_2AlN. *RSC Adv* **2017**, *7*, 37852–37857. [CrossRef]
6. Tang, C.; Grosse, M.K.; Trtik, P.; Steinbrück, M.; Stüber, M.; Seifert, H.J. H_2 Permeation Behavior Of Cr_2AlC And Ti_2AlC MAX Phase Coated Zircaloy-4 By Neutron Radiography. *Acta Polytech.* **2018**, *58*, 69. [CrossRef]

7. Barsoum, M.W. The $M_{n+1}AX_n$ phases: A new class of solids. *Prog. Solid State Chem.* **2000**, *28*, 201–281. [CrossRef]
8. Smialek, J.L. Oxidation of Al_2O_3 Scale-Forming MAX Phases in Turbine Environments. *Met. Mat Trans A* **2017**, *89*, 334. [CrossRef]
9. Zhang, Z.; Jin, H.; Chai, J.; Pan, J.; Seng, H.L.; Goh, G.T.W.; Wong, L.M.; Sullivan, M.B.; Wang, S.J. Temperature-dependent microstructural evolution of Ti_2AlN thin films deposited by reactive magnetron sputtering. *Appl. Surf. Sci.* **2016**, *368*, 88–96. [CrossRef]
10. Sokol, M.; Natu, V.; Kota, S.; Barsoum, M.W. On the Chemical Diversity of the MAX Phases. *Trends Chem.* **2019**, in press. [CrossRef]
11. Laimer, J.; Fink, M.; Mitterer, C.; Störi, H. Plasma CVD of alumina—Unsolved problems. *Vacuum* **2005**, *80*, 141–145. [CrossRef]
12. Gavrilov, N.V.; Kamenetskikh, A.S.; Tretnikov, P.V.; Chukin, A.V. High-rate low-temperature PVD of thick 10 μm α-alumina coatings. *J. Phys. Conf. Ser.* **2019**, *1393*, 12082. [CrossRef]
13. Kyrylov, O.; Kurapov, D.; Schneider, J.M. Effect of ion irradiation during deposition on the structure of alumina thin films grown by plasma assisted chemical vapour deposition. *Appl. Phys. A* **2005**, *80*, 1657–1660. [CrossRef]
14. Checchetto, R.; Bonelli, M.; Gratton, L.M.; Miotello, A.; Sabbioni, A.; Guzman, L.; Horino, Y.; Benamati, G. Analysis of the hydrogen permeation properties of TiN-TiC bilayers deposited on martensitic stainless steel. *Surf. Coat. Technol.* **1996**, *83*, 40–44. [CrossRef]
15. Gröner, L.; Kirste, L.; Oeser, S.; Fromm, A.; Wirth, M.; Meyer, F.; Burmeister, F.; Eberl, C. Microstructural investigations of polycrystalline Ti_2AlN prepared by physical vapor deposition of Ti-AlN multilayers. *Surf. Coat. Technol.* **2018**, 166–171. [CrossRef]
16. Levchuk, D.; Bolt, H.; Döbeli, M.; Eggenberger, S.; Widrig, B.; Ramm, J. Al–Cr–O thin films as an efficient hydrogen barrier. *Surf. Coat. Technol.* **2008**, *202*, 5043–5047. [CrossRef]
17. Chikada, T.; Suzuki, A.; Yao, Z.; Levchuk, D.; Maier, H.; Terai, T.; Muroga, T. Deuterium permeation behavior of erbium oxide coating on austenitic, ferritic, and ferritic/martensitic steels. *Fusion Eng. Des.* **2009**, *84*, 590–592. [CrossRef]
18. Frank, R.C.; Swets, D.E.; Fry, D.L. Mass Spectrometer Measurements of the Diffusion Coefficient of Hydrogen in Steel in the Temperature Range of 25–90 °C. *J. Appl. Phys.* **1958**, *29*, 892–898. [CrossRef]
19. Gorman, J.K.; Nardella, W.R. Hydrogen Permeation through Metals. *Vacuum* **1962**, *12*, 19–24. [CrossRef]
20. Levchuk, D.; Koch, F.; Maier, H.; Bolt, H. Gas-driven Deuterium Permeation through Al_2O_3 Coated Samples. *Phys. Scr.* **2004**, *T108*, 119–123. [CrossRef]
21. Tang, C.; Klimenkov, M.; Jaentsch, U.; Leiste, H.; Rinke, M.; Ulrich, S.; Steinbrück, M.; Seifert, H.J.; Stueber, M. Synthesis and characterization of Ti_2AlC coatings by magnetron sputtering from three elemental targets and ex-situ annealing. *Surf. Coat. Technol.* **2017**, *309*, 445–455. [CrossRef]
22. Ines, D. Raman Spectroscopy Analysis of CVD Hard Coatings Deposited in the $TiC_{1-x}Nx$, $TiB_xC_yN_z$ and Ti-B-N System. Ph.D. Thesis, Eberhard Karls Universität Tübingen, Tübingen, Germany, 2011.
23. Wang, Z.; Li, X.; Li, W.; Ke, P.; Wang, A. Comparative study on oxidation behavior of Ti_2AlN coatings in air and pure steam. *Ceram. Int.* **2019**, *45*, 9260–9270. [CrossRef]
24. Aminzadeh, A. Excitation Frequency Dependence and Fluorescence in the Raman Spectra of Al_2O_3. *Appl. Spectrosc.* **1997**, *51*, 817–819. [CrossRef]
25. Wang, X.H.; Zhou, Y.C. Oxidation behavior of Ti_3AlC_2 powders in flowing air. *J. Mater. Chem.* **2002**, *12*, 2781–2785. [CrossRef]
26. Luo, M.-F.; Fang, P.; He, M.; Xie, Y.-L. In situ XRD, Raman, and TPR studies of CuO/Al_2O_3 catalysts for CO oxidation. *J. Mol. Catal. A Chem.* **2005**, *239*, 243–248. [CrossRef]
27. Zhang, Z.; Jin, H.; Pan, J.; Chai, J.; Wong, L.M.; Sullivan, M.B.; Wang, S.J. Origin of Al Deficient Ti2AlN and Pathways of Vacancy-Assisted Diffusion. *J. Phys. Chem. C* **2015**, *119*, 16606–16613. [CrossRef]

Article

The Effect of Laser Power on the Properties of M_3B_2-Type Boride-Based Cermet Coatings Prepared by Laser Cladding Synthesis

Zhaowei Hu, Wenge Li * and Yuantao Zhao *

Merchant Marine College, Shanghai Maritime University, Shanghai 201306, China; huzw0731@sina.com
* Correspondence: wgli@shmtu.edu.cn (W.L.); zhaoyt@shmtu.edu.cn (Y.Z.); Tel.: +86-1391-799-6912 (W.L.)

Received: 12 March 2020; Accepted: 14 April 2020; Published: 16 April 2020

Abstract: Boride-based cermet can serve as a good protective coating for low-corrosion and wear-resistant materials, such as carbon steels, due to their mechanical and chemical properties. In this study, M_3B_2 (M: Mo, Ni, Fe, and Cr) boride-based cermet coatings were fabricated on Q235 steel with mixed powders of Mo, B, Ni60, and Cr by laser cladding synthesis, and the effects of laser power on the properties of the cermet layer were investigated. Three laser powers (2200, 2500, and 2800 W) were used at the same scanning speed. The X-ray diffraction (XRD), scanning electron microscopy (SEM), and energy-dispersive X-ray spectroscopy (EDS) analysis confirmed that all the coatings were composed of M_3B_2-type borides and {Fe, Ni} alloys. The micro-hardness, corrosion, and frictional experiments showed that the cermet coatings enhanced the corresponding performances of the Q235 steels at the three laser powers. However, the micro-hardness of the coatings decreased as the power increased, and the maximum micro-hardness value was 1166.3 HV (Vickers Hardness). The results of the corrosion and frictional experiments showed that the best performance was obtained at a laser power of 2500 W, followed by 2800 and 2200 W.

Keywords: boride-based cermet; laser cladding synthesis; laser power; microstructure; hardness; corrosion resistance; wear resistance

---

1. Introduction

In recent years, wear- and corrosion-resistant materials have been used in various industrial areas. However, the slurry flow unit breaks easily due to erosive wear and erosion–corrosion, thus a material that has comprehensive properties needs to be used. Boride-based cermet composite coatings consisting of a transition metal base matrix with dispersed hard phases, such as Mo_2FeB_2, WCoB, MoCoB, and Mo_2NiB_2, possess high hardness, high melting points, good wear and corrosion resistance, and good optical and thermal properties [1–6]. These cermet coatings have been applied to injection molding machine parts, cutters for heat sealers, bearings for sea water pumps, offshore engineering parts, and slurry flow units in the coal-mining industry [7,8]. In particular, as emerging and promising materials with superior strength, hardness, wear, and corrosion resistance, M_3B_2-type boride-based cermet coatings, with Mo_2FeB_2, Mo_2NiB_2, and W_2NiB_2 as hard phases, are qualified candidates for applications that require wear and corrosion resistance [9,10].

Kenichi Takagi prepared the ternary boride using the sintering method, and systematically studied the effects of the Mo/B atomic ratio and additional elements on the mechanical properties and structure of boride cermets, and the mechanism of boride formation [11–17]. Cr and V addition resulted in improvement of the mechanical properties, such as TRS (transverse rupture strength) and hardness associated with a simultaneous phase transformation of complex boride from orthorhombic to tetragonal. The addition of Fe, Co, Ti, Mn, Zr, Nb, and W did not result in structural changes, but did degrade the mechanical properties [11]. Adding 3.5 wt.% Cr and 11.5 wt.% V, the TRS of boride increased when the Mo/B atomic ratio increased and then decreased, while the hardness and density increased as the ratio increased.

The maximum TRS and hardness were 2.95 GPa and 90.5 R_A (Rockwell A hardness) [12]. The sintering mechanism of boride was analyzed by studying the microstructure of adding 10 wt.% Cr and 12.5 wt.% V, respectively, at different sintering temperatures, and the mechanical properties were discussed. The results showed that Mo_2NiB_2 was formed in the compact by a reaction of 2MoB + Ni = Mo_2NiB_2, and the addition of V yielded better mechanical properties [13]. The addition of Mn had an effect on the mechanical properties and microstructure of Mo_2NiB_2-based cermets in the powder composition of Mo, Ni, B, V, and x wt.% Mn. The hardness increased with Mn addition, while the TRS increased then decreased; the maximum TRS was 3.5 GPa at 2.5 wt.% Mn [14]. Adding 12.5 wt.% V and 2.5 wt.% Mn, the TRS of boride increased with the increasing Mo/B atomic ratio and then decreased, while the hardness increased as the ratio increased.

The maximum TRS and hardness were 3.25 GPa and 90.8 R_A [15]. Adding V can greatly improve the mechanical properties of cermets. Corrosion tests in the molten fluorocarbon resin revealed that the Mo_2NiB_2 boride cermets had far better corrosion resistance than high-speed steel and SUS 304 [16]. The effects of Cr on the properties of Mo_2NiB_2 ternary boride were studied, while the Mo_2NiB_2 molar formula was assumed as $(Mo_{2-x}Ni_{1-x}Cr_{2x})B_2$ for Cr-substituted Mo and Ni. Cr was doped into Mo_2NiB_2 ternary borides. In particular, 10 and 15 wt.% Cr exhibited high hardness and high elastic moduli, which makes them a suitable alternative material for wear-resistant hard materials, such as WC [17]. Yuan et al. [18] prepared Mo_2NiB_2 cermets and found that the maximum bending strength and hardness of Mo_2NiB_2 cermets reached 1.85 ± 0.04 GPa and 85.7 ± 0.1 R_A, respectively.

In Lei Zhang's study, Mo_2NiB_2-Ni cremets with different Ni contents and ball milling times were fabricated by reaction boronizing sintering [19,20]. The cermets when molar ratio of Ni/B is 1.1 had the best mechanical properties and the lowest wear rates. The cermets with a milling time of 11 h exhibited a maximum hardness and bending strength of 87.6 HRA and 1367.3 MPa, respectively. Further, the high-temperature compressive strength and tribological behaviors of the cermets were investigated [21]. Sevinch [22] prepared Mo_2NiB_2 by self-propagating high-temperature synthesis using a mixture of MoO_3, NiO, B_2O_3, and Al powders.

Most research has focused on the sintering method. Wu et al. [23,24] found that continuous, dense, and adherent Mo_2NiB_2 cermet coatings were obtained by a laser surface cladding technique, which was chosen as laser cladding possesses some advantages over other methods, such as its convenience, high efficiency, high energy, cost effectiveness, and the fact that it is environmentally friendly [25]. However, the influence of laser parameters has not been deeply discussed in laser cladding synthesis. The laser power, as one of the most important variants, should be primarily investigated, as it would increase the scope of applications of Mo_2NiB_2 coatings.

Herein, several various laser powers were studied to reveal the influences of power on the microstructure and properties of M_3B_2 boride-based cermet coatings on Q235 steels. Then, the microstructures of the coatings were analyzed by X-ray diffraction (XRD), scanning electron microscopy (SEM), and energy-dispersive X-ray spectroscopy (EDS). The hardness, corrosion resistance, and wear resistance of the coatings were evaluated. How the laser power affected the microstructure and properties of the coatings was revealed.

2. Materials and Methods

2.1. Laser Cladding Synthesis

Commercially available Mo (Tianjiu, Changsha, China, 99.9% pure, particle size < 48 μm), B (Tianjiu, Changsha, China, 99% pure, particle size < 48 μm), Cr (Tianjiu, Changsha, China, 99% pure, particle size < 48 μm), and Ni60 (Kenna Metal, Pittsburgh, PA, USA, particle size < 45 μm) powders were used as raw materials. The composition of mixed powders is listed in Table 1. The Q235 steel

(C 0.14%–0.22%, Mn 0.3%–0.65%, Si ≤ 0.3%, S ≤ 0.05%, and P ≤ 0.045%) was used as substrate with a size of 100 mm × 80 mm × 10 mm.

Table 1. Composition of the powder.

Element	Mo	Ni	B	Cr	Fe	Si	W	C
wt.%	57.82	17.68	7.58	10	4.56	1.21	0.91	0.24

The raw powders were ball milled for 12 h in a planetary ball mill machine. To remove rust, dust, and oils, the substrates of Q235 steel were sandblasted for 10 min, and then cleaned with 99.8% alcohol in an ultrasonic cleaner for 15 min. As one of the common methods, the pre-placed powder method has been utilized here to investigate the effect of laser power on the properties of coatings. The mixed powders were placed on the surface of the substrate using polyvinyl butyral (PVB) binder. The thickness of the placed layer was about 1 mm. Then, the substrate layer was dried naturally in a drying cabinet. A Rofin DC030 laser (Rofin, Hamburg, Germany) was used as the laser source. The wavelength was 10.6 μm with a rectangular spot of 6 mm × 2 mm. The substrate with the layer was scanned by a laser after drying, as shown in Figure 1. During the laser cladding synthesis process, argon was used as a protective gas to prevent oxidation. As clearly seen in Figure 1a, the argon is blown out of the nozzle shielding oxygen entering the molten pool. The laser scanning sequences are shown in Figure 1b, and the laser spot was overlapped.

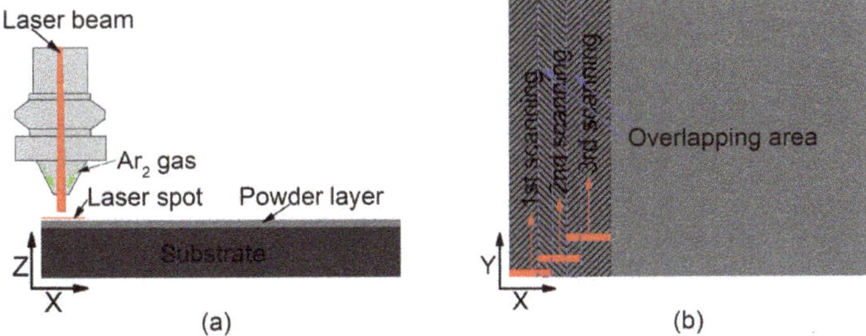

Figure 1. Laser cladding synthesis: (**a**) X–Z direction view; (**b**) X–Y direction view.

Here, the laser powers were 2200, 2500, and 2800 W, respectively, which was decided by good surface conditions in the forgone investigations, while the scanning speed was 1 mm/s, and overlap rate was 33%. The samples of 2200, 2500, and 2800 W were marked as No.1, No.2, and No.3, respectively.

2.2. Microstructure Analysis

The phase compositions of the cermet coatings were analyzed by X-ray diffraction (XRD, Rigaku Ultima IV, Tokyo, Japan) with Cu-Kα radiation (λ = 1.54 A) operated at 40 kV and 30 mA. The detected diffraction angle (2θ) was scanned from 20° to 100° and the scanning speed was 5°/min. The results were obtained using the XRD analysis software X'Pert HighScore Plus (v1.1, Malvern Panalytical, Etten Leur, Netherlands). The microstructure and composition distribution of the cermet coatings were characterized by scanning electron microscopy (SEM, Hitachi TM3030, Tokyo, Japan) equipped with an energy-dispersive X-ray spectroscopy (EDS, Oxford Swift 3000, Oxford, UK) machine.

2.3. Properties Analysis

The microhardness of the cross-sections of the cermet coatings was measured with a Vickers hardness tester (HXD-1000TMC/LCD, Shanghai TaiMing, Shanghai, China) under a load of 0.98 N and a dwell time of 15 s at room temperature. The corrosion resistance of the cermet coatings was evaluated by an electrochemical workstation (Autolab PGSTAT302N, Herisau, Switzerland) in 3.5 wt.%

NaCl solution at room temperature. A standard three-electrode system was used. The reference electrode was a silver chloride electrode (Ag/AgCl), the counter electrode was platinum, and the working electrode was coated with specimens with a surface area of 1 cm^2. Before each corrosion experiment, the cermet coatings were immersed in 3.5 wt.% NaCl solution for 1.5 h. The open circuit potential (OCP) was measured for 1 h.

The electrochemical impedance spectra (EIS) was measured at the OCP with a potential amplitude of 5 mV and a frequency of 0.01 to 100,000 Hz. Potentiodynamic polarization experiments were scanned at 1 mV/s in the range of OCP ± 0.8 V, and the corrosion potential (E_{corr}) and corrosion current (i_{corr}) were obtained by the Tafel extrapolation method [26]. The corrosion results were analyzed by the software Nova 2.1.4 (version 2.1.4, Metrohm Autolab, Utrecht, The Netherlands). The wear resistance of the cermet coatings was measured by tribometer (UMT TriboLab, Bruker, Campbell, CA, USA). The cermet coatings reciprocated sliding against a Si3N4 ball with ϕ8 mm at room temperature under dry sliding conditions. The experiments were continuously applied with a load of 45 N lasting 1 h and a sliding frequency of 5 Hz for a sliding distance of 6 mm. The surface features of the cermet coatings were measured using a 3D optical microscope (ContourGT-1, Bruker, Campbell, CA, USA). After wear testing, the worn morphologies were observed by SEM.

3. Results

3.1. Microstructure and Composition of Cermet Coatings

The XRD patterns of cermet coatings were recorded. The laser power was increased to analyze the change in crystal structures. Figure 2. shows the XRD patterns of the coatings. The $(M_2M)_3B_2$-type boride phase, {Fe, Ni} phase and unknown phase were detected in Sample 1, Sample 2, and Sample 3. The M in the $(M_2M)_3B_2$ phase represents Mo, Ni, Cr, and Fe. They also represent the sorting of the intensity of the M_3B_2 and {Fe, Ni} phases, respectively, in a standard card with the black numbers and green numbers. The intensity of M_3B_2 in Sample 1 was much stronger than in Sample 2, while the intensity of Sample 2 was a little stronger than that of Sample 3. The intensity of {Fe, Ni} in Sample 1 was much stronger than in Sample 2 and 3 at position (1). The intensities of Sample 1 and Sample 2 were approximately the same at position (2) but much stronger than that of Sample 3. Position (1) was the strongest peak and position (2) was the second strongest peak in the standard card. The peak intensities of all samples were similar at the other three positions. The M_3B_2 phase intensity decreased as the laser power increased. The intensity of the {Fe, Ni} phase also decreased, but not sharply, as the laser power increased.

Figure 3a,b show a macro view of the coating after cladding. It can be seen that the overall coating after lapping is relatively flat and there are no visible macro cracks. Figure 3c,d are SEM sample images after aqua regia corrosion is taken from the coating. The light gray is the substrate and the dark gray is the coating. The coating and the substrate are well combined, and the coating can be seen with detailed overlapping patterns.

The cross-section morphological images of cermet coatings were recorded with the increase of laser power, as shown in Figures 4–6. There is an interface between the substrate and the coating in the morphological images. Distributed in cermet coating Sample 1, Sample 2, and Sample 3, there are white and gray phases, which are the M_3B_2 hardness phases and {Fe, Ni} binder phases, respectively. The curve radian of the interface at the molten pool area is evidently larger than that of the overlap area shown in Figure 4a. The white phase is long and uniformly distributed in the coating, as shown in Figure 4, while the laser power was 2200 W. Compared with the gray phase in the coating, the white phase is more abundant. The smaller grain size is about 2.8 μm, the larger grain size is about 10 μm, and the much longer grains will break, as shown in Figure 4b.

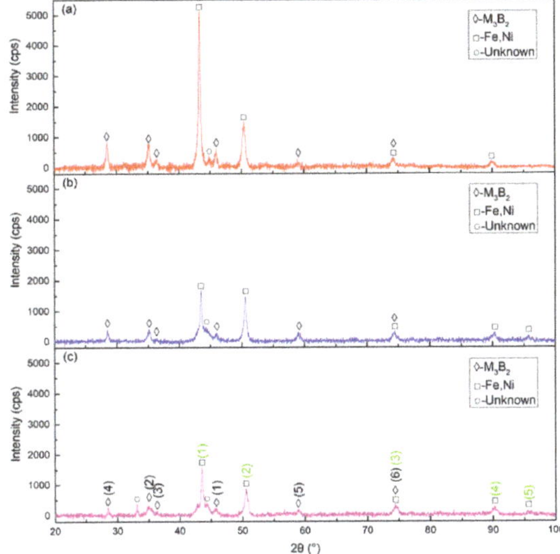

Figure 2. X-ray diffraction (XRD) patterns of the cermet coatings at various laser powers: (**a**) laser power of 2200 W; (**b**) laser power of 2500 W; (**c**) laser power of 2800 W.

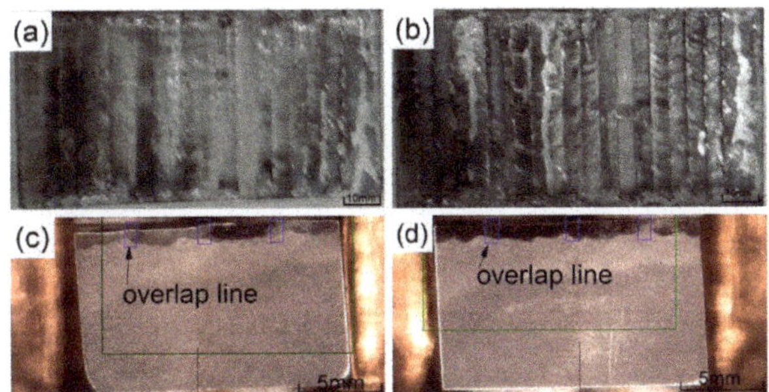

Figure 3. The macro morphology of the coatings: (**a**) 2500 W; (**b**) 2800 W; (**c**) cross section at 2500 W; (**d**) cross section at 2800 W.

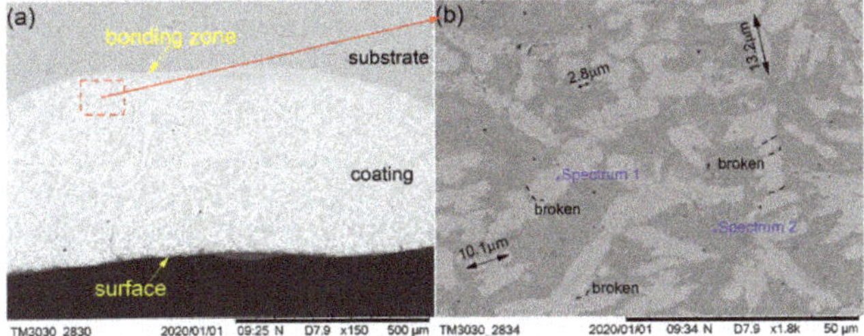

Figure 4. Scanning electron microscopy (SEM) morphology of the cermet coatings at 2200 W: (**a**) low magnification; (**b**) high magnification.

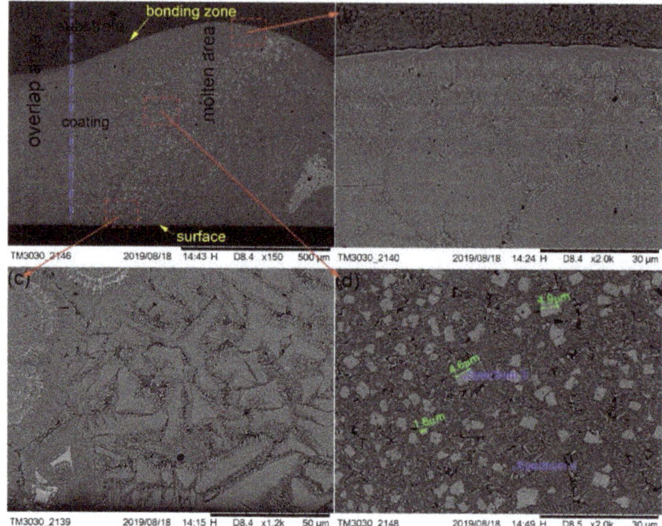

Figure 5. SEM morphology of the cermet coatings at 2500 W: (**a**) low magnification; (**b**) bonding zone at high magnification; (**c**) surface area at high magnification; (**d**) middle area at high magnification.

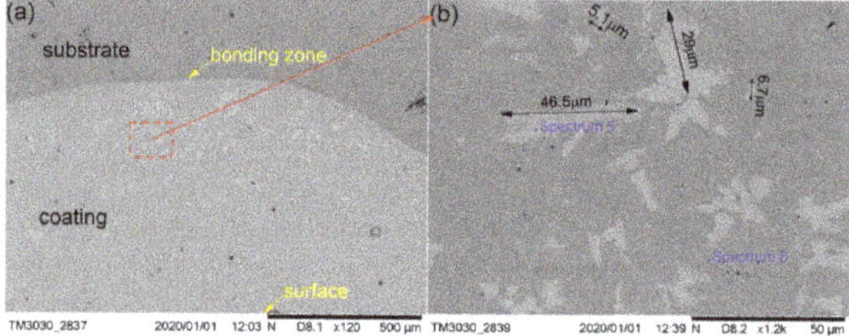

Figure 6. SEM morphology of the cermet coatings at 2800 W: (**a**) low magnification; (**b**) high magnification.

Figure 5 shows the SEM images that depict the cross-sectional morphology of the cermet coatings at the laser power of 2500 W, and shows enlarged views of the bonding zone (Figure 5b), the surface area (Figure 5c), and the middle area (Figure 5d), respectively. The curve radian of the interface at the laser power of 2500 W is larger than that of 2200 W. The curve radian of the bonding zone at the molten pool area is clearly larger than that of the overlap area. Figure 5a shows that the white phase is evenly distributed in the molten pool area, and less distributed in the overlap area and the area near the interface. Compared with the grain at 2200 W, the grain size at 2500 W was significantly smaller and more regular. However, the distribution of grains in the entire cross-section of the coating was more uniform at 2200 W.

It can be seen from Figures 4a and 5a that the white phase was present in a lower quantity at 2500 W than that at 2200 W. The morphology of the gray phase is different in the bonding zone and the surface area. In the surface area, the gray phase is constituted by crystals that are shaped as triangles, squares, bars, and other shapes, and the edges have burrs similar to feathers. In the interface area, most of the gray phases contain crystals that are square-shaped, and some of them contain thin, long white phases. It can be seen from Figure 5d that the shape of crystals in the white phase in the middle area is basically square, the small grain size is approximately 1.8 μm, and the largest grain size is approximately 4.9 μm.

Figure 6 shows the SEM images depicting the cross-sectional morphologies of the cermet coating at the laser power of 2800 W and shows enlarged views of the middle area (Figure 6b). As in Figures

4a and 5a, the curve radian of the interface at the molten pool area is larger than that of the overlap area in Figure 6a. The white phase was present in a lower quantity, and it was unevenly scattered in the coating at 2800 W. There are more white phases distributed at the bottom of molten pool than other areas. The morphology of the white phase is irregular, as shown in Figure 6b. The smaller grain size was approximately 5.1 μm, the larger grain size was approximately 29 μm, and the long grain size could reach up to 50 μm.

The compositions of the white and gray phases—marked as spectrums 1 to 5 in Figures 4–6—were evaluated, as shown in Figure 7, with the increase of laser power. In order to elucidate the effects of laser power on the elements of the white phases, the EDS spectrums 1, 3, and 5 were employed in three kinds of cermet coatings, respectively. Meanwhile, spectrums 2, 4, and 6 were also carried out to depict the variation of gray phases in the cermet coatings. Evidently, the element contents of the white phase and the gray phase are completely different, as shown in Figure 7. The contents of Mo are much higher in the white phases than in the gray phases, whereas the contents of Fe are greatly increased in the gray phase. Thus, the gray phases can be considered as the binder of {Fe, Ni} with few Mo, and the white phases are M_3B_2 with B aggregation, in which M represents Mo, Ni, Cr, and Fe. The high concentration of Fe originates from the dilution of substrate steels, which is a common phenomenon in laser cladding [27]. The detailed element contents are given in Table 2.

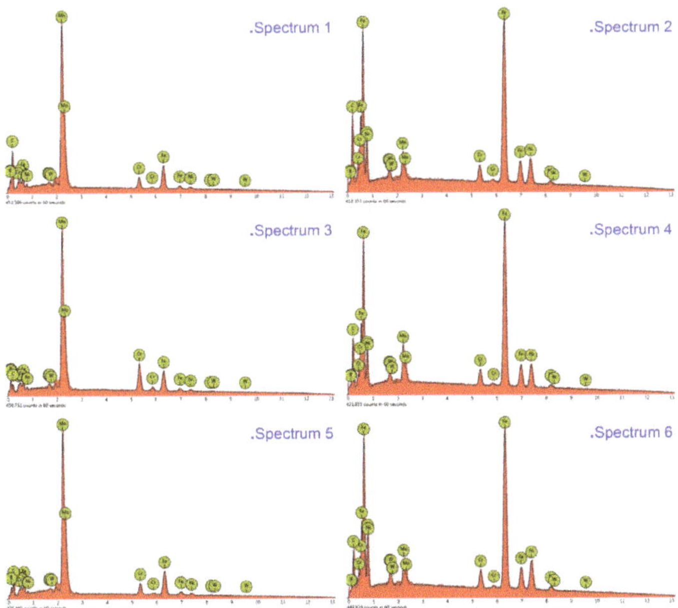

Figure 7. Energy-dispersive X-ray spectroscopy (EDS) analysis of the white and gray phases in the cermet coatings at various laser powers.

Table 2. Element contents (at%) at different locations from EDS.

Laser Power (W)	Location	Mo	B	Ni	Fe	Cr	Si	W
2200	Spectrum1	40.81	28.43	2.32	20.42	6.72	0.76	0.54
	Spectrum2	4.29	0	17.23	71.11	4.38	2.33	0.66
2500	Spectrum3	40.76	28.56	2.11	20.96	6.72	0.62	0.27
	Spectrum4	3.95	0	19.34	68.08	4.72	3.25	0.66
2800	Spectrum5	36.21	30.26	1.48	15.83	14.19	1.2	0.84
	Spectrum6	5.64	0.74	15.71	71.08	4.23	2.22	0.39

3.2. Hardness of M_3B_2-Based Boride Cermet Coatings

The positions at which the micro Vickers hardness tests of the cermet coatings were carried out are shown in Figure 8. Seven points were tested on the cross-section, and the first point was taken as close as possible to the coating surface. The distance between the test points was one fifth of the thickness of the coating. The distance between the test points was different in the three samples as the coating thicknesses of the samples were different.

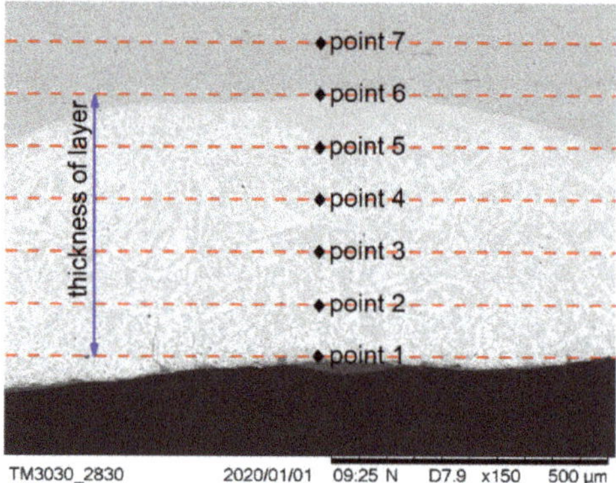

Figure 8. Positions of the hardness measurements.

The results of the hardness test are shown in Figure 9. This indicated that the micro-hardness decreased along the surface of the substrate at different laser powers. The hardness of the coating decreased with increasing laser power. The hardness at 2500 W was slightly greater than the hardness at 2800 W, but significantly less than the hardness at 2200 W. The hardness suddenly changed at a certain position in the middle area of the coating when the power was 2800 W; it was not only higher than the hardness of the two positions before and after it, but also higher than the hardness of the other two samples at the same position. The maximum hardness value in all test points was 1166.3 HV, when the power was 2200 W at the surface position.

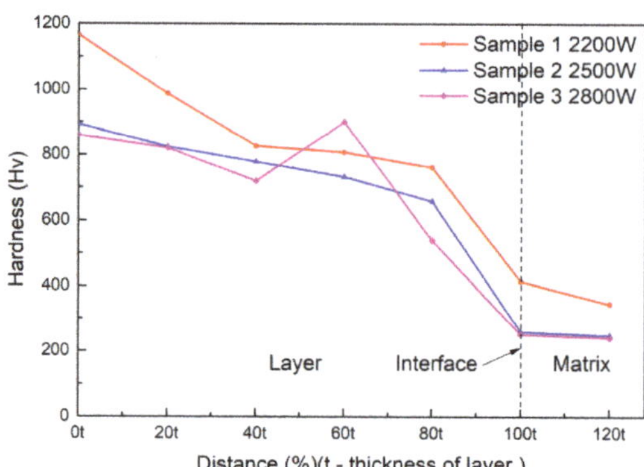

Figure 9. Vickers hardness of the cermet coatings at various laser powers.

3.3. Corrosion Tests of Cermet Coatings

The substrate and all cermet coatings were immersed in 3.5 wt.% NaCl solution for 1.5 h before the OCP test. The results of the OCP were listed in Table 3. Compared to the coatings and substrate, the OCP shifted positively. The difference of the OCP between 2200 and 2800 W was not great, but the OCP reached −0.3161 V at 2500 W with a significant increase.

Table 3. The open circuit potential (OCP) of the substrate and cermet coatings immersed in NaCl solution.

	Substrate	Sample 1	Sample 2	Sample 3
OCP (V)	−0.6438	−0.54443	−0.3161	−0.52393

As is known, the EIS technique has been extensively used to characterize surface coatings in a semiquantitative way [28]. Nyquist plots recorded with samples immersed in NaCl solution are shown in Figure 10. All impedance spectra of the samples displayed depressed semicircles. The diameter of the impedance spectra reflected the rate of the electrochemical reaction [29]. As shown in Figure 10, the diameter of the impedance spectra arc of coatings increased compared to the substrate. The laser power had a great effect on the impedance spectra of the coating, and the diameter of the impedance spectra arc increased first and then decreased with increasing power. When the power was 2500 W, the corrosion rate was minimum.

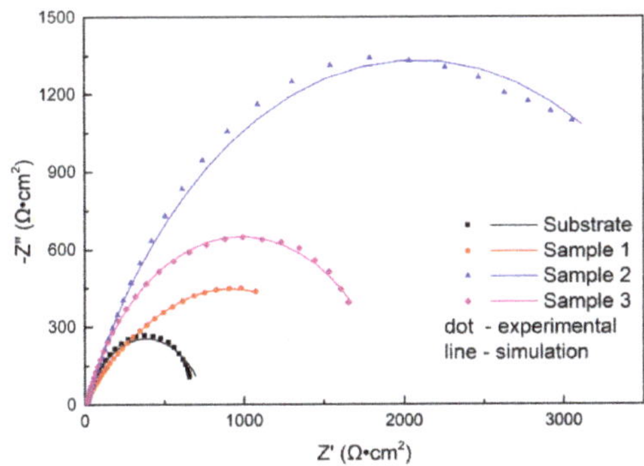

Figure 10. Nyquist diagrams for the substrate and cermet coatings at various laser powers.

Moreover, Figure 11 displays the electrochemical data of the samples as Bode plots. They exhibited the same features as the Nyquist plots. The Z value was the smallest for the substrate and the largest in Sample 2 among the three coating samples at the lowest frequency, which indicated that the coating had better corrosion resistance than the substrate and had the best corrosion resistance at 2500 W.

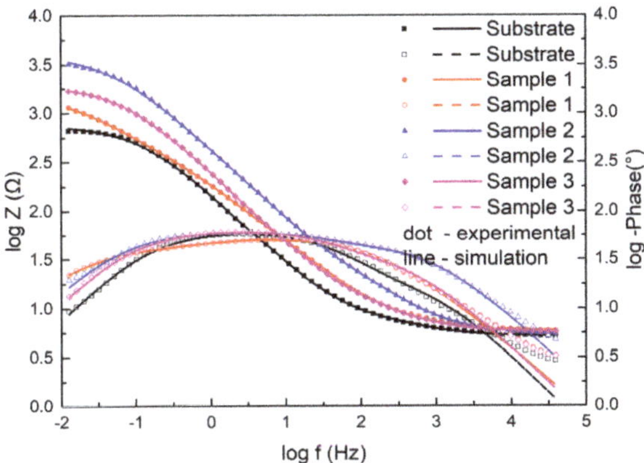

Figure 11. Bode diagrams for the substrate and cermet coatings at various laser powers.

The Bode diagrams with a wide phase angle in Figure 11 indicated that there are at least two time constants [30]. The equivalent circuit of the two time constants in Figure 12 was used to fit the EIS data using Nova software. Here, R_s is the solution resistance from the reference electrode to the working electrode, R_f is the surface coating resistance, and R_{ct} is the constant phase elements of the charge transfer resistance. The impedance of the regular phase-angle element Q is related to the angular frequency (ω) by the following relation: $Z_{CPE} = Y_0^{-1}(j\omega)^{-n}$, where Y_0 is a proportionality factor and n is the deviation parameter, which reflects the roughness of the surface; Q_f is the constant phase angle element (CPE) between the solution and the surface, and Q_{dl} is the CPE of the interface double-layer between the solution and the matrix. The fitting lines are also plotted in Figures 10 and 11, and the fitting results are in good agreement with the experimental results, indicating that the equivalent circuit is feasible.

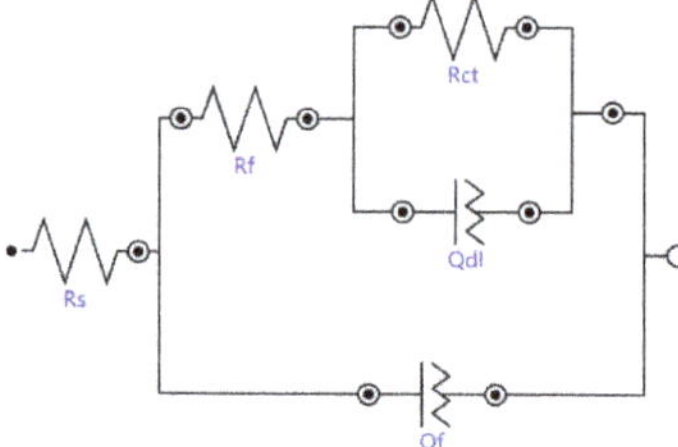

Figure 12. Equivalent circuit model of the substrate and cermet coatings.

The fitting results of the EIS data of all samples immersed in 3.5 wt.% NaCl solution with different processing parameters are listed in Table 4. The solution resistance R_s did not change significantly, indicating that the environmental conductivity was relatively stable. Compared with the substrate, the R_f of coatings increased, and the R_f at 2200 W was much larger than at 2500 W and 2800 W, which indicated that laser power had an effect on the surface coating resistance. The polarization resistance R_p ($R_p = R_f + R_{ct}$) is used to evaluate the corrosion resistance of materials [31]: the greater the R_p value, the better the corrosion resistance. The R_p of Sample 2 was 4115.6 Ωcm^2, more than twice that of Sample 1 and Sample 3, and the R_p of Sample 3 was slightly larger than that of Sample 1. The results are consistent with the previous analysis of the Nyquist and Bode plots.

Table 4. Fitted values of the equivalent circuit of the electrochemical impedance spectra (EIS) diagram.

Sample No	R_s (Ωcm^2)	R_f (Ωcm^2)	Q_f,Y_0 ($m\Omega^{-1}cm^{-2}s^n$)	R_{ct} (Ωcm^2)	Q_{dl},Y_0 ($m^{-1}cm^{-2}s^n$)
Substrate	5.21	5.39	0.675	751	0.952
No.1	5.9	639	1.342	1072	2.165
No.2	5.35	48.6	0.397	4067	0.197
No.3	5.78	20.2	0.659	1928	0.323

The polarization curves of the substrate and cermet coatings immersed in 3.5 wt.% NaCl solution are presented in Figure 13. It is evident from Figure 13 that both the substrate and the coatings have undergone the transition process from the activated state to the passivated state and ultimately, to overpassivation; however, the duration of each process varied. The self-corrosion potentials of Samples 2 and 3 shifted positively compared to the substrate, but the self-corrosion potentials of Sample 1 shifted negatively, while the OCPs of coatings were all positively shifted. The passivation state of the substrate and Sample 3 was short, quickly changing to the overpassivation state, but Sample 3 experienced a brief secondary passivation state. Samples 1 and 2 maintained a long passivation state, and Sample 2 entered a secondary passivation state after overpassivation. The polarization curves show that the coating at 2500 W had very good corrosion resistance.

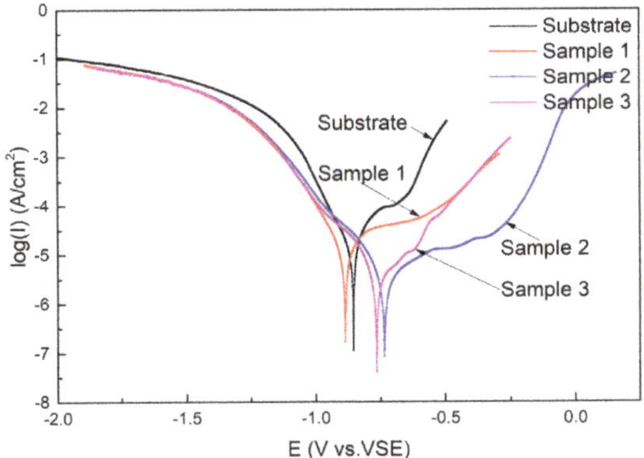

Figure 13. Polarization curves of the substrate and cermet coatings.

According to the relationship between the current density i and the polarization potential E of the Butter–Volmer equation (Equation (1) and (2)), the polarization resistance R_p can be calculated by Equation (3) [32].

$$i = i_{corr}[\exp(2.303\Delta E)/b_a - \exp(-2.303\Delta E)/b_c], \quad (1)$$

$$\Delta E = E - Ecorr \quad (2)$$

$$R_p = b_a b_c/(2.303(b_a + b_c)i_{corr}). \quad (3)$$

The corrosion parameters related to the polarization curves are calculated and listed in Table 5. i_{corr} is the corrosion current density, E_{corr} is the self-corrosion potential, b_a is the anode Tafel constant, and b_c is the cathode Tafel constant. E_{corr},Obs in Table 5 is the observed value, and E_{corr},Calc is the calculated value. The i_{corr} decreased from 16.9 to 2.93 μA, which indicated that the coating on the substrate greatly improved the corrosion performance of the substrate. As the laser power increased, i_{corr} decreased and then increased, and the polarization resistance showed the opposite trend. The results of the analysis are consistent with the previous results.

Table 5. The electrochemical parameters obtained from the polarization curves of the substrate and coatings.

Sample No	E_{corr},Obs (V)	E_{corr},Calc (V)	i_{corr} (A)	$\lvert b_a \rvert$ (V/dec)	$\lvert b_c \rvert$ (V/dec)	Polarization Resistance (Ω)
Substrate	−0.85436	−0.87408	1.6907×10^{-5}	0.17169	0.09951	1618.2
No.1	−0.88434	−0.89038	1.1189×10^{-5}	0.22088	0.10775	2811
No.2	−0.73426	−0.72866	2.9279×10^{-6}	0.24416	0.12023	11949
No.3	−0.76463	−0.74899	3.2315×10^{-6}	0.2036	0.12814	10569

3.4. *Wear Tests of Cermet Coatings*

The frictional coefficient curves of the substrate and cermet coatings are shown in Figure 14. The frictional coefficients of the substrate and coatings rose rapidly at the start of the test, and then the substrate and Samples 2 and 3 entered the steady wear stage after a breaking-in process of approximately 2.5 min, whereas Sample 1 took approximately 7.5 min. After entering the steady state, the frictional coefficient of the substrate and Sample 1 displayed some fluctuations and showed a slight upward trend, while the fluctuations in the frictional coefficient of Samples 2 and 3 were small and showed a downward trend. The order of the frictional coefficients of the substrate and the coating tended to be consistent after sliding for 30 min. The frictional coefficient of the substrate was larger than that of the coatings, and this decreased as the power increased. The frictional coefficient of coatings was the smallest at 2800 W.

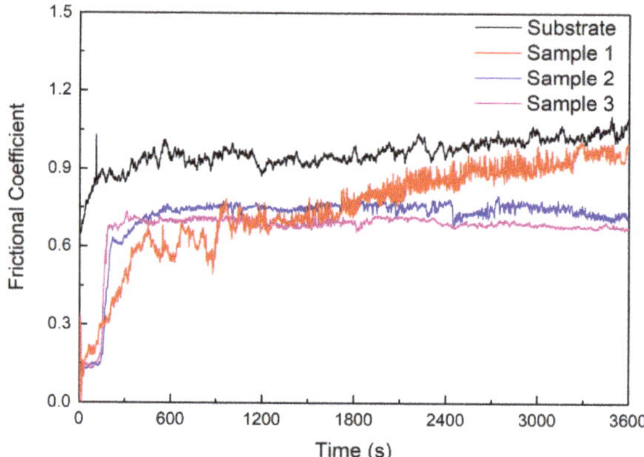

Figure 14. Friction coefficient of the substrate and cermet coatings.

The wear volumes of the substrate and cermet coatings are shown in Figure 15. The volume loss of the substrate was approximately 2.6 to 3.1 times that of the coatings. The influence of laser power on the volume loss of coatings was different from the frictional coefficient. The volume loss increased as the frictional coefficient increased. The frictional coefficient of Sample 2 was larger than that of Sample 3, but the volume loss of Sample 2 was smaller. The coating with the best wear resistance was obtained at a laser power of 2500 W.

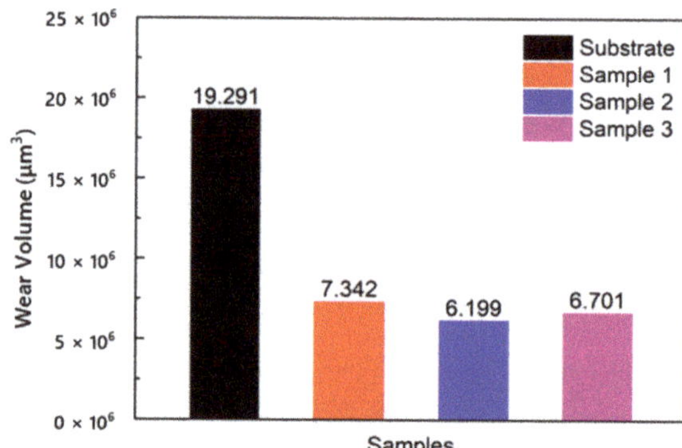

Figure 15. Volume losses of the substrate and cermet coatings at various laser powers.

The three-dimensional morphologies of the wear tracks of the substrate and coatings are shown in Figure 16. The plow-furrows can be clearly identified in the worn area of the substrate, and some pits were found on the surface of Sample 1, while only fine scratches were found on the surfaces of Samples 2 and 3. As seen from Figure 16, the wear tracks of coatings had a lower wear depth than that of the substrate (31.74 μm). The wear depth of Sample 2 (9.41 μm) was similar to that of Sample 3 (9.69 μm), which was much smaller than that of Sample 1 (13.17 μm). This further confirmed that the coating had the best wear resistance at 2500 W.

The worn morphologies of the substrate and cermet coatings were observed by SEM, as shown in Figure 17. The worn surface of the substrate displayed large plastic deformation and furrows, and many adhered materials and microcracks parallel to the ploughing direction (Figure 17a). The worn surface of Sample 1 did not have obvious ploughing furrows, but had distorted plastic deformation, peeling and adhered materials, and micro-cracks appeared in the peeling layer (Figure 17b). The worn surface of Sample 2 was relatively flat, with small plastic deformation and peeling. Micro-cracks were observed in the peeling layer, and no obvious adhered materials were seen (Figure 17c). The worn surface of Sample 3 had shallow furrows and more peeling and adhered materials, and there were micro-cracks in the dark area (Figure 17d). The analysis is consistent with the previous results.

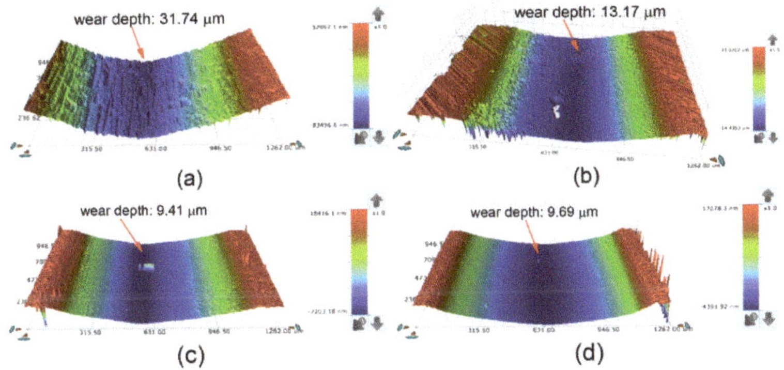

Figure 16. Three-dimensional topography of the wear tracks: (**a**) substrate; (**b**) laser power of 2200 W; (**c**) laser power of 2500 W; (**d**) laser power of 2800 W.

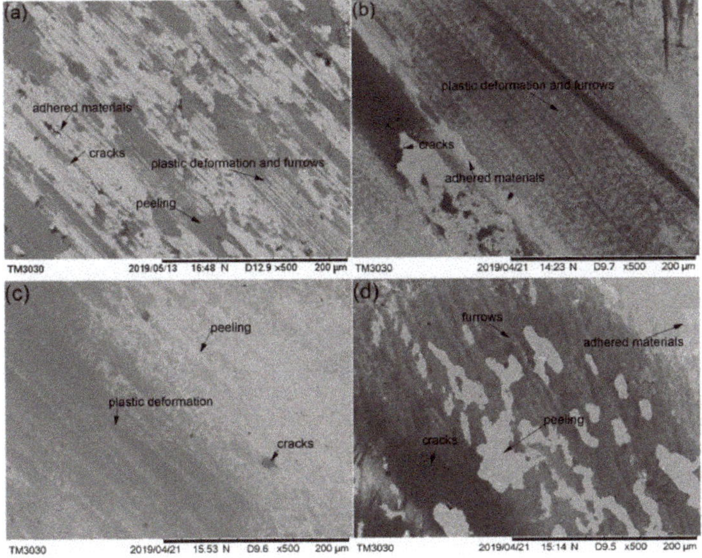

Figure 17. The SEM morphology of the worn substrate and cermet coatings: (**a**) substrate; (**b**) laser power of 2200 W; (**c**) laser power of 2500 W; (**d**) laser power of 2800 W.

4. Discussion

The XRD analysis showed that there were at least two phases in the coating. By comparison with the standard card, the two phases were identified as M_3B_2 and {Fe, Ni}. The absolute intensity of M_3B_2 in Sample 1 was much stronger than in Sample 2, while the intensity of Sample 2 was slightly greater than that of Sample 3. The absolute intensity of {Fe, Ni} in Sample 1 was also much stronger than in Sample 2, but the intensity of Sample 2 was a little stronger than that of Sample 3. From the XRD diffraction peaks of the samples, their integrated intensity changes were the largest at 2200 W, the second largest at 2500 W, and the smallest at 2800 W. Therefore, according to the XRD phase calculation method, the change in the content of the M_3B_2 phase was the largest at 2200 W, followed by 2500 W, and the smallest change was observed at 2800 W. It can be seen from the SEM that the coatings consisted of a white phase and gray phase. The results of EDS analysis of the white and gray phases indicated that the white phase was M_3B_2 and the gray phase was {Fe, Ni}. Fe and Cr occupied the positions of the Ni atoms in the white phase M_3B_2 [21,24].

The M_3B_2 phase was long and uniformly distributed in the coating at 2200 W. Compared with the gray phase in the coating, the white phase was present in a greater quantity. The M_3B_2 phase was evenly distributed in the molten pool area, and less distributed in the overlap area and the area near the interface at 2500 W. The M_3B_2 phase was present in a smaller quantity, and it was unevenly scattered in the coating at 2800 W, while there was a greater amount of M_3B_2 phase distributed in the bottom of the molten pool than other areas. The smaller grain size was approximately 2.8 μm, and the larger grain size was approximately 10 μm at 2200 W. The small grain size was approximately 1.8 μm, and the largest grain size was approximately 4.9 μm at 2500 W. The smaller grain size was approximately 5.1 μm, the larger grain size was approximately 29 μm, and the long grain size could reach 46.5 μm at 2800 W.

When the laser power was changed, the energy obtained per unit time, the depth, and disturbance of the molten pool, and the solidification time changed as a result. When the power was 2500 W, the molten pool became deeper, and Fe in the substrate floated to the top of the molten pool; however, as the disturbance became larger and the cooling time became longer, the grain size became smaller. When the power was 2800 W, a large amount of Fe floated to the top of the molten pool, which changed the atomic ratio of the elements in the molten pool, resulting in a reduction in the quantity of M_3B_2 phase that was formed. The complex spatial microstructures of the coatings including the complex-shaped M_3B_2 and {Fe, Ni} phases were caused by the non-equilibrium solidification behavior of the laser cladding coating in the quick-cooling temperature filed, which has been investigated comprehensively in other works [33,34].

The improvement of the hardness of the cermet coatings can be explained by particle dispersion strengthening [35]. The M_3B_2 boride particles in the composite coatings acted as a strengthening phase that was distributed in the solid solution phase. M_3B_2 is a hard and brittle phase that is difficult to deform. The crystal structure of M_3B_2 is different from the {Fe, Ni} phase; therefore, when the dislocation cut through the particles, the slip resistance increased, which resulted in an increase in the hardness of the coating. It was found from Figure 9 that the hardness of the coating decreased as the laser power increased. The main reason for this result was the different volume fraction of the reinforcing particles in the coating. The volume fraction can be clearly seen in Figure 4a, 5a and 6a. The results of the hardness test show that the hardness gradually decreased from the surface to the bonding zone at 2200 W. Figure 4a shows the microstructure of the coating at 2200 W, the grain size in the surface area was smaller than that in the middle area, and fewer particles were present in the bonding area. The grain refinement can improve the performance according to the Hall–Petch strengthening formula (Equation (4)) [36].

$$\sigma_s \doteq \sigma_0 + Kd^{-1/2}, \quad (4)$$

where σ_s is the yield strength of the materials, σ_0 is the deformation resistance of the grain, K is the influence coefficient of the grain on deformation, and d is the average size of the grain. The hardness of the coating at 2500 W decreased slightly from the surface area to the middle area, and decreased significantly in the bonding zone. Figure 5a shows the microstructure of the coating at 2500 W, the volume fraction and size of the grains did not change greatly in the surface and middle area but occupied a small proportion in the bonding area. Figure 9 shows the hardness of the coating at 2800 W gradually decreased from the surface area to the bonding area, but suddenly increased in the middle area approaching the bonding area. The sudden increase in hardness caused by a large amount of M_3B_2 phase gathered at that area, which can be evidently seen in Figure 6a. Figure 6a shows the microstructure of the coating at 2800 W, the grain size and volume fraction of the second phase in the surface and middle area were small, but they were large at test position 4. The second phase distribution in the coating was different, due to the differences of melting depth, disturbance, and cooling time of the molten pool under different laser powers.

The EIS test results are shown in Figures 10 and 11, which show that the corrosion performances of the coatings were much better than that of the substrate. The corrosion resistance of the coating was different at different laser powers; it was highest at 2500 W, followed by at 2800 W and lowest at 2200 W. The EIS plots were fitted by the Nova software, and the model $R_s(Q_f(R_f(Q_{dl}R_{ct})))$ was applied to analyze the variations in corrosion. Both R_f and R_{ct} increased, indicating that the surface properties of the substrate were changed with the coating, and the corrosion performance was improved. The R_f coating at 2200 W was much larger than that of coatings 2 and 3, due to the difference of the boride ceramic phase volume fraction. The law of R_f at various powers was consistent with the law of the volume fraction of boride ceramic phase M_3B_2 in the coating. In the EIS diagram, the high-frequency part is related to the charge transfer, which is associated with the effect of an electric double-layer [37]. The R_{ct} of the coating at 2500 W was larger than the others, which means that the electric double-layer formed by the structure of the coating at 2500 W was more likely to hinder the electron transfer in 3.5 wt.% NaCl solution.

The polarization curves show that both the substrate and the coatings underwent the transition process from the activated state to the passivated state and ultimately, to overpassivation; however, the duration of each process varied. The corrosion resistance was improved by the coatings compared to the bare substrate. The corrosion current density i_{corr}, the Tafel constant for an anode b_a, and the Tafel constant for a cathode b_c, which reflect the corrosion mechanism and kinetics are considered as electrochemical corrosion parameters [38]. The absolute values of b_a are greater than those of b_c, indicating that the anode is the control step of the whole electrochemical reaction. The corrosion mechanism of substrate and coatings in 3.5 wt.% NaCl solution can be deduced as follows.

Anodic reaction:

$$M = M^{n+} + ne^-, \tag{5}$$

Cathodic reaction:

$$\frac{n}{2}H_2O + \frac{n}{4}O_2 + ne^- = nOH^- \tag{6}$$

where M is the metal elements in the coating (according to the XRD and SEM analyses, M is Fe and Ni), and n^+ is the valence of elements. The formed passivation film can reduce the average speed of the anode dissolution process, which was characterized in Hamadou's study [39]. The polarization curves show that the passivation film formed on the substrate and the coatings, but the electrode reaction of the substrate was faster, and the film was quickly dissolved. The electrode material and surface state had a significant effect on the reaction speed of the electrode. The microstructure of the coatings was different, which caused the differences in the coatings' abilities to hinder electron transfer, and the coating at 2500 W was the strongest according to the EIS results.

The volume fraction of {Fe, Ni} was large in the surface area of Sample 3, and the short maintenance time of the passivation state was similar to that of the substrate. However, the M_3B_2 phase in the coating caused the potential to shift positively, and the corrosion resistance was improved. The volume fraction of M_3B_2 was larger in the surface area of Coatings 1 and 2, the passivation state was maintained in this microstructure. However, the dense M_3B_2 phase was distributed in the {Fe, Ni} phase, which made it easier to form galvanic cells between them, which accelerated the corrosion process.

The tribological properties are not inherent properties of the materials and depend on the mechanical properties and some other factors, such as the tribo-phase, roughness, and toughness [40]. The substrate and coatings were ground under the same process conditions to ensure consistent roughness. The decrease of the frictional coefficient and wear rate of the coatings compared to the substrate due to the hardness improved according to Archard's principle [41].

The substrate underwent abrasive wear and adhesion wear demonstrated by the furrows and adhered materials on the worn track (Figure 17a). The M_3B_2 phase in the coating resisted deformation, so that there were no obvious furrows in the wear track, and the abrasive wear was reduced. The hardness of Sample 1 was greater than that of Samples 2 and 3; however, the frictional coefficient and volume loss were greater, because the volume fraction of the M_3B_2 phase in Sample 1 was large, and a small amount of the bonding phase was insufficient to protect the hard particles during the friction process. Then, the particles were peeled off (Figure 17b).

The frictional coefficient of the coating at 2500 W was greater than at 2800 W in Figure 14, but the volume loss was smaller in Figure 15. Sample 3 displayed a large area of peeling during the friction process, but only a small amount of the peeling layer remained in the wear track, so the frictional coefficient was not large; however, the volume loss was large (Figure 17d). Fine, square-shaped M_3B_2 particles were uniformly distributed in Sample 2, which improved the hardness of the coating, and a sufficient bonding phase protected the M_3B_2 particles, thus, there were no obvious deformations, furrows, or adhesive materials in the wear track, only fewer peeling phenomena (Figure 17c).

5. Conclusions

In this study, M_3B_2 boride-based cermet coatings were developed on Q235 steel by laser cladding synthesis using different laser powers. The laser power had significant effects on the microstructure, and the properties of the coatings were studied. The coating consisted of a white, hard M_3B_2 phase and a gray bonding phase {Fe, Ni}. The laser power increased and the grains in the coating were refined; however, the uniformity of the distribution decreased due to the increase of the molten pool disturbance. When the power was further increased, the dilution rate increased, and a large amount of Fe entered the coating from the steel. The thickness of the coating increased, but the amount of hard M_3B_2 phase in the coating decreased.

The hardness of the coatings demonstrated a linear relationship with the power, that is, the hardness decreased with the increase of power, because the increase in power caused the precipitation of Fe in the steel and then the reduction of the proportion of the hard phase in the coating. The maximum hardness of the coating was 1166.3 HV, which was obtained at 2200 W.

From the EIS and polarization curve analysis, it was shown that the corrosion performance of the coating was much better than that of the substrate. The corrosion current of the coatings was reduced by an order of magnitude compared to the substrate, and the polarization resistance was increased by an order of magnitude. The corrosion performance of the coatings first increased and then decreased with increasing power. The coating with the best corrosion performance was obtained at a power of 2500 W, at which time the corrosion current was 2.93 μA, and the polarization resistance was 11949 Ω.

The frictional coefficient and wear loss of the coatings were lower than those of the substrate. The wear performance was also better. Under the same conditions, the volume loss of the substrate was 19.29 μm^3, and the maximum volume loss of the coating was 7.34 μm^3, which greatly improved the wear resistance. The frictional coefficient decreased with increasing power, while the volume loss was reduced and then increased. The volume loss showed that the coating had the best wear resistance when the power was 2500 W.

The prepared coating greatly improved the overall performance of the substrate. The key process parameter—laser power—should not be too low or too high, but the performance at high power was better than that at low power. At 2500 W, the coating demonstrated not only the best corrosion resistance but also the best wear resistance. These cermet coatings can be used in the manufacture and repair of mining machinery parts, seawater pump parts, offshore engineering parts, and slurry flow units.

Author Contributions: Conceptualization, W.L. and Z.H.; methodology, Z.H.; validation, Z.H., Y.Z., and W.L.; formal analysis, Z.H. and Y.Z.; investigation, Z.H. and Y.Z..; resources, W.L.; data curation, Z.H.; writing—original draft preparation, Z.H.; writing—review and editing, Z.H. and Y.Z.; visualization, Z.H.; supervision, W.L.; project administration, W.L. and Y.Z.; funding acquisition, W.L. All authors have read and agreed to the published version of the manuscript.

Funding: This research was funded by the National Natural Science Foundation of China, grant number 51572168.

Acknowledgments: The authors acknowledge Lei Hong from Logistics Engineering College of Shanghai Maritime University for the laser cladding experiments, Shuxiang Ma from Shanghai Maritime University for the SEM measurements, and Dongsheng Wang from Shanghai Maritime University for the frictional experiments.

Conflicts of Interest: The authors declare no conflicts of interest.

References

1. Wang, H.Q.; Sun, J.S.; Li, C.N.; Geng, S.N.; Sun, H.G.; Wang, G.L. Microstructure and mechanical properties of molybdenum–iron–boron–chromium cladding using argon arc welding. *Mater. Sci. Technol.* **2016**, *32*, 1694–1701. [CrossRef]
2. Zhang, T.; Yin, H.; Zhang, C.; Zhang, R.; Xue, J.; Zheng, Q.; Qu, X. First-principles study on the mechanical properties and electronic structure of V doped WCoB and W_2CoB_2 ternary borides. *Materials* **2019**, *12*, 967. [CrossRef] [PubMed]
3. Li, Q.; Zhou, D.; Zheng, W.; Chen, C. Global structural optimization of tungsten borides. *Phys. Rev. Lett.* **2013**, *110*, 136403. [CrossRef]
4. Bahrami-Karkevandi, M.; Ebrahimi-Kahrizsangi, R.; Nasiri-Tabrizi, B. Formation and stability of tungsten boride nanocomposites in WO_3–B_2O_3–Mg ternary system: Mechanochemical effects. *Int. J. Refract. Met. Hard Mater.* **2014**, *46*, 117–124. [CrossRef]
5. Kadri, M.T.; Heciri, D.; Derradji, N.; Belfarhi, B.; Belkhir, H. Effects of Na, Mg and Al substitution in hypothetical superconducting Be_2B. *Phys. Status Solidi B* **2008**, *245*, 2779–2785. [CrossRef]
6. Togano, K.; Badica, P.; Nakamori, Y.; Orimo, S.; Takeya, H.; Hirata, K. Superconductivity in metal rich Li-Pd-B ternary boride. *Phys. Rev. Lett.* **2004**, *93*, 247004. [CrossRef] [PubMed]
7. Prakash, S.; Karacor, M.; Banerjee, S. Surface modification in microsystems and nanosystems. *Surf. Sci. Rep.* **2009**, *64*, 233–254. [CrossRef]
8. Zheng, X.Q.; Liu, Y. Slurry erosion-corrosion wear behavior in SiC-containing NaOH solution of Mo_2NiB_2 cermets prepared by reactive sintering. *Int. J. Refract. Met. Hard Mater.* **2019**, *78*, 193–200. [CrossRef]
9. Kayhan, M.; Hildebrandt, E.; Frotscher, M.; Senyshyn, A.; Hofmann, K.; Alff, L.; Albert, B. Neutron diffraction and observation of superconductivity for tungsten borides, WB and W_2B_4. *Solid State Sci.* **2012**, *14*, 1656–1659. [CrossRef]
10. Moraes, V.; Riedl, H.; Fuger, C.; Polcik, P.; Bolvardi, H.; Holec, D.; Mayrhofer, P. Ab initio inspired design of ternary boride thin films. *Sci. Rep.* **2018**, *8*, 1–9. [CrossRef]
11. Takagi, K.I.; Yamasaki, Y.; Komai, M. High-strength boride base hard materials. *J. Solid State Chem.* **1997**, *133*, 243–248. [CrossRef]
12. Takagi, K.I.; Yamasaki, Y. Effects of Mo/B atomic ratio on the mechanical properties and structure of Mo_2NiB_2 boride base cermets with Cr and V additions. *J. Solid State Chem.* **2000**, *154*, 263–268. [CrossRef]
13. Takagi, K.I. High tough boride base cermets produced by reaction sintering. *Mater. Chem. Phys.* **2001**, *67*, 214–219. [CrossRef]
14. Takagi, K.I. Effect of Mn on the mechanical properties and microstructure of reaction sintered Mo_2NiB_2 boride-based cermets. *Mater. Mater. Int.* **2003**, *9*, 467–471. [CrossRef]
15. Yamasaki, Y.; Nishi, M.; Takagi, K.I. Development of very high strength Mo_2NiB_2 complex boride base hard alloy. *J. Solid State Chem.* **2004**, *177*, 551–555. [CrossRef]
16. Takagi, K.I. Development and application of high strength ternary boride base cermets. *J. Solid State Chem.* **2006**, *179*, 2809–2818. [CrossRef]
17. Takagi, K.I.; Koike, W.; Momozawa, A.; Fujima, T. Effects of Cr on the properties of Mo_2NiB_2 ternary boride. *Solid State Sci.* **2012**, *14*, 1643–1647. [CrossRef]
18. Yuan, B.; Zhang, G.J.; Kan, Y.M.; Wang, P.L. Reactive synthesis and mechanical properties of Mo_2NiB_2 based hard alloy. *Int. J. Refract. Met. Hard Mater.* **2010**, *28*, 291–296. [CrossRef]
19. Zhang, L.; Huang, Z.; Liu, Y.; Shen, Y.; Li, K.; Cao, Z.; Ren, Z.; Jian, Y. Effect of Ni content on the microstructure mechanical properties and erosive wear of Mo_2NiB_2-Ni cermets. *Ceram. Int.* **2019**, *45*, 19695–19703. [CrossRef]
20. Zhang, L.; Huang, Z.; Liu, Y.; Shen, Y.; Li, K.; Cao, Z.; Ren, Z.; Jian, Y. Effect of mechanical ball milling time on the microstructure and mechanical properties of Mo_2NiB_2-Ni cermets. *Materials* **2019**, *12*, 1926. [CrossRef]
21. Zhang, L.; Huang, Z.F.; Shen, Y.P.; Li, K.M.; Cao, Z.; Jian, Y.X.; Ren, Z.J. High temperature compressive properties and tribological behavior of Mo_2NiB_2-Ni cermets. *Ceram. Int.* **2019**, *45*, 18413–18421. [CrossRef]

22. Moghaddam, S.R.; Derin, B.; Yucel, O.; Sonmez, M.S.; Sezen, M.; Bakan, F.; Sanin, V.N.; Andreev, D.E. Production of Mo_2NiB_2 based hard alloys by self-propagating high temperature synthesis. *High Temp. Mater. Process.* **2019**, *38*, 683–691. [CrossRef]
23. Wu, Q.; Li, W.; Zhong, N.; Wang, G. Microstructure and properties of laser-clad Mo_2NiB_2 cermet coating on steel substrate. *Steel Res. Int.* **2015**, *86*, 293–301. [CrossRef]
24. Hu, Z.; Li, W.; Zhao, Y. Microstructure and properties of M_3B_2-type boride based cermet coatings prepared by laser cladding synthesis. *Coatings* **2019**, *9*, 476. [CrossRef]
25. Sexton, L.; Lavin, S.; Byrne, G.; Kennedy, A. Laser cladding of aerospace materials. *J. Mater. Process. Technol.* **2002**, *122*, 63–68. [CrossRef]
26. Toor, I.U.H. Effect of Mn content and solution annealing temperature on the corrosion resistance of stainless steel alloys. *J. Chem.* **2014**, *2014*, 1–8. [CrossRef]
27. Kim, J.D.; Peng, Y. Melt pool shape and dilution of laser cladding with wire feeding. *J. Mater. Process. Technol.* **2000**, *104*, 284–293. [CrossRef]
28. Schachinger, E.D.; Braidt, R.; Strauß, B.; Hassel, A.W. EIS study of blister formation on coated galvanised steel in oxidising alkaline solutions. *Corros. Sci.* **2015**, *96*, 6–13. [CrossRef]
29. Wang, Y.; Cheng, G.; Wu, W.; Qiao, Q.; Li, Y.; Li, X. Effect of pH and chloride on the micro-mechanism of pitting corrosion for high strength pipeline steel in aerated NaCl solutions. *Appl. Surf. Sci.* **2015**, *349*, 746–756. [CrossRef]
30. Liu, M.; Cheng, X.; Li, X.; Zhou, C.; Tan, H. Effect of carbonation on the electrochemical behavior of corrosion resistance low alloy steel rebars in cement extract solution. *Constr. Build. Mater.* **2017**, *130*, 193–201. [CrossRef]
31. Brytan, Z.; Niagaj, J. Corrosion studies using potentiodynamic and EIS electrochemical techniques of welded lean duplex stainless steel UNSS82441. *Appl. Surf. Sci.* **2016**, *388*, 160–168. [CrossRef]
32. Cáceres, L.; Vargas, T.; Herrera, L. Determination of electrochemical parameters and corrosion rate for carbon steel in un-buffered sodium chloride solutions using a superposition model. *Corros. Sci.* **2007**, *49*, 3168–3184. [CrossRef]
33. Li, C.; Yu, Z.; Gao, J.; Zhao, J.; Han, X. Numerical simulation and experimental study of cladding Fe60 on an ASTM 1045 substrate by laser cladding. *Surf. Coat. Technol.* **2019**, *357*, 965–977. [CrossRef]
34. Liu, C.; Li, C.; Zhang, Z.; Sun, S.; Zeng, M.; Wang, F.; Guo, Y.; Wang, J. Modeling of thermal behavior and microstructure evolution during laser cladding of AlSi10Mg alloys. *Opt. Laser Technol.* **2020**, *123*, 105926. [CrossRef]
35. Munoz-Morris, M.A.; Oca, C.G.; Morris, D.G. An analysis of strengthening mechanisms in a mechanically alloyed, oxide dispersion strengthened iron aluminide intermetallic. *Acta Mater.* **2002**, *50*, 2825–2836. [CrossRef]
36. Hansen, N. Hall-Petch relation and boundary strengthening. *Scr. Mater.* **2004**, *51*, 801–806. [CrossRef]
37. Zhang, K.; Song, R.; Gao, Y. Corrosion behavior of hot-dip galvanized advanced high strength steel sheet in a simulated marine atmospheric environment. *Int. J. Electrochem. Sci.* **2019**, *14*, 1488–1499. [CrossRef]
38. Wang, R.; Luo, S.; Liu, M.; Xue, Y. Electrochemical corrosion performance of Cr and Al alloy steels using a J55 carbon steel as base alloy. *Corros. Sci.* **2014**, *85*, 270–279. [CrossRef]
39. Hamadou, L.; Kadri, A.; Benbrahim, N. Characterisation of passive films formed on low carbon steel in borate buffer solution (pH 9.2) by electrochemical impedance spectroscopy. *Appl. Surf. Sci.* **2005**, *252*, 1510–1519. [CrossRef]
40. Ju, H.; Ding, N.; Xu, J.; Yu, L.; Geng, Y.; Ahmed, F.; Zuo, B.; Shao, L. The influence of crystal structure and the enhancement of mechanical and frictional properties of titanium nitride film by addition of ruthenium. *Appl. Surf. Sci.* **2019**, *489*, 247–254. [CrossRef]
41. Xiang, L.; Shen, Q.; Zhang, Y.; Bai, W.; Nie, C. One-step electrodeposited Ni-graphene composite coating with excellent tribological properties. *Surf. Coat. Technol.* **2019**, *373*, 38–46. [CrossRef]

Article

Investigations on Aging Behavior and Mechanism of Polyurea Coating in Marine Atmosphere

Kaiyuan Che, Ping Lyu *, Fei Wan * and Mingliang Ma

School of Civil Engineering, Qingdao university of technology, Qingdao 266033, China; 18764811850@163.com (K.C.); mamingliang@qut.edu.cn (M.M.)

* Correspondence: lyuping_qut@sina.com (P.L.); shuoyiecool@sina.com (F.W.); Tel.: +86-139-6422-2593 (P.L.); +86-185-6175-2728 (F.W.)

Received: 5 September 2019; Accepted: 1 November 2019; Published: 5 November 2019

Abstract: In this investigation, the aging behaviors of polyurea coating exposed to marine atmosphere for 150 days were studied and the mechanism was analyzed. The influences on surface and mechanical properties, surface morphology, thermal stability behavior, as well as chemical changes evolution of the coating were investigated. By attenuated total reflectance fourier transform infrared spectroscopy (ATR–FTIR) and X-ray photoelectron spectroscopy (XPS), changes in the chemical properties of polyurea coatings before (PCB) and after 150 d (PCA) of aging were analyzed, and emphasis was given to the effect of aging on functional group change, the hydrogen bonding behavior, and phase separated morphology. The results displayed prominent chain scission during aging, such as N–H, C=O, and C–O–C and the hydrogen bonded urea carbonyl content showed a decrease trend. The relative content of soft and hard segments showed a significant change, which increased the degree of phase separation.

Keywords: polyurea; aging mechanism; morphology; chemical properties; phase separation; hydrogen bond

1. Introduction

Polyurea as a thermoset elastomer has excellent performance for its special phase structure and physical crosslinks, which is formed by reacting a diisocyanate with an amine terminated compound by a step growth polymerization process [1–4]. Due to its exceptional mechanical and physical properties, chemical and moisture resistance, polyurea has been extensively applied in a wide number of applications. [1,3] For example, polyurea has been found to change the failure and fragmentation behavior upon impact by hypervelocity projectiles and is applied in military fields such as bulletproof vests [5]. In recent years, polyurea has been increasingly used in marine engineering, such as dock steel piles, sea-crossing bridges, and drilling platforms. Polyurea has also been studied as a binder for marine joint structures exposed to impact loads of varying amplitudes and strain rates. Thus, polyurea's moisture resistance properties and its impact mitigation capacity over a wide range of frequencies make it an ideal candidate for applications that involve exposure to or submergence in sea water [6].

As is known, there are aggressive factors such as ultraviolet radiation, floating and pushing fouling, aggressive chemicals, mechanical stresses, moisture and so on in the ocean atmosphere, which can not only cause failure of coating, but also pose a threat to property and life safety [7,8]. Based on this background, research work on the constitutive properties of polyurea are being given increasingly more interest [9–13]. Youssef [14] investigated the effect of UV radiation on the dynamic mechanical properties of polyurea samples and found that the dynamic creep modulus increased with an increase in the UV exposure duration. Whitten [15] studied the color change from white to deep tan after extended exposure to ultraviolet radiation. Youssef [16] investigated the effect of UV radiation on the ultrasonic properties of polyurea and concluded that the acoustic wave speeds and p-wave attenuation were found to exhibit minimal change, while shear attenuation showed convergence as the temperature and ultraviolet exposure duration increased. Gupta [17] presented a new laser-generated stress-wave-based test method to measure polyurea characteristics under high strain rates and limited strains.

In recent years, many scholars have done a lot of research on the chemical composition of coatings. [18–24]. For example, S. Bhargava et al. [25] have investigated the UV aging mechanism of waterborne polyurethane coatings. The result showed that there exists scission in functional groups such as C–O–C, C=O, C–H, and CO–NH. Chain scission of the polyurethane binder resulted in the appearance of N–H groups. Boubakri [26] irradiated thermoplastic polyurethane (TPU) coatings for 140 hours of ultraviolet light and showed that glass transition temperature (Tg) of TPU coating decreased in the beginning and then increased with UV-exposure time, then drew the conclusion that there were competition between TPU chain breakage and cross-linking. Liu P et al. [27] have characterized the alkyd coating and polyurethane coating after accelerated UV aging. The results showed that the apparent destruction mode of the alkyd system and the polyurethane system were the same, but the failure mechanism was different. Zhu [28] researched the aging behavior of aliphatic polyurethane coatings and acrylic polyurethane coatings under ultraviolet light irradiation, and concluded that the acrylic urethane was mainly C–O bond cleavage, while the aliphatic urethane was mainly C–N bond cleavage. Rossi [29] showed that the main degradation mechanism of the polyurethane under UV exposure was produced by oxidation of the carbon atom at the alpha position of the nitrogen atom of the urethane group. In conclusion, past research mainly focused on the effects of surface morphology, functional group changes, and the thermodynamic properties of coating.

Polyurea has a unique microphase separation structure composed of a soft segment and a hard segment [26,30,31], of which the soft segment is composed of long aliphatic polyether chains and form amorphous domains whereas the hard segment is composed of a urea bond (NH–CO–NH), forming carbonyl to amino hydrogen bonds [32]. Research indicated that the microphase separation structure of polyurea results in the formation of good physical properties, such as high tensile strength, higher elongation, and so on. Iqbal [33] et al. investigated the effect of soft segment length on their mechanical properties and concluded that the relative content of soft and hard segments had a high correlation with its mechanical strength and modulus. Hydrogen bond plays a pronounced role in defining the macroscopic properties of polyurea, which can promote physical cross-linking between macromolecules [34]. As is known, natural exposure has a vital impact on the macro-properties and constitutive properties of polyurea coating [35–37], but there are only a few studies on the effect of exposure in marine atmosphere on the chemical properties of especially hydrogen bond behavior and phase separation morphology.

In this investigation, emphasis was given to the effect of aging on the functional group change, hydrogen bonding behavior, and phase separated morphology in marine atmosphere. The present study evaluated the morphological, chemical, thermal, and mechanical behavior of polyurea during exposure in marine atmosphere and the aging mechanism was analyzed. Spectrometer and contact angle tester, as well as mechanical tests were performed to characterize the degradation. On further studying the SEM, AFM revealed changes in micromorphology. By ATR–FTIR and XPS, changes in the chemical properties of PCB and PCA were analyzed, and emphasis was given to the effect of aging on hydrogen bonding behavior and phase separated morphology.

2. Materials and Methods

2.1. Materials

2.1.1. Raw Materials

The polyurea coating was manufactured by Qingdao Shamu Advanced Material Co., Ltd (Qingdao, China), which was a two-component elastomeric material formed by the reaction of component A and B. Component A was a semi-prepolymer of terminal NCO groups formed from diphenylmethane-4,4′

diisocyanate (MDI) and a polyether polyol and component B consisted of a terminal amino polyether Jeffamine® D2000 and an amine chain extender. The molecular structures are shown in Figure 1. The two components were, respectively, placed in a raw material tank and mixed, and then thoroughly reacted in a ratio of 1:1 before spraying. The spray equipment was PHX-40 proportioner (PMC Global, Inc., Branford, CA, USA) and AP-2 gun (PMC Global, Inc., Branford, CA, USA). The spray temperature and pressure were 65 °C and 2500 psi, respectively. The coating film thickness was 2 mm with 7 days to be fully cured.

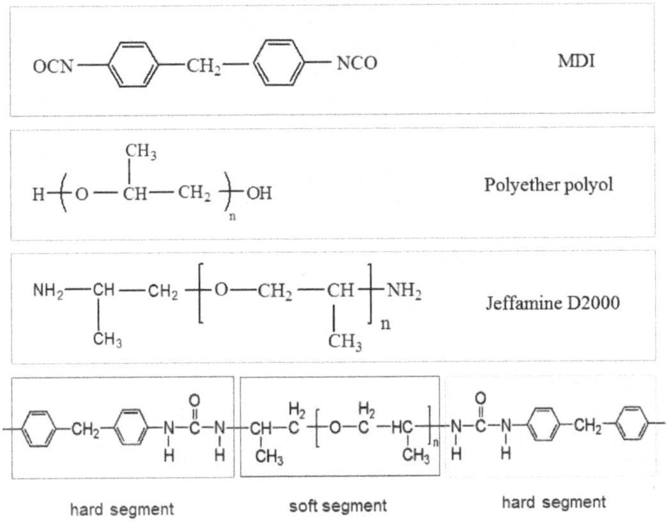

Figure 1. Molecular structures of polyurea precursors.

2.1.2. Properties of Polyurea Coating

The polyurea used in this investigation is pure sheet-like coating without spraying on any substrate, and is mainly applied to marine steel structure protection in engineering. It contains zero Volatile Organic Content, which is extremely friendly to the environment. Due to the fast curing speed and insensitivity to moisture and temperature, it can be continuously sprayed on a sloping or vertical surface without sagging or running.

2.2. *Experimental Method*

The experimental site is located at the Qingdao University of Technology, which is situated 4 kilometers away from the actual coastline. Qingdao is situated in the southeastern part of the Shandong Peninsula, borders the Yellow Sea, and is of a temperate monsoon climate, where the annual average relative humidity, temperature, as well as precipitation are 70%, 12.3 °C, and 680 mm, respectively. The polyurea coating used in the experiment should be hung on the exposed frame in the artificial sea pool in accordance with the requirements of GB/T9276-1996 standard. The exposure frame is oriented 45° to the south, ensuring that it is fully exposed to natural sunlight, as shown in Figure 2. The exposure time is 150 days, from mid-March to mid-August.

Figure 2. Qingdao University of Technology exposure site.

2.3. *Characterization*

2.3.1. Contact Angle Measurements

Static contact angle measurements were carried out at room temperature on an SDC-200 contact angle goniometer (Dongguan Shengding Precision instrument Co., Ltd., Dongguan, China), using a syringe with a needle to drop a droplet of distilled water on the surface of the sample to be tested, which was then stopped for a few seconds, waiting for the drop to stabilize, and then digital images of the droplet silhouette were captured. The contact angle was measured within 30 s. The polar and dispersive components of the surface energy were calculated using the Fowkes method. Before this measurement, the sample surface before and after aging was ultrasonically cleaned with alcohol to remove impurities and oil.

2.3.2. Mechanical Properties

Tensile properties were conducted using MZ-4000D$_1$ universal testing machine (Jiangsu Mingzhu Test Machinery Co., Ltd., Yangzhou, China) at a constant speed of 500 mm/min and at room temperature (25 °C). The dimensions of tensile test samples were taken according to the ASTM D 412 (Standard Test Methods for Vulcanized Rubber and Thermoplastic Elastomers—Tension).

2.3.3. Scanning Electron Microscope

The microstructure of the sample before and after aging was observed by a JSM-7500F scanning electron microscope produced by JEOL (Beijing, China). The resolutions were 1.0 nm (15 kV) and 1.4 nm (1 KV), respectively. The accelerating voltage range was from 0.1KV to 30 KV, and the electron gun used a tungsten filament lamp with magnifications ranging from 25 times to 1,000,000 times. Before the test, the sample was gently wiped with anhydrous ethanol to remove surface dust and was then dried. Then, the sample was sputtered-coated with gold to form a conductive film to improve the image quality and resolution.

2.3.4. Atomic Force Microscopy

Tapping mode atomic force microscopy (AFM) was performed with an Ntegra Prima Scanning Probe system from NT–MDT Prima (NT–MDT Spectrum Instruments Corp., Moscow, Russia). Topographic (height) and phase images were recorded simultaneously under ambient conditions. A silicon cantilever probe with a force constant of 5 N·m^{-1} and a resonance frequency of 70 kHz was used to work on the tapping mode. Before the test, a 15 mm × 15 mm square specimen of PCB and PCA were prepared, and the cleaned samples were obtained by absolute ethanol to remove surface dust.

2.3.5. Thermogravimetric Analysis

Thermogravimetric analysis (TGA) was performed on a STA449C (Netzsch Gerateball, Selb, Germany) under the N_2 atmosphere at 20 mL/min. In the experiment, a sample weighing approximately 8 mg was heated at 10 °C/min from room temperature to 700 °C.

2.3.6. Attenuated Total Reflectance Fourier Transform Infrared Spectroscopy

ATR–FTIR (Attenuated Total Reflectance) spectra of PCB and PCA were investigated with VERTEX 70 FTIR spectrometer (Bruker Optics, Inc., Ettlingen, Germany) with 32 scans and a resolution of 4 cm^{-1}.The spectral region was from 4000 to 500 cm^{-1} and the surface of samples for measurements was cleaned by anhydrous ethanol. The degree of degradation could be detected by detecting changes in the dipole moment associated with the stretching vibration of the functional group.

2.3.7. X-ray Photoelectron Spectroscopic

The X-ray photoelectron spectroscopic (XPS) experiments were performed on Thermo ESCALAB 250XI system (AXIS SUPRA ^{DLD}, Shimadzu Kratos Inc., Kyoto, Japan) with Al K_α (hv = 1486.6 eV) radiation. The binding energy (BE) scale was regulated by setting the C1s transition at 284.8 eV. The accuracy of the BE values was ± 0.2 eV. The surface of the measurement samples was cleaned by anhydrous ethanol and then naturally dried.

3. Results

3.1. Surface and Mechanical Properties

Gloss refers to the ability of a surface to reflect light projected thereon and can be used to indicate the degree of aging of the coating. Figure 3 showed the change of gloss of polyurea coating for 150 days. With extended exposure time, the glossiness value showed a decreasing tendency. As the aging time reached 150 days, the glossiness value decreased 91.95%. Literature [38] stated that the greater the roughness of the coating surface, the lower the reflectivity and gloss value obtained, so the decrease of the coating gloss corresponded well to the cracks of the coating surface.

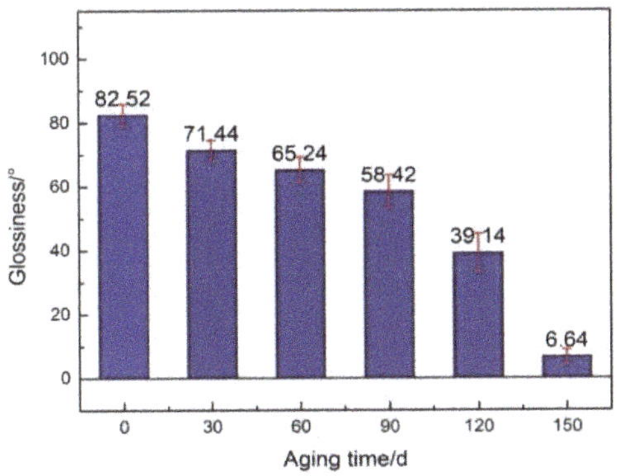

Figure 3. Change in gloss with aging time.

Determining the water contact angle can provide information about the change in the surface topography through the wetting properties [39,40]. The contact angle results and calculated surface energies of the polyurea coatings were presented in Figure 4. As observed, a steady decrease in the value of the contact angle was evident. The contact angle value of the coating before aging was about 78° ± 1.2° and dropped to 37.7° ± 2.1°, as the aging time reached 150 days, which decreased by about 51.28%. The result indicated a significant change in its wettability on the sample. As the value decreased, the levels of hydrophilicity increased, which could increase the adhesion of contaminants [41]. This was one of reasons for the decrease in gloss.

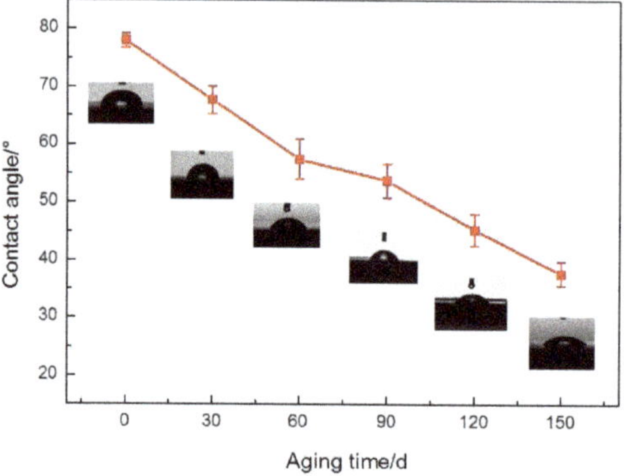

Figure 4. Change in contact angle with aging time.

In this section, the aging impact on the tensile properties of the polyurea coating was investigated. Figure 5 showed the result of a polyurea coating during natural exposure to the ocean atmosphere for 150 days. A slight increase in the tensile strength values was observed in the sample during the early stage of aging. This related to the enhancement of molecular cross-linking in the early aging period, and in this period, the effect of molecular cross-linking was greater than molecular bond rupture [26]. However, the tensile strength presented the descent trend on the whole, in the long term, and the tensile strength reduced by 7.44% after exposure for 150 days.

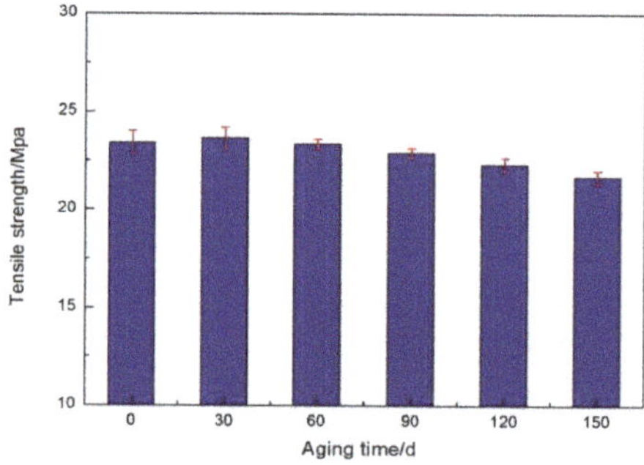

Figure 5. Tensile strength of polyurea coating.

3.2. *Surface Morphology and Topography*

The morphology changes on the surface of samples before and after ocean atmosphere exposure for 30 d, 60 d, 90 d, 120 d, and 150 d were obtained by SEM in Figure 6. As shown in Figure 6, the surface of the unaged coating was relatively smooth and homogeneous, and no obvious holes and cracks appeared. With an increase in the aging time, the coating surface generated more cracks, which resulted in the increase of the surface roughness. The defect area calculated by the image J software is shown in Figure 7. The increase in area defects correlated to the exposure duration, which was consistent with research conducted by Youssef [14]. The contact angle decreased with an increase in the defect area, and the two were basically negatively correlated. The appearance of cracks on the coating was related to the breakage of molecular bonds. As found in the literature [38], under the combined action of violet radiation and other influencing factors, the molecules in the coating could degrade to form micropores, and then microcracks had developed. The degradation products might also leave the surface, creating a rough surface.

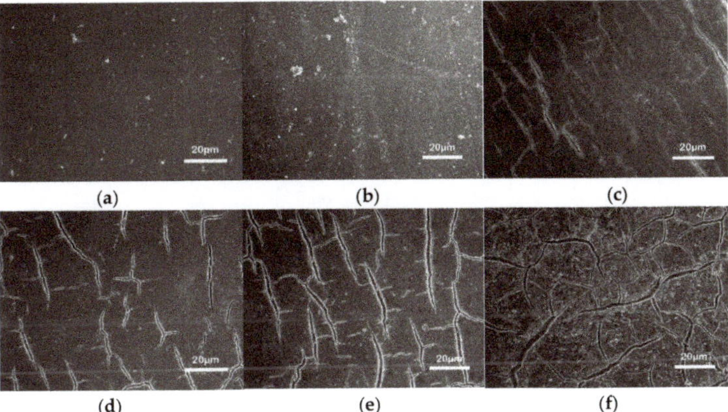

Figure 6. SEM of polyurea—(**a**) 0-day aged; (**b**) 30-days aged; (**c**) 60-days aged; (**d**) 90-days aged; (**e**) 120-days aged; and (**f**) 150-days aged.

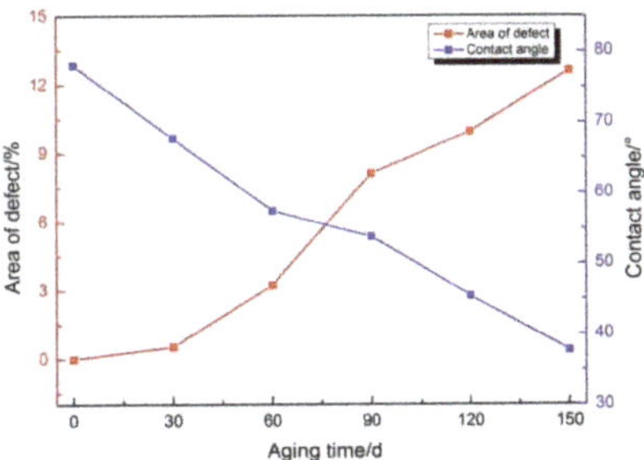

Figure 7. Correlation between the defect area and the contact angle.

The surface morphology of PCB and PCA were probed using tapping mode AFM, and the phase contrast, and topographic images were shown in Figures 8 and 9. Figure 8 provided direct visual evidence of two distinct phases—the darker areas correspond to the soft segment domain and the lighter areas correspond to the hard segment domain [42], the hard segment dispersed randomly within the soft segment, forming a microphase separation structure. Compared to the PCB, the PBA had a higher ordinate, indicating a greater degree of microphase separation. Figure 9 showed AFM 2D (Figure 9a,b) and 3D (Figure 9c,d) topographic image representations of PCB and PCA. We can clearly observe in Figure 9c,d, using a qualitative approach, that an increase in surface roughness was achieved after exposure in the marine atmosphere. According to the NOVA software analysis, the root mean square roughness (R_{rms}) value for the original sample was 6.03 nm and the value of average roughness (R_a) was 3.24 nm. The surface topography was relatively flat. As the exposure time reach 150 days, an obvious increase in the roughness degree values could be observed, the R_{rms} value reached 49.50 nm while the R_a was close to 36.3 nm. Surface roughness was the main reason for reducing the contact angle [41] which explains the reason for the above-mentioned contact angle reduction. The surface height of PCB and PCA was analyzed by the NOVA software, and the results are shown in Figure 9e. As observed, there are two different roughness profile zones—some very rough zones with a surface height from 50 to 300 nm and other zones had very low roughness values in the range of less than 50 nm.

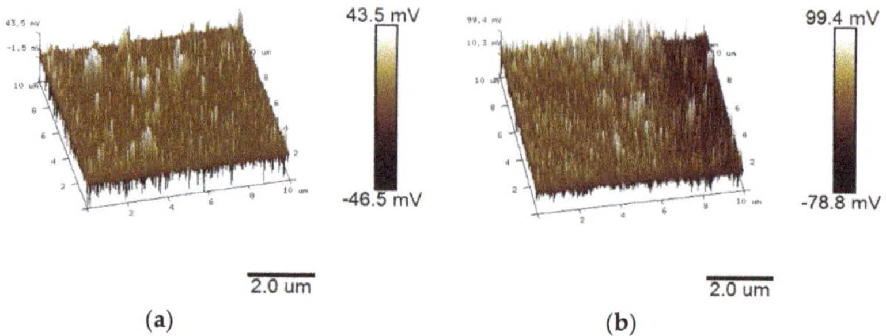

(a) (b)

Figure 8. Atomic force microscopy (AFM) of polyurea coating (**a**) 3D phase contrast image of PCB; and (**b**) 3D phase contrast image of PCA.

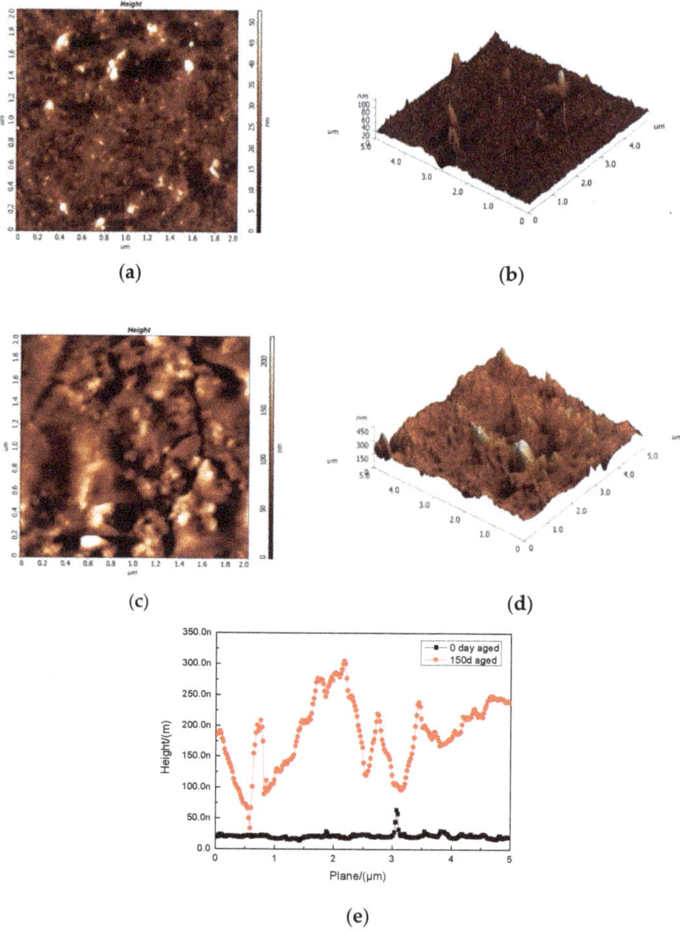

(a) (b) (c) (d) (e)

Figure 9. AFM of polyurea coating (**a**) 2D topographic image of PCB; (**b**) 3D topographic image of PCB; (**c**) 2D topographic image of PCA; (**d**) 3D topographic image of PCA; and (**e**) the height of surface before and after 150 d aging.

3.3. Thermal Stabilities

Thermal gravity (TG) weight loss curves and the differential thermal gravity (DTG) for PCB and PCA are shown in Figure 10. The onset degradation temperatures are defined by the temperatures of 5% weight loss in TGA curves, whereas temperatures of the maximum degradation rate were evaluated by the peaks in the DTG curves. It can be seen that PCB and PCA both exhibited only one degradation step in nitrogen atmosphere during 250–500 °C. Compared to PCB, PCA had a better thermal stability, for it started to degrade at as high as 297 °C and the maximum weight loss rate occurred at 413 °C, while the coating before aging started to degrade and the maximum weight loss rate occurred at the temperature of 272 °C and 403 °C, respectively. The results of this experiment had the following implications. First, it is possible for PCA to degrade into small molecules, which has a better heat resistance than macromolecules. Second, the soft segment content is reduced, and the coating with a high hard segment content exhibits better heat resistance [42]. Practical reasons need further experimentation.

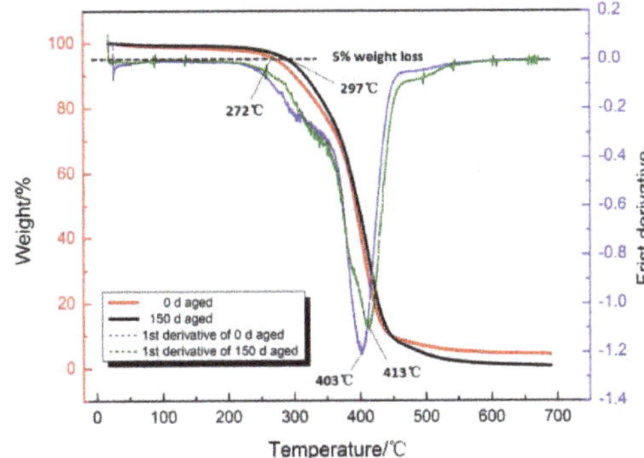

Figure 10. TG and DTG curves for polyurea before and after aging.

3.4. Chemical Changes

The FTIR spectra of PCB and PCA are shown in Figure 11. The spectral bands corresponding to the hard segment domains were observed at 3284–3287 cm^{-1} for N–H stretching vibrations and at 1600–1700 cm^{-1} attributed to C=O stretching vibrations. The absorption bands pertaining to the C–N stretching vibrations were at around 1540 cm^{-1}, all above were associated with the existence of urea bond in the interior of the composites. The Figure 11 shows that the major difference between the spectra corresponding to PCB and PCA was their intensity of bands. A decrease in the intensity signal of the aged sample indicated the beginning of chain scission or the disappearance of a certain group. For example, the absorbance peaks at 1020–1100 cm^{-1} decreased, confirming the cleavage of the C–O–C bond at exposure aging, and at the same time, the peak of C=O absorption decreased because of the scission of C=O band. The secondary amino N–H stretching vibration peak near 3287 cm^{-1} was weakened and broadened, and shifted towards a lower frequency, which indicated the existence of N–H bond cleavage and the reduction in the degree of hydrogen bonding. The C–H stretching vibration in the 2860–2970 cm^{-1} range was obviously weakened, probably the C–H bond in the D2000 side methyl and methylene had fractured [43]. The results coincided perfectly with the evolution of the surface energy and the tensile properties.

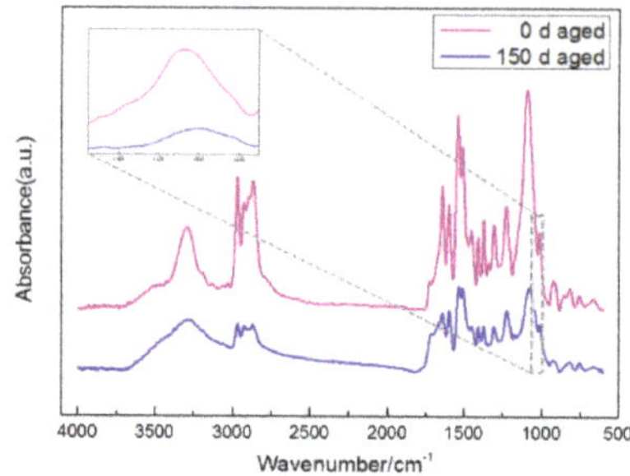

Figure 11. FTIR spectra at the region of 4000–500 cm^{-1} for unaged and aged polyurea.

Hydrogen bonding characteristic was one of the most notable features of the polyurea materials, especially polyurea materials with a high hard segment content [44]. As major objectives of hydrogen bonding behavior research, amino, and carbonyl largely determine the hard segment structure and affect the structure and properties of the material [30]. In order to investigate the effect of aging on hydrogen bonding degree, emphasis was given to the amino region and the urea carbonyl region of the spectrum.

The peak at about 3440 cm^{-1} was assigned to the amino that had not formed a hydrogen bond. In the N–H region, the stretching vibration band of N–H of PCB and PCA showed in the frequency of 3287.28 cm^{-1} and 3284.46 cm^{-1}, respectively. The magnitude of the shift of the infrared absorption peak number was the indicator of the strength of the hydrogen bond, thus, the length of hydrogen bond can be calculated through the following Equation (1) [45].

$$R = 3.21 - \frac{\Delta V}{0.548 \times 10^3} \tag{1}$$

where ΔV represented the shift distance from the amino band without hydrogen bonding to the existing amino band, while R denote the length of hydrogen bond. From the above formula, the lengths of the amino hydrogen bonds before and after aging were both calculated to be about 2.93 Å, thus, aging has little effect on the hydrogen bond length in the N–H region.

A detailed analysis of hydrogen bonding characteristics was carried out by curve fitting the broad band in the carbonyl region of PCB and PCA. The band deconvolution analysis was performed on the carbonyl region at 1620–1690cm^{-1} using the OMNIC 8.2 software (Thermo nicolet, Madison, WI, USA). Hydrogen bonded carbonyl ureas were characterized by the presence of disordered and ordered urea carbonyls, consistent with previous reports [46,47]. The peaks due to the urea domain were characterized by bands at 1626–1645 cm^{-1} and 1651–1669 cm^{-1}, attributed to the ordered and disordered hydrogen-bonded urea carbonyls, respectively [43,47,48]. The absorption bands arising around 1673–1685 cm^{-1} corresponded to free hydrogen-bonded urea carbonyl. The wavenumber assignment areas as well as the hydrogen bonding degree of the deconvoluted bands are shown in Table 1. The degree of hydrogen bonding corresponding to the urethane carbonyl was calculated using Equation (2) [47]:

$$\begin{gathered} X_o = \frac{A_o}{\left(A_o + A_{diso} + A_{free}\right)} \\ X_{diso} = A_{diso}/\left(A_o + A_{diso} + A_{free}\right) \\ X_b = X_o + X_{diso} \end{gathered} \tag{2}$$

where A_o, A_{diso}, and A_{free} denote the area of ordered, disordered, and free hydrogen bonded urea carbonyl, respectively. Whereas, X_o, X_{diso}, and X_b showed the percentage of ordered, disordered, and free hydrogen bonding. A comparative analysis of the relative trends in the hydrogen bonding degree of PCB and PCA needed to be considered separately. According to the data in Table 1, major differences between the two sets of samples towards the hydrogen bonding behavior were the degree of ordered, disordered, and total urea hydrogen bonding.

Table 1. Degree of hydrogen bonding in the polyurea obtained from the deconvolution of the carbonyl stretching region.

Aging Time/d	Wavenumber (cm^{-1})	Assignment	Area	Percentage of Hydrogen Bonding (%)		
				Xo	Xdiso	Xb
0	1629–1644	Ordered hydrogen bonding	6.9697	57.66		83.41
	1651–1669	Disordered hydrogen bonding	3.1131		25.75	
	1681	free	2.0055			
150	1626–1645	Ordered hydrogen bonding	3.1411	46.13		74.56
	1651–1666	Disordered hydrogen bonding	1.9355		28.43	
	1673–1685	free	1.7321			

The X_o value of PCB and PCA were 57.66% and 46.13%, respectively, whereas the X_b values were 83.41% and 74.56%. In contrast to PCB, a significant decrease in the X_o and X_b values of PCA was observed. The X_o value showed a slight increase tendency compared with PCB. Based on the observations above, the ordered hydrogenated urea carbonyl content in PCA was less than PCB, which was related to the intermolecular force, for a reduction in the degree of hydrogen bonding can result in the weakening of the aggregation between hard segments and a decrease in the order of the hard segment structure [49]. Thus, results clearly demonstrated that the intermolecular force of the polyurea after aging was reduced. Literature pointed out [46], the hydrogen bonding in the region of carbonyl can limit the phase separation, as compared to the cases with less hydrogen bonding. The results showed the reduction of hydrogen bonding urea carbonyl content, which improved the degree of microphase separation in the soft and hard sections, in good agreement with the results of AFM.

In the XPS spectrum shown as Figure 12, three obvious signals were detected at 285.3 eV (C 1s), 399.8 eV (N 1s), and 532.7eV (O 1s). It can be seen that relative content of N, C, and O elements was slightly changed and no new elements were produced. The contents of N, C, and O elements and the ratio of elements of PCB and PCA were shown in Table 2. After aging, the relative content of the N element increased more obviously, while the O element and the C element decreased slightly. The oxygen–to–carbon (O/C) and the nitrogen–to–carbon (N/C) ratio of PCB was 0.2598 and 0.0565, respectively, whereas that of PCA was 0.2605 and 0.0765, respectively. The increase of O/C and N/C ratio indicated that the coating degraded during exposure in the marine atmosphere, and the degradation caused the inefficiency of coating.

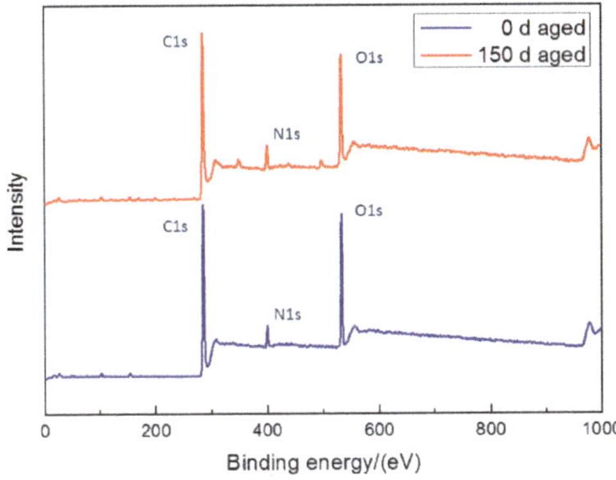

Figure 12. XPS full spectrum analysis before and after polyurea aging.

Table 2. Elemental composition and atomic ratios of polyurea before and after aging.

Coating Treatment	Atomic Percentage/%			Atomic Ratio	
	C	N	O	O/C	N/C
Before the aging	75.97	4.29	19.74	0.2589	0.0565
After aging 150 d	74.81	5.69	19.49	0.2605	0.0761

In order to elucidate the changes in chemical groups, the $C1_S$ spectrum was chosen and deconvolved with the XPSPEAK software as shown in Figure 11. The $C1_S$ spectrum of PCB can be resolved into four contributions (Figure 13a). The peak with lowest BE, 284.8 eV, was assigned to carbon atoms which were linked to hydrogen (C–H) or carbon (C–C). The peak observed at 285.7 was assigned to the carbon atom bonded to the nitrogen atom (C–N). The peak located at 286.3 eV corresponded to carbon atoms single-bonded to oxygen atoms (C–O–C) [50]. According to the data in Table 3, the relative content of C–O–C before aging was estimated to be 39.860%, while C=O was about 1.859%. After 150 days of aging, the percentage of C–O–C and C=O content changed greatly, the C–O–C content fell to 24.626%, while the C=O content increased to 15.963%. This phenomenon can be explained as the bond breakage degree of the C–O–C in the soft segment after aging, which was much higher than that of the hard segment urea bond –NHCONH–, and the soft segment phase was more damaged than the hard segment. This indicated the different sensitiveness of the hard and soft segment to the attack of the marine atmosphere exposure, and that the soft segments were more susceptible to erosion, which explained the reason for a greater degree of microphase separation of PCA.

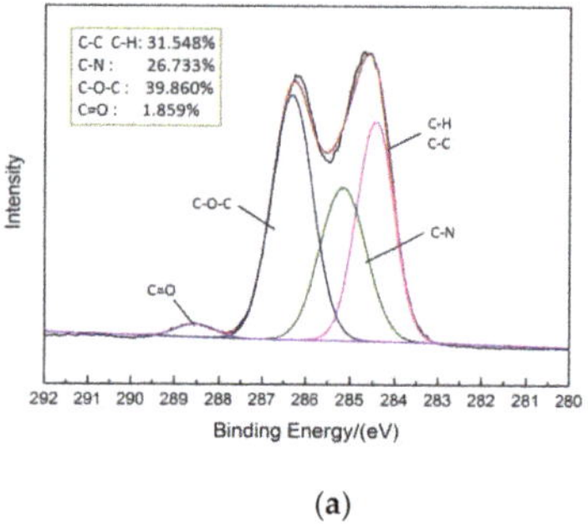

(a)

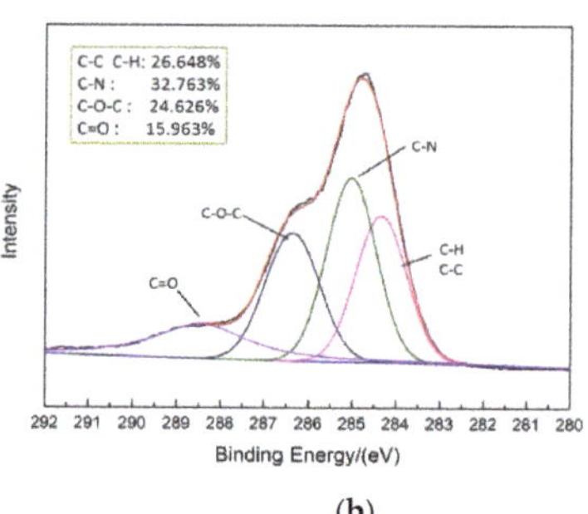

(b)

Figure 13. Component-fitted XPS spectra of the C1s peaks of polyurea coating (**a**) before and (**b**) after aging.

Table 3. Component-Fitted C1s Spectra of polyurea before and after aging.

Element	C–C C–H	C–N	C–O–C	C=O
BE(eV)	284.8	285.7	286.3	288.5
PCB/%	31.548	26.733	39.860	1.859
PCA/%	26.648	32.763	24.626	15.963

Through the above analysis results, we could draw up the partial aging mechanism shown in Scheme 1. As observed, aging had a great influence on the hydrogen bonding of the hard segment, especially the carbonyl region. After aging, the degree of hydrogen bonding in the carbonyl region was significantly reduced. The decrease of hydrogen bonding degree could weaken the aggregation between hard segments and reduce the order of the hard segment structure, which resulted in the decrease in the mechanical properties. Further studies found that the sensitivity of polyurea between soft and hard segment were significantly different, and the degree of C–O–C bond breakage in the soft segment was higher than that in the hard segment, which caused the soft segment to be etched, and then resulted in the increase of roughness and the microphase separation degree. Macroscopically, the gloss of the material was decreased, and the micro-cracks were generated on the surface of the coating on the microscopic surface, which was one of the important reasons for the decline in the mechanical properties.

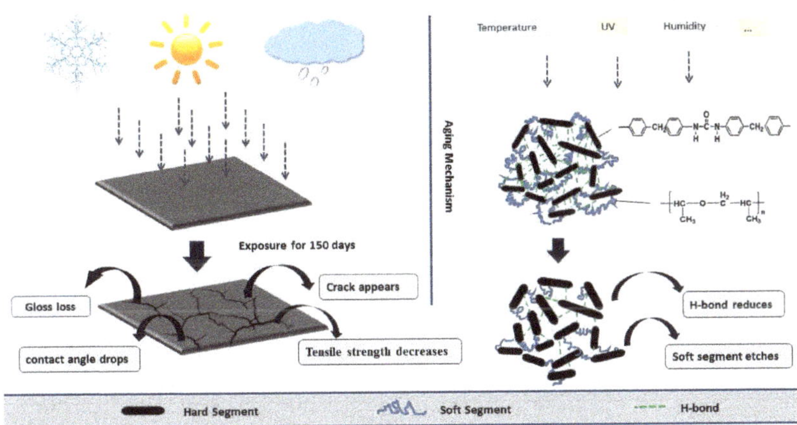

Scheme 1. Schematic illustration of the aging mechanism of polyurea coating.

4. Conclusions

1. An experimental investigation was designed and conducted to determine the aging behavior and mechanism of polyurea coating in marine atmosphere. From the appearance and morphology, PCA exhibits macroscopic phenomena such as loss of light and pulverization while the gloss and contact angle decreased by 91.95% and 51.28%, respectively. It was found that the cracks increased with exposure duration, and the area of defects was basically negatively correlated with the contact angle. The surface roughness of PCA increased significantly, the R_{rms} value reached 49.50 nm, while the R_a was close to 36.3 nm.
2. FTIR showed that there was a lot of decrease in functional groups such as C–O–C, C=O and C–N bond. The length of the hydrogen bond in the amide-to-amino region remained stable, via calculation, PCB and PCA were both about 2.93 Å. The total hydrogen bonding degree of the urea carbonyl group decreased from 83.41% to 74.56%, indicating that the interaction between the polyurea molecules was weakened.
3. XPS showed the percentage of C–O–C and C=O content changed greatly, the C–O–C content fell to 24.626%, while the C=O content increased to 15.963%. The soft segment of the PCA was etched more than the hard segment, which increased the degree of microphase separation.

Author Contributions: Conceptualization, K.C., P.L., and F.W.; Data curation, K.C. and P.L.; Formal analysis, K.C., P.L., and F.W.; Funding acquisition, P.L. and M.M.; Investigation, K.C., P.L., F.W., and M.M.; Methodology, K.C., P.L., and F.W.; Project administration, K.C., P.L., and F.W.; Resources, K.C. and P.L.; Supervision, K.C., P.L. and F.W.; Validation, K.C., P.L., and F.W.; Writing—original draft, K.C. and P.L.; Writing—review & editing, K.C., P.L., and F.W.

Funding: National Natural Science Foundation of China (No. 51503116 and No.51578298).

Conflicts of Interest: The authors declare no conflict of interest.

References

1. Weibo, H. *Spray Polyurea Elastomer Technology*; Chemical Industry Press: Beijing, China, 2005.
2. Iqbal, N.; Tripathi, M.; Parthasarathy, S.; Kumar, D.; Roy, P.K. Polyurea coatings for enhanced blast-mitigation: A review. *RSC Adv.* **2016**, *6*, 109706–109717. [CrossRef]

3. Ma, S.; Van, E.H.; Noordover, B.; Sablong, R.; Van, R.B.; Koning, C. Isocyanate-free approach to water-borne polyurea dispersions and coatings. *ChemSusChem* **2018**, *11*, 149–158. [CrossRef] [PubMed]
4. Chattopadhyay, D.K.; Raju, K.V.S.N. Structural engineering of polyurethane coatings for high performance applications. *Prog. Polym. Sci.* **2007**, *32*, 352–418. [CrossRef]
5. Grujicic, M.; Pandurangan, B.; He, T.; Cheeseman, B.A.; Yen, C.F.; Randow, C.L. Computational investigation of impact energy absorption capability of polyurea coatings via deformation-induced glass transition. *Mater. Sci. Eng.* **2010**, *527*, 7741–7751. [CrossRef]
6. Gupta, V.; Youssef, G. Orientation-Dependent Impact Behavior of Polymer/EVA Bilayer Specimens at Long Wavelengths. *Exp. Mech.* **2014**, *54*, 1133–1137. [CrossRef]
7. Davies, P.; Evrard, G. Accelerated ageing of polyurethanes for marine applications. *Polym. Degrad. Stab.* **2007**, *92*, 1455–1464. [CrossRef]
8. Olajire, A.A. Recent advances on organic coating system technologies for corrosion protection of offshore metallic structures. *J. Mol. Liq.* **2018**, *269*, 572–606. [CrossRef]
9. Youssef, G.; Gupta, V. ARTICLES Dynamic tensile strength of polyurea. *J. Mater. Res.* **2012**, *27*, 494–499. [CrossRef]
10. Grujicic, M.; He, T.; Pandurangan, B. Development and parameterization of an equilibrium material model for segmented polyurea. *Multidiscip. Model. Mater. Struct.* **2011**, *7*, 96–114. [CrossRef]
11. Jain, A.; Youssef, G.; Gupta, V. Dynamic tensile strength of polyurea-bonded steel/E-glass composite joints. *J. Adhes. Sci. Technol.* **2013**, *27*, 403–412. [CrossRef]
12. Youssef, G.; Brinson, J.; Whitten, I. The Effect of Ultraviolet Radiation on the Hyperelastic Behavior of Polyurea. *J. Polym. Environ.* **2017**, *26*, 183–190. [CrossRef]
13. Sarva, S.S.; Deschanel, S.; Boyce, M.C.; Chen, W. Stress—Strain behavior of a polyurea and a polyurethane from low to high strain rates. *Polymer* **2007**, *48*, 2208–2213. [CrossRef]
14. Youssef, G.; Whitten, I. Dynamic properties of ultraviolet-exposed polyurea. *Mech. Time Depend. Mater.* **2016**, *21*, 351–363. [CrossRef]
15. Whitten, I.; Youssef, G. The effect of ultraviolet radiation on ultrasonic properties of polyurea. *Polym. Degrad. Stab.* **2016**, *123*, 88–93. [CrossRef]
16. Youssef, G.H. Dynamic Properties of Polyurea. Ph.D. Thesis, University of California, Los Angeles, CA, USA, 2011.
17. Youssef, G.; Gupta, V. Dynamic response of polyurea subjected to nanosecond rise-time stress waves. *Mech. Time Depend. Mater.* **2012**, *16*, 317–328. [CrossRef]
18. Amrollahi, M.; Sadeghi, G.M.M. Assessment of adhesion and surface properties of polyurethane coatings based on non-polar and hydrophobic soft segment. *Prog. Org. Coat.* **2016**, *93*, 23–33. [CrossRef]
19. Malviya, A.K.; Tambe, S.P.; Malviya, A.K.; Tambe, S.P. Accelerated test methods for evaluation of WB coating system comprising of epoxy primer and polyurethane top coat. *Prog. Color Color. Coat.* **2017**, *10*, 93–104.
20. Yang, X.F.; Tallman, D.E.; Bierwagen, G.P.; Croll, S.G.; Rohlik, S. Blistering and degradation of polyurethane coatings under different accelerated weathering tests. *Polym. Degrad. Stab.* **2002**, *77*, 103–109. [CrossRef]
21. Liu, K.; Su, Z.; Miao, S.; Ma, G.; Zhang, S. UV-curable enzymatic antibacterial waterborne polyurethane coating. *Biochem. Eng. J.* **2016**, *113*, 107–113. [CrossRef]
22. Wang, M.; Xu, X.; Ji, J.; Yang, Y.; Shen, J.; Ye, M. The hygrothermal aging process and mechanism of the novolac epoxy resin. *Compos. Part B* **2016**, *107*, 1–8. [CrossRef]
23. Aglan, H.; Calhoun, M.; Allie, L. Effect of UV and hygrothermal aging on the mechanical performance of polyurethane elastomers. *J. Appl. Polym. Sci.* **2008**, *108*, 558–564. [CrossRef]
24. Guo, T.X.; Weng, X.Z. Evaluation of the freeze-thaw durability of surface-treated airport pavement concrete under adverse conditions. *Constr. Build. Mater.* **2019**, *206*, 519–530. [CrossRef]
25. Bhargava, S.; Kubota, M.; Lewis, R.D.; Advani, S.G.; Prasad, A.K.; Deitzel, J.M. Ultraviolet, water, and thermal aging studies of a waterborne polyurethane elastomer-based high reflectivity coating. *Prog. Org. Coat.* **2015**, *79*, 75–82. [CrossRef]
26. Boubakri, A.; Guermazi, N.; Elleuch, K.; Ayedi, H.F. Study of UV-aging of thermoplastic polyurethane material. *Mater. Sci. Eng.* **2010**, *527*, 1649–1654. [CrossRef]
27. Liu, P. Artificial Aging Test Analysis of Alkyd and Polyurethane Coating System. *Adv. Mater. Res.* **2012**, *496*, 116–120. [CrossRef]

28. Zhu, Y.; Xiong, J.; Tang, Y.; Zuo, Y. EIS study on failure process of two polyurethane composite coatings. *Prog. Org. Coat.* **2010**, *69*, 7–11. [CrossRef]
29. Rossi, S.; Fedel, M.; Petrolli, S.; Deflorian, F. Accelerated weathering and chemical resistance of polyurethane powder coatings. *J. Coat. Technol. Res.* **2016**, *13*, 427–437. [CrossRef]
30. Yılgör, E.; Yılgör, I.; Yurtsever, E. Hydrogen bonding and polyurethane morphology. I. Quantum mechanical calculations of hydrogen bond energies and vibrational spectroscopy of model compounds. *Polymer* **2002**, *43*, 6551–6559. [CrossRef]
31. Boubakri, A.; Elleuch, K.; Guermazi, N.; Ayedi, H.F. Investigations on hygrothermal aging of thermoplastic polyurethane material. *Mater. Des.* **2009**, *30*, 3958–3965. [CrossRef]
32. Iqbal, N.; Kumar, D.; Roy, P.K. Emergence of time-dependent material properties in chain extended polyureas. *J. Appl. Polym. Sci.* **2018**, *135*, 46730. [CrossRef]
33. Iqbal, N.; Tripathi, M.; Parthasarathy, S.; Kumar, D.; Roy, P.K. Tuning the properties of segmented polyurea by regulating soft-segment length. *J. Appl. Polym. Sci.* **2018**, *135*, 46284. [CrossRef]
34. Iqbal, N.; Tripathi, M.; Parthasarathy, S.; Kumar, D.; Roy, P.K. Aromatic versus Aliphatic: Hydrogen Bonding Pattern in Chain-Extended High-Performance Polyurea. *ChemistrySelect* **2018**, *3*, 1976–1982. [CrossRef]
35. Zhang, X.; Wang, J.; Guo, W.; Zou, R. A bilinear constitutive response for polyureas as a function of temperature, strain rate and pressure. *J. Appl. Polym. Sci.* **2017**, *134*, 45256. [CrossRef]
36. Jia, Z.; Amirkhizi, A.V.; Nantasetphong, W.; Nemat-Nasser, S. Experimentally-based relaxation modulus of polyurea and its composites. *Mech. Time Depend. Mater.* **2016**, *20*, 155–174. [CrossRef]
37. Chao, L.; Ma, C.; Xie, Q.; Zhang, G. Self-repairing silicone coating for marine anti-biofouling. *J. Mater. Chem. A* **2017**, *5*, 15855–15861.
38. Zhang, H.; Dun, Y.; Tang, Y.; Zuo, Y.; Zhao, X. Correlation between natural exposure and artificial ageing test for typical marine coating systems. *J. Appl. Polym. Sci.* **2016**, *133*. [CrossRef]
39. Zhang, Y.; Qi, Y.; Zhang, Z. Synthesis of PPG-TDI-BDO polyurethane and the influence of hard segment content on its structure and antifouling properties. *Prog. Org. Coat.* **2016**, *97*, 115–121. [CrossRef]
40. Meiron, T.S.; Marmur, A.; Saguy, I.S. Contact angle measurement on rough surfaces. *J. Colloid Interface Sci.* **2004**, *274*, 637–644. [CrossRef]
41. Huang, F.; Wei, Q.; Wang, X.; Xu, W. Dynamic contact angles and morphology of PP fibres treated with plasma. *Polym. Test.* **2006**, *27*, 22–27. [CrossRef]
42. Awad, W.H.; Wilkie, C.A. Investigation of the thermal degradation of polyurea: The effect of ammonium polyphosphate and expandable graphite. *Polymer* **2010**, *51*, 2277–2285. [CrossRef]
43. Ping, L. Studies on the Novel Polyaspartic Ester Based Polyurea Coatings for Marine Concrete Protection. Ph.D. Thesis, Ocean University of China, Shandong, China, 2007.
44. Lu, P.; Chen, G.H.; Huang, W.B. Degradation of Polyaspartic Polyurea Coating under Different Accelerated Weathering Tests. *J. Sichuan Univ. Eng. Sci. Ed.* **2007**, *39*, 92–97.
45. Lu, P.; Chen, G.; Huang, W. Study on Synthesis, Morphology and Properties of Novel Polyureas Based on Polyaspartic Esters. *J. Chem. Eng. Chin. Univ.* **2008**, *22*, 106–112.
46. Jiang, B.; Tsavalas, J.G.; Sundberg, D.C. Morphology control in surfactant free polyurethane/acrylic hybrid lattices—The special role of hydrogen bonding. *Polymer* **2018**, *139*, 107–122. [CrossRef]
47. Chavan, J.G.; Rath, S.K.; Praveen, S.; Kalletla, S.; Patri, M. Hydrogen bonding and thermomechanical properties of model polydimethylsiloxane based poly(urethane-urea) copolymers: Effect of hard segment content. *Prog. Org. Coat.* **2016**, *90*, 350–358. [CrossRef]
48. Rakshit, P.; Joshi, N.; Jain, R.; Shah, S. Synthesis and Characterization of Polyurea Resin for Dielectric Coating Applications. *Polym. Plast. Technol. Eng.* **2016**, *55*, 1683–1692. [CrossRef]
49. Yang, J.; Wang, G.; Hu, C. Effect of Chain Extenders on Structures and Properties of Aliphatic Polyurethaneureas and Polyureas. *Acta Chim. Sin.* **2006**, *64*, 1737–1742.
50. Chen, T.F.; Siow, K.S.; Ng, P.Y.; Nai, M.H.; Lim, C.T.; Yeop Majlis, B. Ageing properties of polyurethane methacrylate and off-stoichiometry thiol-ene polymers after nitrogen and argon plasma treatment. *J. Appl. Polym. Sci.* **2016**, *133*. [CrossRef]

Article

Material Analysis and Molecular Dynamics Simulation for Cavitation Erosion and Corrosion Suppression in Water Hydraulic Valves

Masoud Kamoleka Mlela [1], He Xu [1,*], Feng Sun [1], Haihang Wang [1] and Gabriel Donald Madenge [2]

1 College of Mechanical and Electrical Engineering, Harbin Engineering University, Harbin 150001, China; masoudkamoleka@gmail.com (M.K.M.); sunfeng@hrbeu.edu.cn (F.S.); wanghaihang@hrbeu.edu.cn (H.W.)
2 College of Aerospace and Civil Engineering, Harbin Engineering University, Harbin 150001, China; gmadenge@icloud.com
* Correspondence: railway_dragon@sohu.com; Tel.: +86-133-5111-7608

Received: 29 December 2019; Accepted: 13 January 2020; Published: 17 January 2020

Abstract: In the milestone of straggling to make water hydraulics more advantageous, the choice of coating polymer for water hydraulics valves plays an essential role in alleviating the impact of cavitation erosion and corrosion, and this is a critical task for designers. Fulfilling the appropriate selection, we conflicted properties that are vital for erosion and corrosion inhibitors, as well as the tribology in the sense of coefficient of friction. This article aimed to choose the best alternative polymer for coating on the selected substrate, that is, Cr_2O_3, Al_2O_3, $T_{i2}O_3$. By applying PROMETHEE (Preference Ranking Organization Method for Enrichment Evaluations), the best polymer obtained with an analyzed performance attribute is Polytetrafluoroethylene (PTFE) that comes up with higher outranking (0.5932052). A Molecular Dynamics (MD) simulation was conducted to identify the stronger bonding with the regards of the better cleave plane between Polytetrafluoroethylene (PTFE) and the selected substrate. Polytetrafluoroethylene (PTFE)/Al_2O_3 cleaved in (010) plane was observed to be the strongest bond in terms of binding energy (3188 kJ/mol) suitable for further studies.

Keywords: polymer; water hydraulic valve; cavitation; erosion; corrosion; coating selection; molecular dynamics

1. Introduction

Today, water hydraulics still face some major adversity to expand their application. Generally, the initial cost of hydraulic fluid components is higher than the hydraulic oil components. This property can be reduced using a deficient cost pressure medium (water), much lower insurance, and disposal costs, if the time used is long enough [1]. Specifically, water hydraulics have a more rapid response and better efficiency compared to oil hydraulics. They are also more stable (in terms of flow velocity and efficiency) over a wide range of operating temperatures due to the higher volume module of water (about 50% higher than that of mineral oil). They also have a lower viscosity (less than 1/30 at 50 °C mineral oil) and higher specific heat capacity (multiple of 2.2 higher than mineral oil). All of the above benefits make water hydraulics particularly interesting for robotics for high-performance actuation techniques, in addition to improving people's awareness of environmental protection and sustainable development requirements [2–5]. Hydraulic systems that utilize water as a pressure medium could be the correct solution for the environmental and safety problems of most oil hydraulic systems [6]. Water hydraulic systems have been widely used in the fields of steel and glass production, ocean exploration, food and medicine processing, and coal mining [3,7]. Water-based systems are highly profoundly vulnerable to cavitation. The imploding of vapour cavities inside the flow motivates pressure pulsations, which may conjointly cause intense noise, part vibration, energy loss, erosion and corrosion of internal surfaces and eventually, scale back the performance of the system or failure of the part [8,9]. In rivers and seawater, several works studied the mechanism of cavitation, erosion abrasion, and corrosion. Cavitation bubbles are created when the fluid's static pressure drops below the vapour pressure at a specific temperature. [3,8]. The formation and collapse of bubbles of vapor would result in high local temperatures and high pressure [10]. The valve is one of the critical hydraulic components and due to the sharp drop in pressure caused by throttling, the occurrence of cavitation is normal [10]. Significantly, fluid valves can be classified as non-continuous valves (e.g., shut-off and change valves) and continuous valves (e.g., servo and proportional valves) [11]. There are already numerous water hydraulic power control systems on the market. Usually, their components are made of stainless steel to ensure satisfactory performance in harsh, conventional operating conditions. They do not, however, provide the necessary quality and long-term, low-friction, and low-wear efficiency [6]. Tensile properties are made up of the materials' reaction to resist when tension forces are applied. It is vital to establish tensile properties because it provides information on the elasticity module, elastic limit, elongation, proportional limit, area reduction, tensile strength, yield point, yield strength, and other tensile properties [12]. Fairfield in [13] declares that the studied erosion resistance was a substitute for a primarily unknown combination of other properties, including fracture resistance, strength, impact resistance, surface roughness, hardness, and service temperature limitation. He additionally clarifies that material properties with the largest contribution to jets resistance are compressive, tensile, and flexural, without forgetting surface roughness, thickness, peak service temperature, and thermal conductivity. Due to polymer significance in many applications, the characterization and study of superhydrophobic surfaces are of significant interest. Many material science studies have focused on surfaces with thrilling behaviors of wettability, such as superhydrophobic and superhydrophilic. The surface must have hydrophobic chemistry for the coating of polymer with other metal oxides of choice. The angle of contact is the angle that a liquid and solid surface create when they come into contact. This angle depends on both the material properties and the attractive and repulsive interaction between the materials. When the liquid spreads over the surface, it forms a small contact angle, whereas the contact angles are higher when the contact area between the surface and the liquid is smaller. The surface properties were analyzed and showed a good anti-corrosion ability [14]. The estimation of polymer solubility in solvents is one of the most important applications of solubility parameters [15]. Joel H. Hildebrand in [16] (who performed definitive work on the solubility of non-electrolytes in 1916 and lay the groundwork for solubility theory) recommended the square root of the cohesive energy density as a numerical value showing a particular solvent's solvency behavior [15]. The cohesive energy density is a numerical value describing the vaporizing energy in calories per cubic centimeter and a direct representation of the severity of van der Waals forces binding the substance's well-organized molecules [17]. Nevertheless, water's low viscosity, strong corrosion, and inadequate lubricating capacity have created significant challenges for this component's development and application. It is, therefore, an urgent task to screen materials suitable for the hydraulic water valve, which is the purpose of this research [18]. Hardness has long been considered a key criterion for the measurement of wear resistance. The erosion rate is correlated with these material properties, such as dynamic hardness, critical failure strain, and toughness [19].

2. Materials and Methods

2.1. Polymer Selection for Water Hydraulics Valves

In the engineering design and manufacturing process, the material plays a vital role. The appropriate selection of materials for a specific function is one of the designers' vital tasks. To meet the end specifications of the consumer, developers need to evaluate, with specific functionalities, the quality of various materials and find appropriate materials. The selection of materials is a challenging and time-consuming task because of the presence of a large number of materials with different attributes. [20]. The advancement of MCDA methods was not only inspired by a variety of real-life issues that need to address several considerations but also by experts' ability to incorporate stronger decision-making methods using recent developments in mathematical modeling, statistical analysis, and computer technology. To date, many computational approaches have been developed and applied to solve problems of material selection arising from various fields of engineering [20]. The PROMETHEE (Preference Ranking Organization Method for Enrichment Evaluations) approach is one of the latest MCDA methods developed by Brans [21] and further extended by Vincke and Brans [22].

PROMETHEE is an outranking method for a finite set of alternative actions to be ranked and selected among the often widely divergent criteria. PROMETHEE is, indeed, a straightforward design and application ranking method compared to other multi-criteria analysis methods [23]. This part will deal with the polymer coating selection and particularly utilizing the PROMETHEE (Preference Ranking Organization Method for Enrichment Evaluations) method. Figure 1 shows a flowchart for coating selection that uses the PROMETHEE technique, and Table 1 shows the used criteria used for polymer selection.

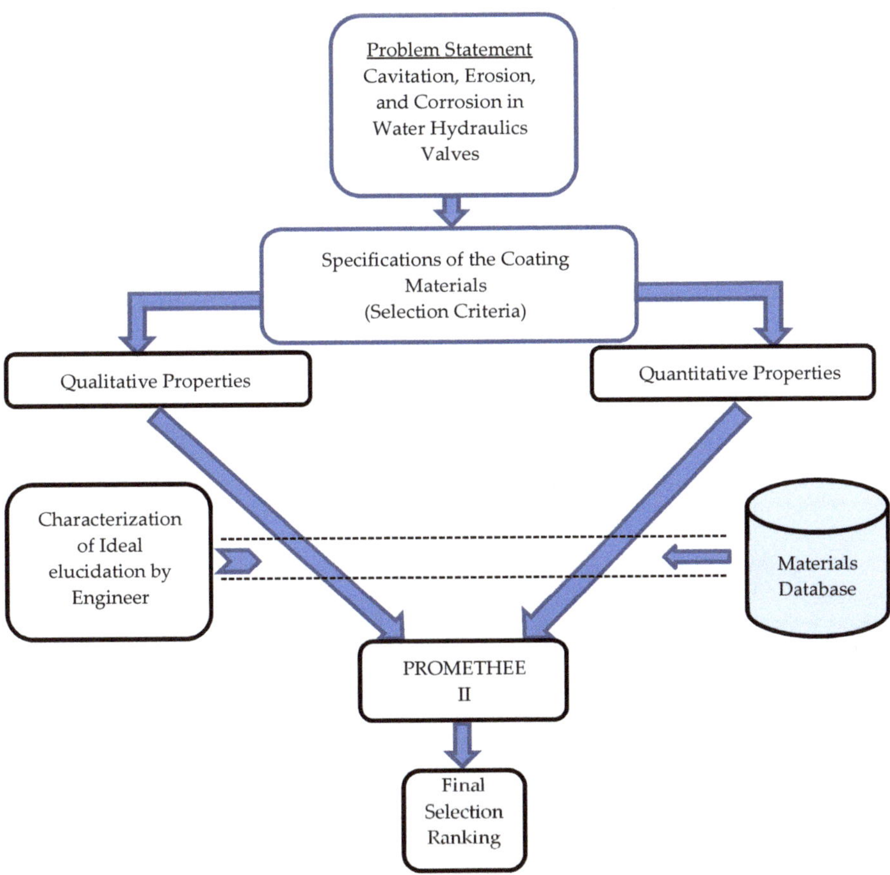

Figure 1. Flowchart for polymer coating selection.

2.1.1. Polyvinylchloride (PVC)

PVC was examined as amongst the most chemically antagonistic polymers and hydrophilicity [24]. Generally, PVC as seen in Figure 2a dissolves into polar solvents but is very resistant to hydrocarbons, alcohols, esters, acids, bases, and salts. Its chemical defiance can be predicted using solubility parameters [25]. When warmed exceedingly its T_g (about 87 °C), unstabilized PVC encounters dehydrochlorination [26]. PVC is the third most widely produced polymer in the world, with its outstanding mechanical strength, flame choke off, thermal stability, and insulation peculiarity, and has been mostly used as a contentedness resin in many products [27]. Table 1 gives some useful PVC properties for water hydraulics valves.

Table 1. Selected polymers and properties that are suitable for water hydraulic valves.

	Attribute or Criteria	Units	Polymer					Reference
			P1	P2	P3	P4	P5	
Non-beneficial	A1	%	0.4	0	0.17	0.3	0.5	[28,29]
	A2	($MPa^{0.5}$)	19.1	12.7	14.9	21.3	22.8	[28]
	A3	-	0.8	0.06	0.8	0.539	0.4	[28,30,31]
Beneficial	A4	(degree)	91.9	122	110	74.7	90	[28]
	A5	MPa	51.7	35	9.7	72.4	103	[32–34]
	A6	D	25	50	70	78	88	[28,35,36]
	A7	J/m	200	188	22	20	80	[17,28,37]
	A8	-	Satisfactory	Excellent	Good	Poor	Very Good	[28]

Annotation: P1 is the polyvinylchloride, P2 is the polytetrafluoroethylene, P3 is the polydimethylsiloxane, P4 is the Polymethylmethacrylate, and P5 is the Polyaryletheretherketone. While non-beneficial criteria are water absorption or equilibrium in the water at 23° C (A1), Hildebrand Solubility(A2), and coefficient of friction(A3). The beneficial criteria are Contact angle (A4), Tensile strength (A5), Hardness Shore D (A6), Impact Strength (A7) and Chemical resistance (A8) [24–27].

2.1.2. Polytetrafluoroethylene (PTFE)

Polytetrafluoroethylene (PTFE) in Figure 2b also recognized as Teflon. This material notably used in spacecraft design, automotive industries, and semiconductor design [38]. Disclosed utilization of the PTFE is divulging with the fact that it has meritorious chemical inactivity and high thermal stability. The coefficient of sliding friction between PTFE and many engineering materials is extremely low and has the self-lubrication capability [39,40]. It is insoluble in all common solvents and is resistant to almost all acidic and caustic substances [40]. PTFE is resistant to attack even by corrosive solutions and is practically unaffected by water (hydrophobicity) [26]. PTFE is subjected and considered as superplastic. Table 1 provides some useful PTFE properties suitable for water hydraulic valves. The essential characteristic of a superplastic material is its high strain rate sensitivity of flow stress that entertains high resistance [41]. Although PTFE has a high impact strength, its tensile strength and resistance to wear and creep are low compared to other polymers in the materials' database. Polytetrafluoroethylene is primarily used in applications needing extreme strength, exceptional chemical and heat tolerance, good electrical properties, low friction, or a blend of these properties [26].

2.1.3. Polydimethylsiloxane (PDMS)

The Figure 2c shows a silicon-based elastomer Polydimethylsiloxane which has several interesting properties, along with biological and chemical inertness, extremely low T_g, high temperature, and oxidation ability to resist, and vapor permeability [42]. After preparing the PDMS solid, the hydrophilic uniqueness of the PDMS was lowered so that it can be covered with metals for electrode and microchannel applications [43]. Table 1 provides some vital water hydraulic valve PDMS properties.

2.1.4. Polymethylmethacrylate (PMMA)

Poly (methyl methacrylate) as shown in Figure 2d, is the most significant member of acrylic polymers, distinguished by its thrilling weatherability, strong crystallinity, and extreme glassiness to visible light. It is mechanically robust with excellent insulating properties, making this thermoplastic polymer a choice for many mechanical, microelectronic and electrical uses, and good chemical and electrical and thermal tolerance [26,44]. In the arena of mechanical strength, PMMA has a low elongation at breakage and a high Young's Modulus. Therefore, it does not splinter upon rupture and happens to be one of the hardest thermoplastics with high scratch resistance [45]. Table 1 presents some useful PMMA properties for valves for water hydraulics.

2.1.5. Polyaryletheretherketone (PEEK)

Polyetheretherketone (PEEK) is a tough semi-crystalline thermoplastic polymer with excellent mechanical and dielectric properties, as well as good chemical resistance [39,46]. Polyetheretherketone (PEEK) as seen in Figure 2e is a member of the Polyaryletherketone group that exhibit a strong-performance polymer (HPP) that is commonly used in relatively movement applications. It has extreme chemical resistance; it is suitable for tribological applications due to its high strength and wears resistance. Polyaryletheretherketone (PEEK) has outstanding properties due to its semi-crystalline structure and the molecular strength of its repeating units [46].

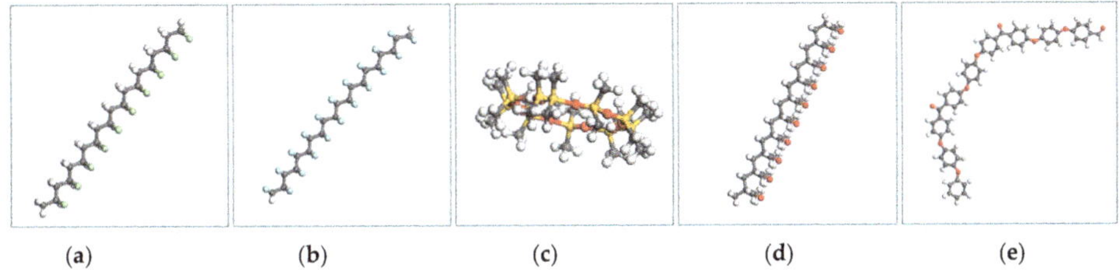

Figure 2. Shows (**a**) Polyvinylchloride; (**b**) Polytetrafluoroethylene; (**c**) Polydimethylsiloxane; (**d**) Polymethylmethacrylate; (**e**) Polyaryletheretherketone.

2.2. PROMETHEE (Preference Ranking Organization Method for Enrichment Evaluations) II

Material selection for engineering design is a multicriteria decision-making model (MCDM) problem that requires consideration of several available materials and conflicting evaluation criteria. The materials screened to be under selection are designated as P1 (Polyvinylchloride), P2 (Polytetrafluoroethylene), P3 (Polydimethylsiloxane), P4 (Polymethylmethacrylate) and P5 (Polyaryletheretherketone), as seen in Table 2. In addition, the beneficial and non-beneficial attributes or criteria to perform well in the water-logged condition of water hydraulic valves under the impact of water cavitation were chosen.

Figure 3 shows the hierarchy of the Material Selection based on the non-beneficial and beneficial criteria. Among the non-beneficial criteria are water absorption or equilibrium in the water at 23 °C (A1), Hildebrand Solubility (A2), and Coefficient of friction (A3). The beneficial criteria are Contact angle (A4), Tensile strength (A5), Hardness Shore D (A6), Impact Strength (A7) and Chemical resistance (A8). The criterion must fit the requirement of the system to survive from the prone to hydrodynamic cavitation surge pressure and erosion wears due to massive pressurized fluid flow in the valve chamber. Corrosion of the valve's internal wall must be prevented to sustain the longer life of the valves and to resist the erosion notch. Water absorption or equilibrium in the water at 23 °C is based on the superhydrophobic surface. However, as a result, the low viscosity and poor lubrication properties of water lead to potential risks of higher wear and friction, especially in the proportional spool valves [6].

For some materials, hardness (macro and micro) is a good cavitation erosion resistance indicator [47]. In terms of hydrophobicity, water absorption is commonly inferred as a measure of a solute's relative tendency to prefer a non-aqueous environment rather than an aqueous one. Hydrophobicity plays an essential role in the biological and physicochemical behaviour of numerous types of organic compounds [48]. Simply, hydrophobicity composite material, remarkable mechanochemical robustness, stain repellency, oil-water separation. Figure 4 presents a stepwise procedure for implementing PROMETHEE II [49,50]. Based on the requirement of the valve for water hydraulics and considering the flow of most significant selected attributes, engineering and material fields are integrated by aggregating the significant weights of criteria and the ratings of alternatives. Table 6 applies the weight distribution of the selected attribute.

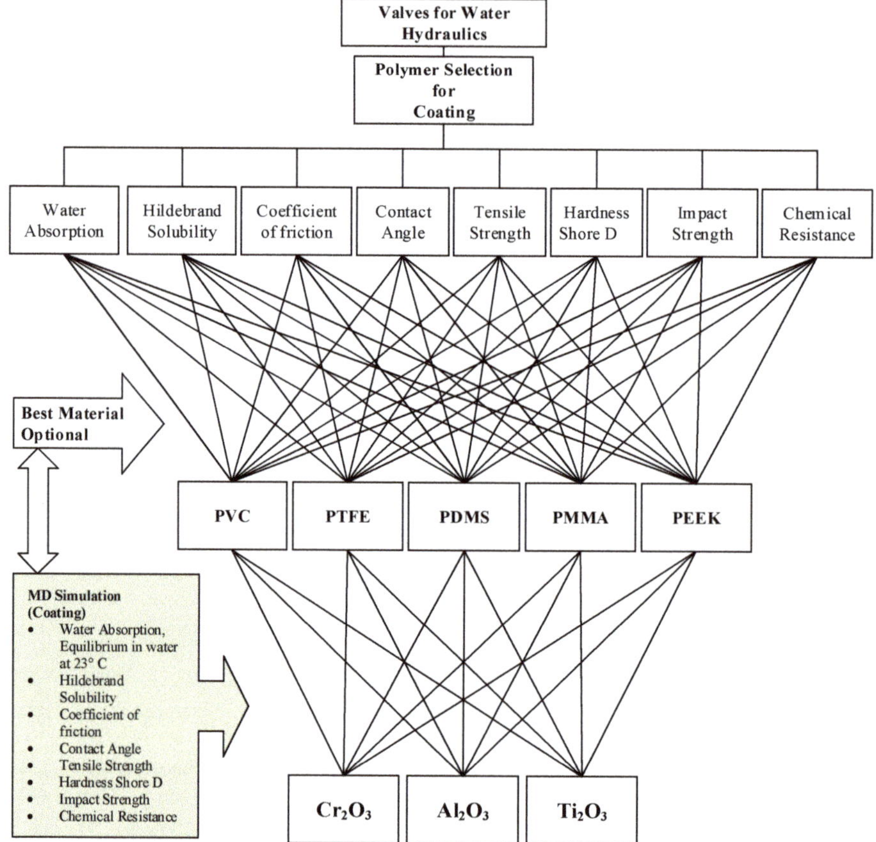

Figure 3. The hierarchy of the Material Selection.

Table 2. Polymer properties and introducing the 5-point scale.

Polymer	Attributes							
	A1	**A2**	**A3**	**A4**	**A5**	**A6**	**A7**	**A8**
P1	0.4	19.1	0.4	91.9	51.7	25	200	2
P2	0	12.7	0.06	122	35	50	188	5
P3	0.17	14.9	0.8	110	9.7	70	22	3
P4	0.3	21.3	0.539	74.7	72.4	78	20	1
P5	0.5	22.8	0.4	90	103	88	80	4

Annotation: Apply the 5-point scale for Chemical Resistance attribute (A8) in Table 1. Poor = 1, Satisfactory = 2, Good = 3, Very Good = 4 and Excellent = 5.

Step 1: Determine the objective and the evaluation criteria or attributes. Also determine / Select the weights for different attributes by relative comparison of attributes.

Step 2: Normalize the evaluation criteria to dissolve the difference in units

For beneficial criteria, the properties that its small value is the prominent for consideration

$$R_{ij} = \frac{[(X_{ij}) - Min(X_{ij})]}{[Max(X_{ij}) - Min(X_{ij})]} \quad (i = 1, 2, .., m; j = 1, 2, ..n) \tag{1}$$

For non-beneficial criteria, the properties that its large value is prominent to consideration

$$R_{ij} = \frac{[Max(X_{ij}) - (X_{ij})]}{[Max(X_{ij}) - Min(X_{ij})]} \quad (i = 1, 2, .., m; j = 1, 2, ..n) \tag{2}$$

Where X_{ij} denote, the specific value of criteria, $Max(X_{ij})$ denote maximum value of criteria, $Min(X_{ij})$ denote the minimum vale of the criteria

Step 3: Calculate the evaluative differences of ith alternative with respect to other alternatives

$$d_j(a,b) = g_j(a) - g_j(b) \tag{3}$$

Where $d_j(a,b)$ denotes the difference between the evaluations of a and b on each criterion

Step 4: Calculate the evaluative differences of ith alternative with respect to other alternatives

$$P_j(a,b) = F_j[d_j(a,b)] \quad (j = 1,2,..k) \tag{4}$$

where $P_j(a, b)$ denotes the reference of alternative a with regard to alternative b against the jth criterion, as a function of $d_j(a, b)$. F_j is a preference function, which translates the difference between the evaluations of alternatives a and b on the jth criterion into a preference degree ranging from 0 to 1.

Step 5: Calculate the overall or global preference index

$$\pi(a,b) = \sum_{j=1}^{n} P_j(a,b) w_j \tag{5}$$

where, w_j is the weight of the jth criterion, and $\pi(a, b)$ of a over b is defined as the weighted sum of $P(a, b)$ for each criterion.

Step 6: Calculated the positive and negative outranking flows

$$\phi^+(a) = \frac{1}{m-1} \sum_{X \in A} \pi(a, X) \tag{6}$$

$$\phi^-(a) = \frac{1}{m-1} \sum_{X \in A} \pi(x, a) \tag{7}$$

Where $\phi^+(a)$ and $\phi^-(a)$ denote the positive outranking flow and negative outranking flow for each alternative respectively.

Step 7: Calculate the net outranking flow and the complete ranking

$$\phi(net) = \phi^+(a) - \phi^-(a) \tag{8}$$

Figure 4. Stepwise complete ranking PROMETHEE II.

2.3. Molecular Dynamics Simulation to Predict the Substrate and Suitable Coating Plane

MD simulation is a modeling technique based on physics that has been commonly used to represent different material systems as it can provide detailed information on variations and conformational changes in the structure and behavior of material systems at the molecular level [51,52]. Figure 5 below shows the Molecular dynamics algorithms. Here, the MD Simulation was conducted to predict a polymeric coating on the selected substrates (Cr_2O_3, Al_2O_3, and Ti_2O_3) along the planes (100), (010) and (001). The simulation computation was created using the material science simulation software developed by Accelrys Software Inc. in San Diego, CA, USA, named Material Studio ® Software, after several phases of model design and calculation. The tasks were to create engineered Cr_2O_3, Al_2O_3, and Ti_2O_3, slabs with an equal thickness of 13.233×10^{-10} m each in a cleaved plane along (100), (010) and (001) each. Using a smarter algorithm with a maximum iteration of 100,000 steps, the surfaces were relaxed at minimum energy using model typing for force-field and condensed-phase optimized molecular potentials (COMPASS) to ensure bonding between metal oxide (Cr_2O_3, Al_2O_3, and Ti_2O_3). The supercell was designed by growing the surface to the lattice vector of $U \times V = 6 \times 6$ and then constructing the slab without the vacuum in order to change the periodicity from 2D to 3D. The steps of making a slab are the same as all-metal oxides (Cr_2O_3, Al_2O_3, and Ti_2O_3) picked. The periodic layer of the amorphous polytetrafluoroethylene cell (PTFE) was constructed and the structure was geometrically refined from the atactic polymer used in this work. This was pre-constructed, manufactured from the 50-chain length repeat-unit tetrafluoroethylene with a number of configurations set to 1.0, the final design goal density agreed to 2 g/cm^3. The next step was to build a layer with a vacuum of 70×10^{-10} m. Minimized polytetrafluoroethylene was deposited on the matched metal oxide slabs of Cr_2O_3, Al_2O_3, and Ti_2O_3.

A molecular dynamics simulation was conducted in each coating composite layer. The temperature control method of Andersen was used, ensembles of a constant number of particles, constant volume, and constant temperature (NVPT). Each arrangement of molecular dynamics (MD) simulation was done with an interval of 1 femtosecond (fs) and each frame of 5000. COMPASS forcefield was determined in all the arrangements. Simulation steps were set to 200,000 with a total dynamic time of 200 picoseconds (ps).

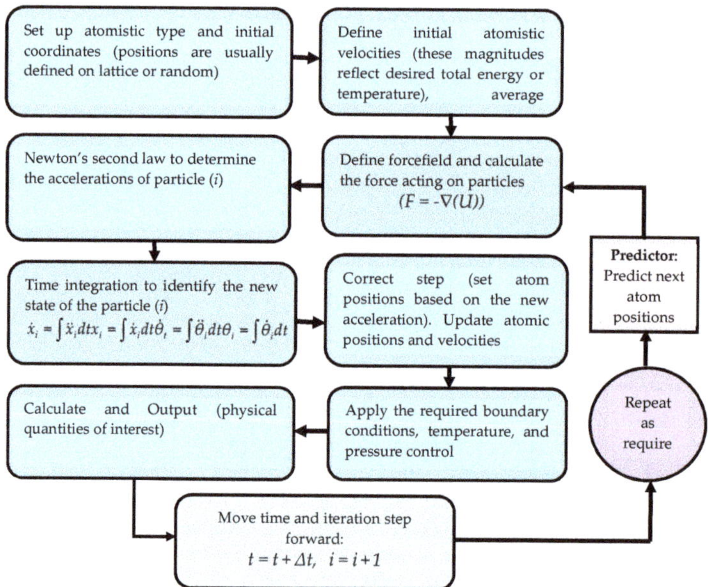

Figure 5. Molecular dynamics simulation algorithm.

3. Results and Discussion

PROMETHEE was used so solve the polymer selection problem based on the Equations (1)–(8) as shown in the Tables 3–8. The PROMETHEE analysis was done by following all the essential steps of the analysis. Finally, the PTFE was the best polymer based on the selected significant properties for coating in the valve in water hydraulics. Table 9 shows the net flow of polymer alternatives. The PTFE has a higher value of 0.5932052, as seen in Figure 6 of the PROMETHEE I-II partial and complete ranking that the (Φnet values) of alternative P2 is the better coating polymer for water hydraulics valves to resist the cavitation erosion and corrosion. Moreover, PDMS, PVC, PEEK, and PMMA, as shown in Figure 7 of the ranking sequence, follow PTFE.

Table 3. Applying normalization to the evaluation criteria to obtain R_{ij} (beneficial and non-beneficial) using Equations (1) and (2).

Polymer	Attributes							
	A1	A2	A3	A4	A5	A6	A7	A8
P1	0.2	0.46835	0	0.36364	0.45016	0	1	0.25
P2	1	1.2848	1	1	0.37513	0.39683	0.93333	1
P3	0.66	1	0	0.7463	0	0.71429	0.01111	0.5
P4	0.4	0.18987	0.3527	0	0.67202	0.84127	0	0
P5	0	0	0.54054	0.32347	1	1	0.33333	0.75
Max (X_{ij})	0.5	22.8	0.8	122	103	88	200	5
Min (X_{ij})	0	14.9	0.06	74.7	9.7	25	20	1

Table 4. Applying the calculation of the evaluative differences of ith alternative with respect to other alternatives using Equation (3).

d_j (a,b)	Attributes							
	A1	A2	A3	A4	A5	A6	A7	A8
d(P1-P2)	−0.80	−0.81645	−1	−0.63636	0.07503	−0.39683	0.06667	−0.75
d(P1-P3)	−0.46	−0.53165	0	−0.38266	0.45016	−0.71429	0.98889	−0.25
d(P1-P4)	−0.2	0.27848	−0.3527	0.36364	−0.22186	−0.84127	1	0.25
d(P1-P5)	0.2	0.46835	−0.54054	0.04017	−0.54984	−1	0.66667	−0.5
d(P2-P1)	0.8	0.81645	1	0.63636	−0.07503	0.39683	−0.06667	0.75
d(P2-P3)	0.34	0.2848	1	0.2537	0.37513	−0.31746	0.92222	0.5
d(P2-P4)	0.6	1.09493	0.6473	1	−0.29689	−0.44444	0.93333	1
d(P2-P5)	1	1.2848	0.45946	0.67653	−0.62487	−0.60317	0.6	0.25
d(P3-P1)	0.46	0.53165	0	0.38266	−0.45016	0.71429	−0.98889	0.25
d(P3-P2)	−0.34	−0.2848	−1	−0.2537	−0.37513	0.31746	−0.92222	−0.5
d(P3-P4)	0.26	0.81013	−0.3527	0.7463	−0.67202	−0.12698	0.01111	0.5
d(P3-P5)	0.66	1	−0.54054	0.42283	−1	−0.28571	−0.32222	−0.25
d(P4-P1)	0.2	−0.27848	0.3527	−0.36364	0.22186	0.84127	−1	−0.25
d(P4-P2)	−0.6	−1.09493	−0.6473	−1	0.29689	0.44444	−0.93333	−1
d(P4-P3)	−0.26	−0.81013	0.3527	−0.7463	0.67202	0.12698	−0.01111	−0.5
d(P4-P5)	0.4	0.18987	−0.18784	−0.32347	−0.32798	−0.15873	−0.33333	−0.75
d(P5-P1)	−0.2	−0.46835	0.54054	−0.04017	0.54984	1	−0.66667	0.5
d(P5-P2)	−1	−1.2848	−0.45946	−0.67653	0.62487	0.60317	−0.6	−0.25
d(P5-P3)	−0.66	−1	0.54054	−0.42283	1	0.28571	0.32222	0.25
d(P5-P4)	−0.4	−0.18987	0.18784	0.32347	0.32798	0.15873	0.33333	0.75

Table 5. Application of the preference function using Equation (4).

$P_j(a,b)$	Attributes							
	A1	**A2**	**A3**	**A4**	**A5**	**A6**	**A7**	**A8**
d(P1-P2)	0	0	0	0	0.0750	0	0.0667	0
d(P1-P3)	0	0	0	0	0.4502	0	0.9889	0
d(P1-P4)	0	0.2785	0	0.3636	0	0	1	0.25
d(P1-P5)	0.2	0.4684	0	0.0402	0	0	0.6667	0
d(P2-P1)	0.8	0.8165	1	0.6364	0	0.3968	0	0.75
d(P2-P3)	0.34	0.2848	1	0.2537	0.3751	0	0.9222	0.5
d(P2-P4)	0.6	1.0949	0.6473	1	0	0	0.9333	1
d(P2-P5)	1	1.2848	0.4595	0.6765	0	0	0.6	0.25
d(P3-P1)	0.46	0.5317	0	0.3827	0	0.7143	0	0.25
d(P3-P2)	0	0	0	0	0	0.3175	0	0
d(P3-P4)	0.26	0.8101	0	0.7463	0	0	0.0111	0.5
d(P3-P5)	0.66	1	0	0.4228	0	0	0	0
d(P4-P1)	0.2	0	0.3527	0	0.2219	0.8413	0	0
d(P4-P2)	0	0	0	0	0.2969	0.4444	0	0
d(P4-P3)	0	0	0.3527	0	0.6720	0.1270	0	0
d(P4-P5)	0.4	0.1899	0	0	0	0	0	0
d(P5-P1)	0	0	0.5405	0	0.5498	1	0	0.5
d(P5-P2)	0	0	0	0	0.6249	0.6032	0	0
d(P5-P3)	0	0	0.5405	0	1	0.2857	0.3222	0.25
d(P5-P4)	0	0	0.1878	0.3235	0.3280	0.1587	0.3333	0.75

Table 6. Applying the calculation of the overall or global preference index, using Equation (5).

	Attributes								
Weight →	**A1**	**A2**	**A3**	**A4**	**A5**	**A6**	**A7**	**A8**	$\pi(a,b)$
	0.2	**0.18**	**0.08**	**0.16**	**0.06**	**0.06**	**0.14**	**0.12**	
$w_j \times$ d(P1-P2)	0	0	0	0	0.0045	0	0.0093	0	0.0138356
$w_j \times$ d(P1-P3)	0	0	0	0	0.0270	0	0.1384	0	0.1654542
$w_j \times$ d(P1-P4)	0	0.0501	0	0.0582	0	0	0.14	0.03	0.2783088
$w_j \times$ d(P1-P5)	0.04	0.0843	0	0.0064	0	0	0.0933	0	0.224064
$w_j \times$ d(P2-P1)	0.16	0.1470	0.08	0.1018	0	0.0238	0	0.09	0.6025884
$w_j \times$ d(P2-P3)	0.068	0.0512	0.08	0.0406	0.0225	0	0.1291	0.06	0.4514746
$w_j \times$ d(P2-P4)	0.12	0.1971	0.0518	0.16	0	0	0.1307	0.12	0.7795376
$w_j \times$ d(P2-P5)	0.2	0.2313	0.0368	0.1082	0	0	0.084	0.03	0.6902656
$w_j \times$ d(P3-P1)	0.092	0.0957	0	0.0612	0	0.0429	0	0.03	0.32178
$w_j \times$ d(P3-P2)	0	0	0	0	0	0.0191	0	0	0.0190476
$w_j \times$ d(P3-P4)	0.052	0.1458	0	0.1194	0	0	0.0016	0.06	0.3787868
$w_j \times$ d(P3-P5)	0.132	0.18	0	0.0677	0	0	0	0	0.3796528
$w_j \times$ d(P4-P1)	0.04	0	0.0282	0	0.0133	0.0505	0	0	0.1320038
$w_j \times$ d(P4-P2)	0	0	0	0	0.0178	0.0267	0	0	0.0444798
$w_j \times$ d(P4-P3)	0	0	0.0282	0	0.0403	0.0076	0	0	0.076156
$w_j \times$ d(P4-P5)	0.08	0.0342	0	0	0	0	0	0	0.1141766
$w_j \times$ d(P5-P1)	0	0	0.0432	0	0.0330	0.06	0	0.06	0.1962336
$w_j \times$ d(P5-P2)	0	0	0	0	0.0375	0.0362	0	0	0.0736824
$w_j \times$ d(P5-P3)	0	0	0.0432	0	0.06	0.0171	0.0451	0.03	0.1954966
$w_j \times$ d(P5-P4)	0	0	0.0150	0.0518	0.0197	0.0095	0.0467	0.09	0.2326512

Table 7. Applying the calculation of the positive and negative outranking flows, using Equations (6) and (7).

Aggregate Preference Function	P1	P2	P3	P4	P5	Leaving Flow $\phi^{+}(a)$
P1	-	0.0138356	0.1654542	0.2783088	0.224064	0.17041565
P2	0.6025884	-	0.4514746	0.7795376	0.6902656	0.63096655
P3	0.32178	0.0190476	-	0.3787868	0.3796528	0.2748168
P4	0.1320038	0.0444798	0.076156	-	0.1141766	0.09170405
P5	0.1962336	0.0736824	0.1954966	0.2326512	-	0.17451595
Entering Flow $\phi^{-}(a)$	0.31315145	0.03776135	0.22214535	0.4173211	0.35203975	

Table 8. Applying the Equation (8) to calculate the net outranking flow and the complete ranking.

Polymer	Leaving Flow	Entering Flow	Net Flow	Ranking
P1	0.17041565	0.31315145	−0.1427358	3
P2	0.63096655	0.03776135	0.5932052	1
P3	0.2748168	0.22214535	0.05267145	2
P4	0.09170405	0.4173211	−0.32561705	5
P5	0.17451595	0.35203975	−0.1775238	4

Table 9. The value of the net flow of polymer alternatives.

Alternative Polymer	P1	P2	P3	P4	P5
ϕ_{net}	−0.1427358	0.5932052	0.05267145	−0.32561705	−0.1775238

Water hydraulics valves are subjected to cavitation erosion, wear, and corrosion due to a high bubbles implosion impact; selected properties of PTFE fit well with the coating requirements. Furthermore, the molecular dynamics results show that the PTFE coated with Al_2O_3 in the cleaved plane of (010) gives more considerable binding energy compared to other composites in all the cleaved planes. This may be due to difference in the chemical reactivity functions within the frameworks of the density functional theory (DFT), the high rate of transfer of electrons to or from the reactants in the plane (010) of PTFE/Al_2O_3, or due to the change in the geometric structure of the molecules (equivalent in external potential). In addition, the chemisorption takes place due to the strong electrostatic (ionic) bond.

These results show the sign of promising the strength of the composite over the effect of the impairment of water hydraulics valves due to the sudden surge pressure caused by the imploding of cavitation bubbles. Figure 8 below shows the pictorial presentation and data obtained after simulation.

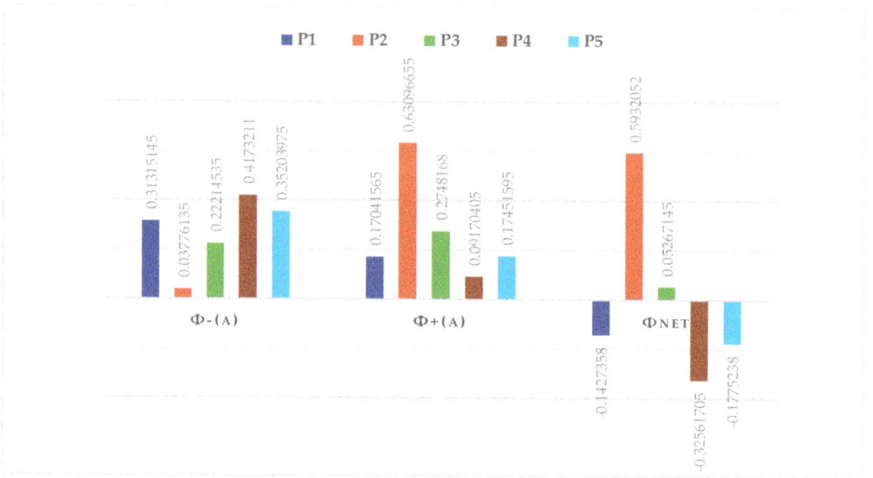

Figure 6. PROMETHEE I-II partial and complete ranking.

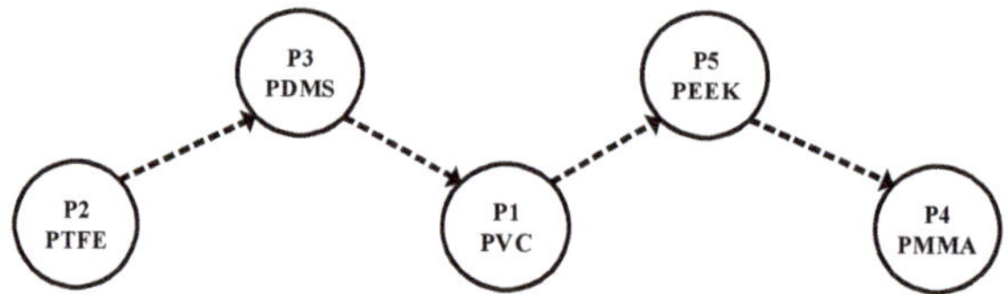

Figure 7. Ranking sequence.

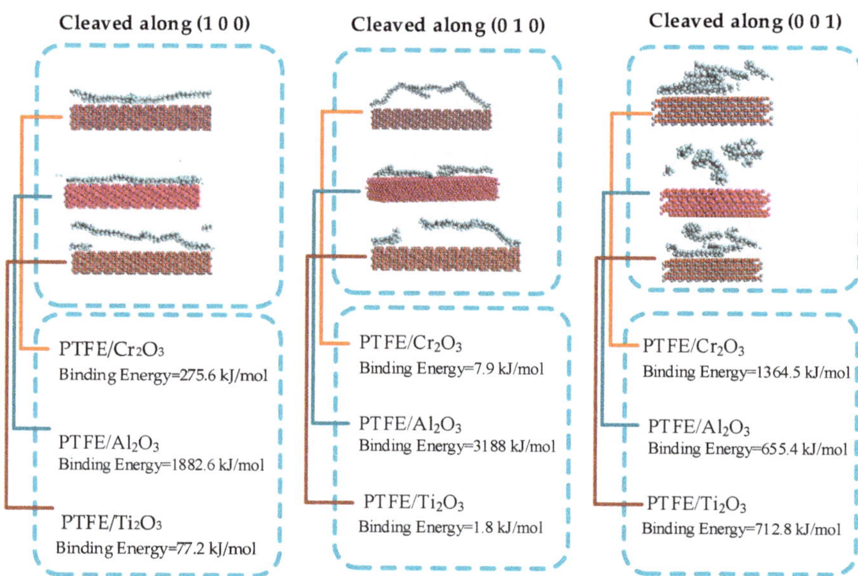

Figure 8. Molecular dynamics Polytetrafluoroethylene coating on Metal Oxides (Ti_2O_3, Al_2O_3, and Cr_2O_3).

4. Conclusions

The selection and screening of polymer coating on metal oxides were studied for the design of liquid hydraulic valves. For this reason, PROMETHEE's decision-making methods were used. The results show the most suitable polymer materials being simulated. The improvement in the cavitation erosion resistance can mainly be linked to the increase in hardness and the elastic response of the PTFE/Al_2O_3 enhancement. Allison and Tong, in [53], using the density functional theory, depicted that binding energy is proportional to hardness. Moreover, simulation of molecular dynamics found the best substrate and cleaved plane in further studies should be considered by the characterization of a different PTFE/Al_2O_3 composition ratio.

Author Contributions: M.K.M. carried out the critical study and did all the comprehensive procedure and wrote the paper; H.X. provided instructions for this research and checked the findings of the analysis; F.S. and H.W. reviewed, revised the manuscript and looked for the literature; G.D.M. involved during the polymer attribute weighing. All authors have read and agreed to the published version of the manuscript.

Funding: The Natural Science Foundation of China supported this work under Grant 5187511, Natural Science Foundation of the Heilongjiang Province of China under Grant F2016003, 'Jinshan Talent' Zhenjiang Manufacture 2025 Leading Talent Project, "Jiangyan Planning" Project in Yangzhong City.

Conflicts of Interest: No conflict of interest declared by the authors.

References

1. Pham, N.P.; Ito, K.; Ikeo, S. Energy Saving for Water Hydraulic Pushing Cylinder in Meat Slicer. *JFPS Int. J. Fluid Power Syst.* **2017**, *10*, 24–29. [CrossRef]
2. Zhang, H. Cavitation Effect to the Hydraulic Piston Pump Flow Pulsation. *Appl. Mech. Mater.* **2014**, *599–601*, 230–236. [CrossRef]
3. Han, M.; Liu, Y.; Wu, D.; Zhao, X.; Tan, H. A numerical investigation in characteristics of flow force under cavitation state inside the water hydraulic poppet valves. *Int. J. Heat Mass Transf.* **2017**, *111*, 1–16. [CrossRef]
4. Xia, Q.; Pan, G. Flow Field Analysis and Structure Optimization of Water Hydraulic Poppet Valve Based on the CFD. *Adv. Mater. Res.* **2013**, *791–793*, 734–737. [CrossRef]
5. Yang, Y.S.; Semini, C.; Tsagarakis, N.G.; Caldwell, D.G.; Zhu, Y. Water Hydraulics—A Novel Design of Spool-type valves for Enhanced Dynamic Performance. In Proceedings of the International Conference on Advanced Intelligent Mechatronics, Xi'an, China, 4–7 June 2008. [CrossRef]
6. Majdič, F.; Velkavrh, I.; Kalin, M. Improving the performance of a proportional 4/3 water–hydraulic valve by using a diamond-like-carbon coating. *Wear* **2013**, *297*, 1016–1024. [CrossRef]
7. Han, M.; Liu, Y.; Wu, D.; Tan, H.; Li, C. Numerical Analysis and Optimisation of the Flow Forces in a Water Hydraulic Proportional Cartridge Valve for Injection System. *IEEE Access* **2018**, *6*, 10392–10401. [CrossRef]
8. Trostmann, E. *Water Hydraulics Control Technology*; CRC Press: Boca Raton, FL, USA, 1996.
9. Kozák, J.; Rudolf, P.; Hudec, M.; Štefan, D.; Forman, M. Numerical and experimental investigation of the cavitating flow within venturi tube. *J. Fluids Eng. Trans. ASME* **2019**, *141*, 041101. [CrossRef]
10. Liu, Y.S.; Huang, Y.; Li, Z.Y. Experimental investigation of flow and cavitation characteristics of a two-step throttle in water hydraulic valves. *Proc. Inst. Mech. Eng. Part A* **2002**. [CrossRef]
11. Herakovič, N. Flow-Force Analysis in a Hydraulic Sliding-Spool Valve. *CODEN STJSAO* **2009**, *51*, 555–564.
12. Rahman, R.; Zhafer, S. Firdaus Syed Putra, 5—Tensile properties of natural and synthetic fiber-reinforced polymer composites. In *Mechanical and Physical Testing of Biocomposites, Fibre-Reinforced Composites and Hybrid Composites*; Jawaid, M., Thariq, M., Saba, N., Eds.; Woodhead Publishing: Cambridge, UK, 2019; pp. 81–102. [CrossRef]
13. Fairfield, C.A. Cavitation erosion resistance of sewer pipe materials. *Proc. ICE Constr. Mater.* **2015**, *168*, 1–15. [CrossRef]
14. Jayadev, D. Chapter 5—Characterization of superhydrophobic polymer coating. In *Superhydrophobic Polymer Coatings*; Samal, S.K., Mohanty, S., Nayak, S.K., Eds.; Elsevier: Amsterdam, The Netherlands, 2019; pp. 91–121. [CrossRef]
15. Jensen, C.W. The Book and Paper Group Annual. Available online: https://cool.conservation-us.org/coolaic/sg/bpg/annual/v03/bp03-04.html (accessed on 28 November 2019).

16. Hildebrand, J.H. Dipole attraction and hydrogen bond formation in their relation to solubility. *Science* **1936**, *83*, 21–24. [CrossRef] [PubMed]
17. Mark, J.E. *Physical Properties of Polymers Handbook*, 2nd ed.; Springer: Berlin/Heidelberg, Germany, 2007; p. 1038. [CrossRef]
18. Zhu, B.; He, X.; Zhao, T. Friction and wear characteristics of natural bovine bone lubricated with water. *Wear* **2015**, *322–323*, 91–100. [CrossRef]
19. Ji, X. Erosive wear resistance evaluation with the hardness after strain-hardening and its application for a high-entropy alloy. *Wear* **2018**, *398–399*, 178–182. [CrossRef]
20. Kumar, R.; Jagadish; Ray, A. Selection of Material for Optimal Design Using Multi-criteria Decision Making. *Procedia Mater. Sci.* **2014**, *6*, 590–596. [CrossRef]
21. Brans, J.P. L'ingéniérie de la décision; Elaboration d'instruments d'aide à la décision. La méthode PROMETHEE. In *L'aide à la décision: Nature, Instruments et Perspectives d'Avenir*; Presses de l'Université Laval: Quebec City, QC, Canada, 1982; pp. 183–213.
22. Brans, J.P.; Vincke, P.; Mareschal, B. How to select and how to rank projects: The Promethee method. *Eur. J. Oper. Res.* **1986**, *24*, 228–238. [CrossRef]
23. Behzadian, M.; Kazemzadeh, R.B.; Albadvi, A.; Aghdasi, M. PROMETHEE: A comprehensive literature review on methodologies and applications. *Eur. J. Oper. Res.* **2010**, *200*, 198–215. [CrossRef]
24. Yong, M.; Zhang, Y.; Sun, S.; Liu, W. Properties of polyvinyl chloride (PVC) ultrafiltration membrane improved by lignin: Hydrophilicity and antifouling. *J. Membr. Sci.* **2019**, *575*, 50–59. [CrossRef]
25. Wypych, G. 2-PVC PROPERTIES. In *PVC Formulary*, 2nd ed.; Wypych, G., Ed.; ChemTec Publishing: San Diego, CA, USA, 2015; pp. 5–44. [CrossRef]
26. Ebewele, R.O. *Polymer Science and Technology*; CRC Press: Boca Raton, FL, USA, 2000. [CrossRef]
27. Jia, P.; Hu, L.; Feng, G.; Bo, C.; Zhang, M.; Zhou, Y. PVC materials without migration obtained by chemical modification of azide-functionalized PVC and triethyl citrate plasticizer. *Mater. Chem. Phys.* **2017**, *190*, 25–30. [CrossRef]
28. Wypych, G. *Handbook of Polymers*, 2nd ed.; ChemTec Publishing: Toronto, ON, Canada, 2016.
29. Stanton, M.M. Super-hydrophobic, highly adhesive, polydimethylsiloxane (PDMS) surfaces. *J. Colloid Interface Sci.* **2012**, *367*, 502–508. [CrossRef]
30. Penskiy, I.; Gerratt, P.A.; Bergbreiter, S. Friction, adhesion, and wear properties of PDMS coatings in MEMS devices. In Proceedings of the 2011 IEEE 24th International Conference on Micro Electro Mechanical Systems, Cancun, Mexico, 23–27 January 2011. [CrossRef]
31. Win, K.; Loong, P.Y.; Liu, E.; Li, L. Enhancing electrical and tribological properties of poly (methyl methacrylate) matrix nanocomposite films by co-incorporation of multiwalled carbon nanotubes and silicon dioxide microparticles. *J. Polym. Eng.* **2016**, *36*, 23–30. [CrossRef]
32. William, D.; Callister, D.G.R., Jr. *Materials Science and Engineering: An Introduction*, 9th ed.; John Wiley & Sons, Inc.: Hoboken, NJ, USA, 2009.
33. Van Krevelen, D.W.; Nijenhuis, K.T. *Properties of Polymers*, 4th ed.; van Krevelen, D.W., Nijenhuis, K.T., Eds.; Elsevier: Amsterdam, The Netherlands, 2009. [CrossRef]
34. Gilbert, M.; Patrick, S. Chapter 13—Poly (Vinyl Chloride). In *Brydson's Plastics Materials*, 8th ed.; Gilbert, M., Ed.; Butterworth-Heinemann: Oxford, UK, 2017; pp. 329–388. [CrossRef]
35. Handbook, P. Teflon PTFE. Available online: http://www.rjchase.com/ptfe_handbook.pdf (accessed on 28 November 2019).
36. Kumar, M.; Arun, S.; Upadhyaya, P.; Pugazhenthi, G. Properties of PMMA/clay nanocomposites prepared using various compatibilizers. *Int. J. Mech. Mater. Eng.* **2015**, *10*, 7. [CrossRef]
37. Vilčáková, J.; Kutějová, L.; Jurča, M.; Moučka, R. Enhanced Charpy impact strength of epoxy resin modified with vinyl-terminated polydimethylsiloxane: Research Article. *J. Appl. Polym. Sci.* **2017**, *135*, 45720. [CrossRef]
38. Wyszkowska, E.; Leśniak, M.; Kurpaska, L.; Prokopowicz, R.; Jozwik, I.; Sitarz, M.; Jagielski, J. Functional properties of poly(tetrafluoroethylene) (PTFE) gasket working in nuclear reactor conditions. *J. Mol. Struct.* **2018**, *1157*, 306–311. [CrossRef]
39. Mizobe, K. Effect of PTFE Retainer on Friction Coefficient in Polymer Thrust Bearings under Dry Contact. *Adv. Mater. Res.* **2013**, *683*, 90–93. [CrossRef]

40. Rae, P.; Dattelbaum, D.J.P. The properties of poly (tetrafluoroethylene) (PTFE) in compression. *Polymer* **2004**, *45*, 7615–7625. [CrossRef]
41. Nunes, L.C.S.; Dias, F.W.R.; Mattos, H.S.D. Mechanical behavior of polytetrafluoroethylene in tensile loading under different strain rates. *Polym. Test.* **2011**, *30*, 791–796. [CrossRef]
42. Charitidis, C.A. Influence of accelerated aging on nanomechanical properties, creep behaviour and adhesive forces of PDMS. *Plast. Rubber Compos.* **2012**, *41*, 94–99. [CrossRef]
43. Yadhuraj, S.R.; Gandla, S. Preparation and Study of PDMS Material. *Mater. Today Proc.* **2018**, *5*, 21406–21412. [CrossRef]
44. Saha, B. A study on frictional behavior of PMMA against FDTS coated silicon as a function of load, velocity and temperature. *Tribol. Int.* **2016**, *102*, 44–51. [CrossRef]
45. Ali, U.; Karim, K.J.B.A.; Buang, N.A. A Review of the Properties and Applications of Poly (Methyl Methacrylate) (PMMA). *Polym. Rev.* **2015**, *55*, 678–705. [CrossRef]
46. Chandrasena, L. Investigating the compatibility of PEEK polymer for the fabrication of sample cells for use in muon spin spectroscopy. *J. Phys. Conf. Ser.* **2014**, *551*, 012038. [CrossRef]
47. Taillon, G. Cavitation erosion mechanisms in stainless steels and in composite metal–ceramic HVOF coatings. *Wear* **2016**, *364–365*, 201–210. [CrossRef]
48. García, M.A.; Marina, M.L.; Ros, A.; Valcrcel, M. Separation modes in capillary electrophoresis. In *Comprehensive Analytical Chemistry*; Elsevier Science: Madrid, Spain, 2005; Volume 45, pp. 31–134. [CrossRef]
49. Lee, P.T.-W.; Yang, Z. *Multi-Criteria Decision Making in Maritime Studies and Logistics: Applications and Cases*; Palgrave Macmillan: Cham, Switzerland, 2018; pp. 19–21. [CrossRef]
50. Kittur, J. Using the PROMETHEE and TOPSIS Multi-Criteria Decision Making Methods to Evaluate Optimal Generation. In Proceedings of the 2015 International Conference on Power and Advanced Control Engineering (ICPACE), Bangalore, India, 12–14 August 2015; pp. 80–85. [CrossRef]
51. Shi, L.; Han, Q. Molecular dynamics study of deformation mechanisms of poly (vinyl alcohol) hydrogel. *Mol. Simul.* **2018**, *44*, 1363–1370. [CrossRef]
52. Epa, V.; Winkler, D.; Tran, L. Chapter 5—Computational Approaches. In *Adverse Effects of Engineered Nanomaterials*; Fadeel, B., Pietroiusti, A., Shvedova, A.A., Eds.; Academic Press: Cambridge, MA, USA, 2012; pp. 85–96.
53. Allison, T.C.; Tong, Y.J. Application of the condensed Fukui function to predict reactivity in core–shell transition metal nanoparticles. *Electrochim. Acta* **2013**, *101*, 334–340. [CrossRef]

Article

Mechanically Robust and Repairable Superhydrophobic Zinc Coating via a Fast and Facile Method for Corrosion Resisting

Junfei Ou *, Wenhui Zhu, Chan Xie and Mingshan Xue *

School of Materials Science and Engineering, Nanchang Hangkong University, Nanchang 330063, China; 70299@nchu.edu.cn (W.Z.); 70362@nchu.edu.cn (C.X.)
* Correspondence: oujunfei_1982@163.com (J.O.); xuems04@mails.ucas.ac.cn (M.X.)

Received: 6 May 2019; Accepted: 28 May 2019; Published: 31 May 2019

Abstract: Zinc coatings and superhydrophobic surfaces have their own characteristics in terms of metal corrosion resistance. Herein, we have prepared a robust and repairable superhydrophobic zinc coating (SZC) based on a widely commercially available cold galvanized paint via a fast (within 10 min) and facile process for corrosion resistance. Specifically, the cold galvanized paint was sprayed onto the iron substrate, followed by acetic acid (HAc) etching and stearic acid (STA) hydrophobizing. The as-obtained sample was coded as Fe-Zn-HAc-STA and possessed an apparent contact angle of 168.4 ± 1.5° as well as a sliding angle of 3.5 ± 1.2°. The Fe-Zn-HAc-STA sample was mechanically durable and easily repairable. After being ultrasonicated in ethanol for 100 min, the superhydrophobicity was still retained. The Fe-Zn-HAc-STA sample lost its superhydrophobicity after being abraded against sandpaper with a load of 100 g and regained its superhydrophobicity after HAc etching and subsequent STA hydrophobizing. The corrosion resistance of the SZC was investigated by immersing the Fe-Zn-HAc-STA sample into the static or dynamic aqueous solution of NaCl (3.5 wt.%) and the lasting life of the entrapped underwater air layer (EUAL) was roughly determined by the turning point at the variation curve of surface wettability against immersion time. The lasting life of the EUAL iwas 8 to 10 days for the SZC in the static NaCl solution and it decreased sharply to 12 h in a dynamic one with the flow rate of 2 and 4 m/s. This suggests that the superhydrophobic surface provided extra corrosion protection of 8 to 10 days or 12 h to the zinc coating. We hope that the SZC may find its practical application due to the facile and fast fabrication procedure, the good mechanical durability, the easy repairability, and the good corrosion protection.

Keywords: superhydrophobic; zinc-rich coating; cold galvanized coating; durability

1. Introduction

Iron and steel are the most common metals used in industry. However, the corrosion and damage of iron are serious; therefore, the anti-corrosion is an important issue in the iron industry [1]. People are using conversion coatings, such as chromate conversion coating and phosphate conversion coating, to improve its corrosion resistance. These conversion coatings have found wide practical application, however, also possess certain inherent disadvantages such as the carcinogenicity caused by the hexavalent chromium [2] and the high energy costs caused by the high treating temperature [3].

The cold zinc-spraying coating is an environment-friendly alternative with the standard electrode potential of −0.76 V, which is more active than iron (−0.44 V) [4]. When the coating is attacked, the zinc powder is corroded as the anode first, and the base iron is protected as the cathode, so the corrosion rate of iron can be slowed down significantly.

Superhydrophobic surface (SHS) with an apparent contact angle above 150° and a sliding angle below 10° is another newly-developed corrosion resistance strategy [5–15]. Once the superhydrophobic sample is immersed into water, a layer of air will be entrapped at the solid/liquid interface, which hinders the penetration of corrosive species (such as the Cl^- ions dissolved in water) to reach the metallic substrate and improve the corrosion resistance greatly [5].

To combine the sacrificing anode effect of zinc coating with the air barrier effect of SHS, Zhang et al. [16] and Brassard et al. [17] have fabricated superhydrophobic zinc coatings (SZC) via electrochemical zinc deposition and subsequent surface passivation by polypropylene or silicone polymer. Such SZC samples improve the anti-corrosion performance of iron or steel significantly. However, the zinc deposits are obtained via an electrochemical deposition, which makes the fabrication of the SZC sample instrument-dependent.

Herein, we aimed to fabricate an SZC sample via a more facile method. The metallic zinc layer was obtained by spraying the cold galvanized paint onto the iron substrate, followed by acetic acid etching and subsequent stearic acid hydrophobizing. The procedure was time-saving (within 10 min) and easy to perform. The mechanical durability was assessed by ultrasonication in ethanol and abrasion against sandpaper. The corrosion resistance was investigated by immersing the Fe-Zn-HAc-STA sample into the static or dynamic aqueous solution of NaCl (3.5 wt.%), and the variation in surface wettability, surface morphology, and surface composition was monitored.

2. Experimental

2.1. Sample Fabrication

The cold galvanized coating (Shanghai Roval Zinc-rich Coating Co, Ltd., Shanghai, China) was shaken evenly and sprayed directly onto a clean iron substrate (2.5 mm × 2.5 mm × 1 mm). Then, the sample was dried at room temperature for 12 h to obtain a uniform cold galvanized coating. The sample was immersed into the aqueous solution of acetic acid (HAc, 18 wt.%) followed by rinsing with ultrapure water and blown dry with nitrogen. Finally, the sample was immersed into the ethanol solution of stearic acid (STA, 0.01 M, chemically pure, Shantou Xilong Chemical Co. Ltd., Shantou, China) for 30 s and blown dry with hot air. To obtain the optimum superhydrophobicity, the immersion in STA was repeated twice. For convenience, the sample after spraying, etching, and hydrophobizing was coded as Fe-Zn, Fe-Zn-HAc, and Fe-Zn-HAc-STA, respectively.

2.2. Characterization

The surface morphology of the sample was characterized by field emission scanning electron microscopy (FE-SEM, Nova, Nano-SEM 450, FEI, Hillsboro, OR, USA; at 1 kV) with attached energy dispersive spectroscopy (EDX, FEI Quanta 200, USA; at 20 kV) under a vacuum environment. The chemical compositions and valence states of the samples were measured using X-ray photoelectron spectroscopy (XPS, Physical Electronics, Chanhassen, MN, USA, PHI-5702) with the Al Kα X-ray source (hν = 1486.6 eV) at a base pressure of 10^{-7} Pa. The surveyed (enlarged) spectra were recorded with a pass energy of 20 (5) eV and an energy step of 0.5 (0.05) eV. The binding energy of adventitious carbon (C1s: 284.8 eV) was used as a basic reference. The crystal orientation of the samples was characterized by X-ray diffraction (XRD, Bruker D8 ADVANCE, Karlsruhe, Germany) using CuKα radiation from 10 to 80° (2 theta) at 4 °/min. The powders scraped from the surface of the samples were recorded by Fourier Transform Infrared spectrometer (FTIR Nicolet IS10, Thermo Scientific, Waltham, MA, USA) by the KBr pellet methods with air as the reference. A resolution of 4 cm^{-1} and 32 scans were chosen. The surface wettability was measured by a contact angle meter (CAM, DSA 20, Krüss, Hamburg, Germany) with a computer-controlled liquid dispensing system and a motorized tilting stage at ambient temperature (20 °C). For apparent contact-angle measurements, a 10 µL water droplet was used with the Laplace-Young fitting. Advancing and receding angles were measured by the addition and subtraction of 5 µL of water to/from the 10 µL droplets sitting on the surfaces. For the sliding angle measurement, the water droplet (10 µL) was dripped onto the sample, which was then rotated until the water drop slid away. The critical angle where a water droplet began to slide down the inclined plate was measured to be the sliding angle. Each reported datum was an average of at least three measurements on the surface.

The adhesion of the coating to the iron substrate was assessed according to the standard of GB9286-98 by a crosscut tester (QFH-A, Dongguan Sanhe Instrument Manufacturing Factory, Dongguan, China). After the crosscut pattern at 90° was formed on the coating, the sample was brushed lightly with a soft brush to remove excess debris from the surface. A transparent tape was applied to the cut surface and rubbed with a rubber to ensure good contact with the coating and then removed after 90 s. Samples were evaluated under a magnifying glass (7×) and a bright light, and rated according to the American Society for Testing Materials (ASTM) rating scheme.

The variation of the surface wettability of the sample against ultrasonication (40 kHz, 100 W) in ethanol or against abrasion under a load of 100 g in contact with sandpaper (600 grits) was used to test the mechanical durability. The corrosion resistance was measured by immersing the sample into the static or dynamic aqueous solution of NaCl (3.5 wt.%) and the variation of surface morphology, chemical composition, and surface wettability was measured by FE-SEM, EDX, and CAM, respectively.

3. Results and Discussion

3.1. Wettability, Surface Morphology, and Surface Chemistry

The surface wettability of the Fe-Zn-HAc-STA sample varies against the etching time in the aqueous solution of HAc. The apparent contact angle or sliding angle levels off after rising or falling steeply for the first 6 min to a value of 161.8 ± 3.0° or 13.0 ± 1.5°, respectively (Figure 1a). As shown in Figure 1(b1), the so-sprayed cold galvanized paint was composed of smooth zinc micro-particles (MPs). After for 6 min, the MPs became rough with nano-wrinkles (Figure 1(b2)). As the etching time increased to 12 min, the nano-wrinkles became much denser (Figure 1(b3)) and the superhydrophobicity was optimum (apparent contact angle of 168.4 ± 1.5° and sliding angle of 3.5 ± 1.2°). The formation mechanism for these nano-wrinkles was due to the un-uniform etching between HAc and zinc as observed by Qian et al. [18]. It is explained that there are numerous dislocation defects in metals, which possess relatively higher energy and are prone to be attacked by chemical etchants such as HAc. Therefore, the etching rate in these dislocation sites was quicker and consequently, micro/nano-hierarchical surface structures were generated (Figure 1c).

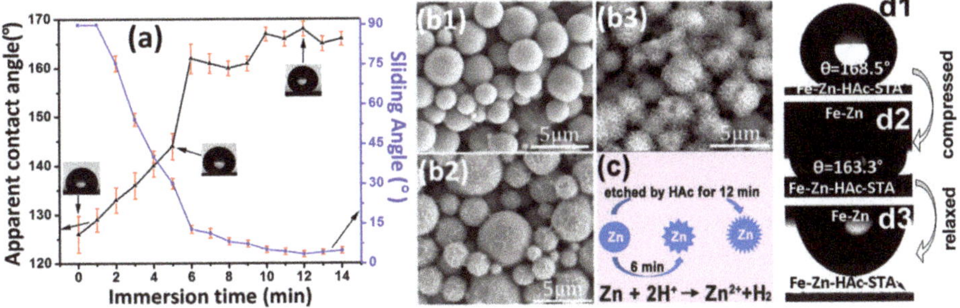

Figure 1. Variation of the surface wettability for the Fe-Zn-HAc-STA sample against immersion time in the aqueous solution of HAc (**a**). Scanning electron microscopic images of the Fe-Zn sample (**b1**), the Fe-Zn-HAc with an etching time of 6 min (**b2**) and 12 min (**b3**). Representative for zinc particles in images (**b1–b3**) and the etching reaction (**c**). Compression of a water droplet on the Fe-Zn-HAc-STA sample with the etching time of 6 min in HAc by a Fe-Zn sample (**d1–d3**).

The wetting state of the Fe-Zn-HAc-STA sample with an etching time in HAc for 6 min was further assessed by measuring the contact angle hysteresis (CAH), which was approximately 4° and equivalent to the difference between the advancing angle (approximately 169°) and the receding angle (approximately 165°). The Fe-Zn sample possessed an advancing angle of approximately 143° and a receding angle of about approximately 35°. A water droplet on the Fe-Zn-HAc-STA sample was compressed by the Fe-Zn sample and the typical images are shown in Figure 1d. After compression, the contact angle decreased slightly from 168.5° to 163.3°; after relaxation, the water droplet adhered to the upper surface and there was no residual on the lower superhydrophobic surface. These analyses suggest that the water droplet on the Fe-Zn-HAc-STA sample with an etching time in HAc of 6 min was in the Cassie state. The hierarchical surface roughness was helpful to resist the destabilization from Cassie wetting state to Wenzel state [15].

The XRD pattern (Figure 2a) for the Fe-Zn sample shows the strongest diffraction peak at 2θ of 43.25°, and weak ones at 36.34°, 39.04°, 54.41°, 70.21°, 70.72°, and 77.19°. These peaks are consistent with the standard PDF card of zinc (JCPDS 04-0831). Therefore, the main component of the Fe-Zn sample was zinc powder. Figure 2b shows the FTIR spectra of the Fe-Zn sample and the Fe-Zn-HAc-STA sample. The peaks at 1536 cm^{-1} and 1455 cm^{-1} in curve (i) for the Fe-Zn sample were attributed to the skeleton vibration of the benzene ring [19]. This was most likely attributed to the benzene ring in the bisphenol A epoxy resin, which was a routine ingredient for the cold galvanized paint to enhance the adhesion to the substrate. The peaks at 2919 cm^{-1} and 2851 cm^{-1} in curve (ii) were attributed to the C–H in epoxy resin [19], which became much stronger for the Fe-Zn-HAc-STA sample (curve ii). These stronger peaks were due to the anchored STA molecules abundant with C–H bonds. Compared with curve i for the Fe-Zn sample, two peaks emerged in curve ii for the Fe-Zn-HAc-STA sample, viz., a weak peak at 1752 cm^{-1} attributed to the carboxyl groups (COOH) [20] and a strong peak at 1400 cm^{-1} attributed to the deprotonated carboxyl groups (COO^-) [17]. The stronger signal at 1400 cm^{-1} suggests that most of the carboxyl groups are deprotonated to facilitate the anchoring of STA to the Zn MPs via coordination bonding (Figure 2(c1)) or ion bonding (Figure 2(c2)) [21].

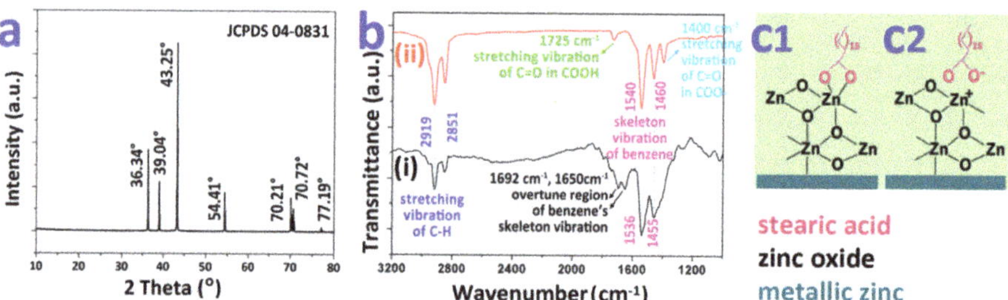

Figure 2. X-ray diffraction pattern of the Fe-Zn sample (**a**). Fourier transform infrared spectra of the Fe-Zn sample (curve i) and the Fe-Zn-HAc-STA sample (curve ii) (**b**). A possible mechanism for stearic acid anchored to the oxidized zinc: coordination (**c1**) or ionic (**c2**) bonding.

Figure 3 shows the XP spectra of the Fe-Zn-HAc-STA sample. In the survey spectrum (Figure 3a), peaks were attributed to the elements of oxygen, carbon, and zinc emerge. The Zn 2p peaks were located at 1021.9 eV and 1044.8 eV, which were higher than the binding energy of Zn (0) and ascribed to the Zn (II) species [22]. This suggests that, due to its high reactivity, the outermost surface of Zn MPs was oxidized. The as-formed oxide layer facilitated the anchoring of STA to the Fe-Zn-HAc sample just as shown in Figure 2c. The C 1s spectrum was deconvoluted to three components, viz., the C–H bonding at 284.8 eV, the C–O bonding at 286.2 eV, and the C=O bonding at 288.3 eV [23]. The deconvoluted peak at 531.8 eV for O 1s was due to the O=C group and the one at 530.6 eV was due to the O–C bond, both of which once again prove the presence of STA on the sample surface.

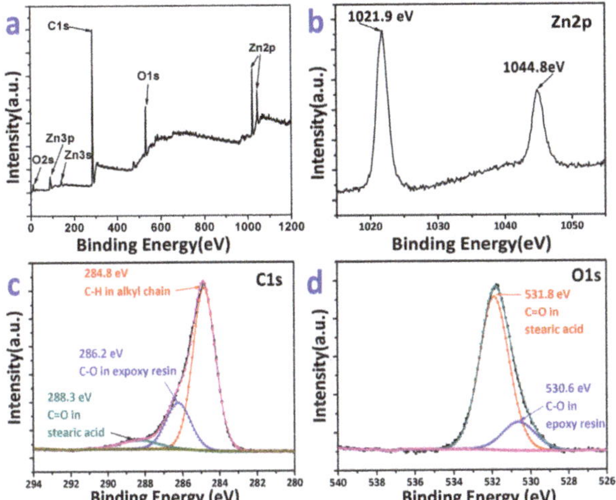

Figure 3. XP spectra for the Fe-Zn-HAc-STA sample: survey spectrum (**a**), Zn 2p spectrum (**b**), C 1s spectrum (**c**), and O 1s spectrum (**d**).

3.2. Mechanical Durability and Repairability

Ultrasonication has been widely used to evaluate the mechanical durability of the superhydrophobic samples [24–28]. Bubble implosions by quick forming and violent collapsing generate shock waves and exert mechanical forces on the surface. Herein, the Fe-Zn-HAc-STA sample was immersed in ethanol, and the variation of surface wettability was plotted (Figure 4a). After 120 min, the apparent contact angle decreased slightly from 168.4 ± 1.5° to 150.6 ± 1.1° and the sliding angle increased from 3.5 ± 1.2° to 13.9 ± 1.3°. The most durable sample in the references was reported by Peng et al., which retained its superhydrophobicity after ultrasonication in ethanol for 14 h [21]. As reported by Liu et al. [25] and Huang et al. [26], the superhydrophobic samples can withstand ultrasonic treatment of 1 and 2 h, respectively. As reported by Liu et al. [27], after ultrasonication of 30 min, the apparent contact angle decreased from 167° to 141°. As reported by He et al. [28], after ultrasonication of 10 min, the apparent contact angle decreased from approximately 153° to approximately 136°. Comparing with these References [25–28], we can say that the Fe-Zn-HAc-STA sample was mechanically robust.

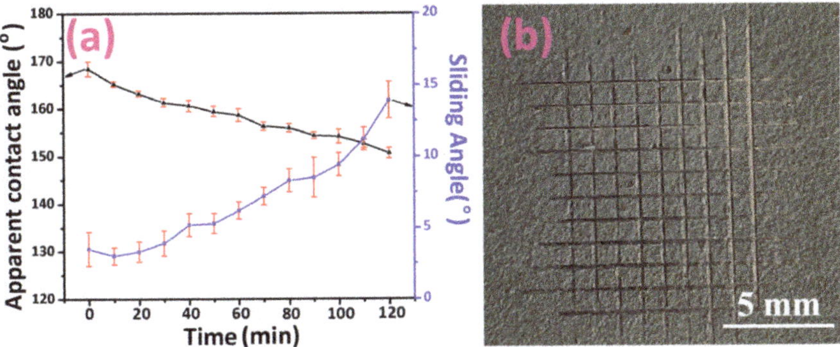

Figure 4. Variation of surface wettability for the Fe-Zn-HAc-STA sample against ultrasonication time in ethanol (**a**). The surface appearance after cross-cut testing (**b**).

The good mechanical durability was due to the strong interlayer interaction. To evaluate its adhesion strength to the iron substrate, the Fe-Zn-HAc-STA sample was crosscut and peeled off by Scotch tape according to the standardized coating adhesion test method (ASTM D 3359) [29]. As shown in Figure 4b, the edges of the cuts were completely smooth and none of the squares of the lattice were detached. Therefore, the Zn-HAc-STA coating adhered to the iron substrate in the level of 5B. The high adhesion strength was due to the epoxy resin, which was a routine ingredient in cold galvanized paint as confirmed by FTIR (Figure 2b). Moreover, as discussed earlier, the STA molecules were supposed to be anchored to the zinc coating firmly via coordination or ionic bonding (Figure 2c).

An abrading test was also performed to further evaluate the mechanical durability of the Fe-Zn-HAc-STA sample, which was subjected to a weight of 100 g while in contact with a 600 grit sandpaper at a length of 20 cm (Figure 5a). After 10 cycles, the micro/nanostructures on the Fe-Zn-HAc-STA sample were nearly destroyed (Figure 5(c1,c2)) and the apparent contact angle decreased to approximately 120°. This suggests that the outermost hydrophobic STA layer and the rough structures on the Fe-Zn-HAc-STA sample were abraded away and consequently the sample becomes hydrophobic just as the Fe-Zn sample. However, this does not mean that the coating will fail. After re-impregnation in an aqueous solution of HAc and ethanol solution of STA, the micro/nanostructures appeared on the surface (Figure 5(c3,c4)) and the coating regained the superhydrophobicity (Figure 5b). The abrading not only caused the change in the surface morphology but also decreased the thickness. As shown in Figure 5d, the thickness of the Fe-Zn-HAc-STA sample was reduced by 49 μm after 100 cycles.

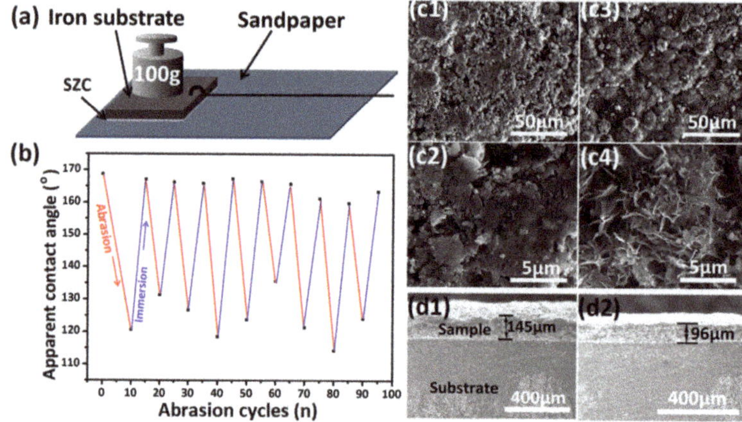

Figure 5. Schematic view of the setup for wearing test (**a**). Variation of apparent contact angle of the Fe-Zn-HAc-STA sample against wearing cycles (**b**). Scanning electron images for the Fe-Zn-HAc-STA sample after 10 cycles of wearing (**c1**,**c2**) and re-impregnation in acetic acid solution and ethanol solution of stearic acid (**c3**,**c4**). Scanning electron images for the cross-section of the Fe-Zn-HAc-STA sample before (**d1**) and after (**d2**) wearing for 100 cycles.

3.3. Corrosion Resistance

Figure 6a shows the variation of the surface wettability for the Fe-Zn-HAc-STA sample (curves i, iii) and the Fe-Zn sample (curve ii) in the aqueous solution of NaCl for 20 days. After 6 days, the apparent contact angle of the Fe-Zn sample decreased sharply to approximately 0°; however, the Fe-Zn-HAc-STA sample still possessed good hydrophobicity (apparent contact angle of 150.7 ± 0.9° and sliding angle of 16.0 ± 0.3°). The much slower deterioration rate of hydrophobicity for the Fe-Zn-HAc-STA sample was due to the entrapped underwater air layer (EUAL) at the solid/liquid interface, which was reflected by the "silver mirror" as shown in Figure 6(b1). In other words, once the superhydrophobic Fe-Zn-HAc-STA samplewas immersed into the aqueous solution of NaCl, most of the solid surface (up to 90%) was in contact with the EUAL rather than water [25]. The EUAL blocked the diffusion of corrosive ions to the solid substrate [5] and consequently, the surface hydrophobicity of the Fe-Zn-HAc-STA sample deteriorated much more slowly as compared with the hydrophobic Fe-Zn sample. We can speculate the lasting life of the EUAL from the curves in Figure 6a. Specifically, the apparent contact angle (curve i) and the sliding angle (curve iii) for the Fe-Zn-HAc-STA sample varied very gently before 8 days and 10 days, respectively; thereafter, the variation was much faster. Moreover, the underwater "silver mirror" after 8 days as shown in Figure 6(b2) was not so obvious and became discontinuous. These data hint that the lasting life of EUAL was about 8 to 10 days. After that, statistically speaking, the STA layer rather than the EUAL was in contact with water. The surface micro-morphology of the Fe-Zn sample after immersion is shown in Figure 6c. Some area was covered by the corrosion products. In contrast, the micro/nano- hierarchical surface structures (Figure 6d) for the Fe-Zn-HAc-STA sample varied a little but were still observable, which is the morphology requirement for the superhydrophobicity.

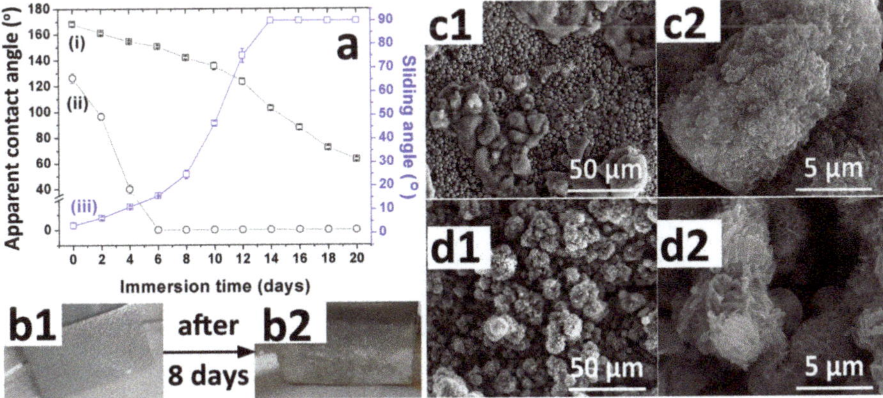

Figure 6. Variation of surface wettability for the Fe-Zn-HAc-STA sample (i, iii) and the Fe-Zn sample (ii) against the immersion time in the aqueous solution of NaCl (3.5 wt.%) (**a**). Digital pictures to show the underwater mirror for the Fe-Zn-HAc-STA sample just immersed into the NaCl solution (**b1**) and after 8 days (**b2**). Surface micro-morphology for the Fe-Zn sample (**c1**,**c2**) and the Fe-Zn-HAc-STA sample (**d1**,**d2**) after immersed in the aqueous solution of NaCl for 8 days.

The element content of the samples before and after immersion was measured by EDX and is summarized in Table 1. As confirmed by FTIR, XPS, and XRD earlier, the Fe-Zn sample was mainly composed of Zn MPs and epoxy resin; for the sample of Fe-Zn-HAc-STA, STA molecules were attached. Therefore, the main elements for these two samples were carbon, oxygen, and zinc. After immersion, the signal attributed to the element of chlorine emerged, which was attributed to the adsorbed NaCl or the chlorine contained corrosion products. For the Fe-Zn sample, the element content of carbon decreased sharply from 61.83 ± 6.23% to 17.48 ± 4.45%, and the oxygen increased from 6.14 ± 0.75% to 53.77 ± 7.42%. This suggests that the carbon-rich organic molecules were detached severely from the Fe-Zn sample and lots of oxygen-rich corrosion products such as $Zn(OH)_2$ were formed via the following electrochemical reactions.

$$\text{anodereaction} : \text{Zn} - 2\text{e} \rightarrow \text{Zn}^{2+} \tag{1}$$

$$\text{cathodereaction} : \text{O}_2 + 2\text{H}_2\text{O} + 4\text{e} \rightarrow 4\text{OH}^- \tag{2}$$

$$\text{overallreaction} : 2\text{Zn} + \text{O}_2 + 2\text{H}_2\text{O} \rightarrow 2\text{Zn}(\text{OH})_2 \downarrow \tag{3}$$

However, for the Fe-Zn-HAc-STA sample, the content change of carbon and oxygen was not so obvious. This is due to the barrier effect of the EUAL with a lasting life of about 8 to 10 days as discussed earlier.

Table 1. Atomic element content of the Fe-Zn sample and the Fe-Zn-HAc-STA sample before and after immersing in the aqueous solution of NaCl for 8 days.

Element	Fe-Zn		Fe-Zn-HAc-STA	
	Before Corrosion	After Corrosion	Before Corrosion	After Corrosion
C	61.83 ± 6.23%	17.48 ± 4.45%	78.07 ± 4.29%	80.63 ± 8.53%
O	6.14 ± 0.75%	53.77 ± 7.42%	8.46 ± 3.20%	7.61± 1.58%
Zn	32.03 ± 4.15%	19.58 ± 3.68%	13.47 ± 1.47%	9.24 ± 1.08%
Cl	0.00%	9.16 ± 1.51%	0.00%	2.53 ± 0.89%

We monitored the variation of the surface wettability in the dynamic aqueous solution of NaCl for 48 h (Figure 7). The flow rate affected the variation of the surface wettability significantly. Curves in Figure 7a (static solution) and Figure 7b (1 m/s) vary smoothly. However, as the flow rate increased to 2 or 4 m/s, similar turning points as shown in Figure 6a at 8 to 10 days appeared after a short immersion of approximately 12 h. This suggests that the lasting life of the EUAL was shortened significantly under dynamic solution with the high flow rate of 2 to 4 m/s.

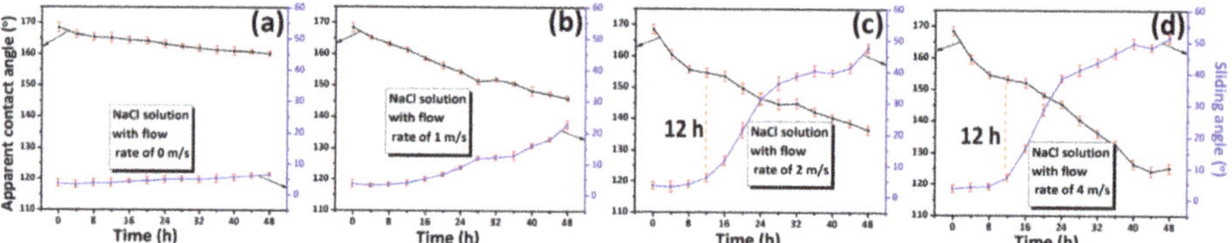

Figure 7. Variation of the surface wettability for the Fe-Zn-HAc-STA sample immersed in 3.5 wt.% NaCl solution for 48 h at a flow rate of 0 m/s (**a**), 1 m/s (**b**), 2 m/s (**c**) and 4 m/s (**d**).

Figure 8 shows the micro-morphological images for the Fe-Zn-HAc-STA sample after immersed in the aqueous solution of NaCl for 48 h. The sample the underwent static corrosion still possessed a hierarchical surface microstructure (Figure 8a). As the flow rate increased to 1 or 2 m/s, some area (the circle in Figure 8(b1)) of the surface became compact and the MPs were covered by the corrosion products (Figure 8(b2,c2)). At 4 m/s, the shape of Zn MPs was difficult to recognize.

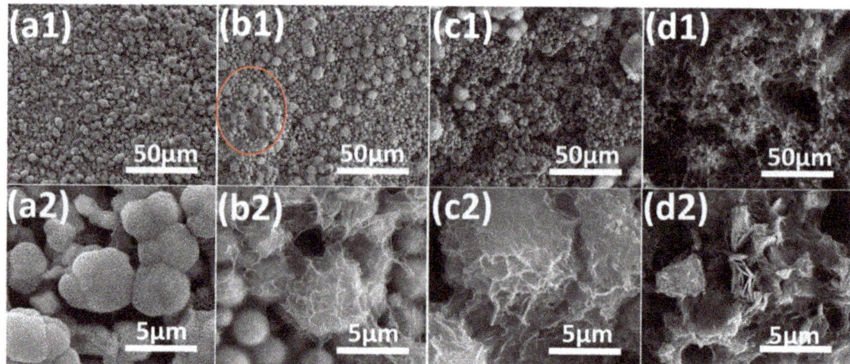

Figure 8. FE-SEM images for the Fe-Zn-HAc-STA samples in the aqueous solution of NaCl (3.5 wt.%) for 48 h at a flow rate of 0 m/s (**a1**,**a2**), 1 m/s (**b1**,**b2**), 2 m/s (**c1**,**c2**) and 4 m/s (**d1**,**d2**).

The element content of the Fe-Zn-HAc-STA sample after dynamic corrosion at different flow rates was measured by EDX (Table 2). The oxygen content increased from 8.46 ± 3.20% to 13.63 ± 3.89% at 1 m/s or to 56.47 ± 5.89% at 4 m/s after 48 h. The carbon content remained almost unchanged at 1 m/s and decreased sharply to 10.50 ± 2.14% at 4 m/s. As discussed earlier, the decreasing carbon content was due to the detachment of STA molecules and the increasing of oxygen content is due to the formation of corrosion products. These data suggest that the flow rate affects the corrosion of the superhydrophobic sample significantly. The reason is that the flowing liquid made it difficult to maintain the EUAL at the submerged superhydrophobic surface.

Table 2. Atomic element content of the Fe-Zn-HAc-STA sample before and after immersion in the aqueous solution of NaCl (3.5 wt.%) for 48 h at different flow rates.

Element	Before Corrosion	After Corrosion at Different Flow Rate	
		1 m/s	4 m/s
C	78.07 ± 4.29%	80.04 ± 5.53%	10.50 ± 2.14%
O	8.46 ± 3.20%	13.63 ± 3.89%	56.47 ± 5.89%
Zn	13.47 ± 1.47%	6.17 ± 1.20%	30.82 ± 4.18%
Cl	0.00%	0.16 ± 0.02%	2.21 ± 3.55%

4. Conclusions

A superhydrophobic zinc coating coded as Fe-Zn-HAc-STA was prepared on the iron substrate via a facial multi-step procedure including (a) the spraying of the widely commercially available cold galvanized paint, (b) the etching by acetic acid (HAc), and (c) the hydrophobizing by stearic acid (STA). The Fe-Zn-HAc-STA sample possessed good mechanical durability and its superhydrophobicity remained after being ultrasonicated in ethanol for 100 min. After being abraded against the sandpaper for 10 cycles, the surface superhydrophobicity was lost due to the destruction of the hierarchical surface structures and the detachment of the hydrophobic STA molecules. However, after quick HAc etching and STA hydrophobizing, the sample regained the superhydrophobicity easily. The Fe-Zn-HAc-STA sample was immersed into the aqueous solution of NaCl to evaluate the corrosion resistance, and an entrapped underwater air layer (EUAL) was formed at the solid/liquid interface. The lasting life of EUAL in the static solution was 8 to 10 days and decreased sharply to 12 h as the flow rate of the corrosive solution increased to 2 or 4 m/s. We can say that the superhydrophobic surface provided extra corrosion protection of 8 to 10 days or 12 h to the zinc coating. We hope that this research may be helpful for the design of anti-corrosion superhydrophobic coatings due to the facile and fast fabrication procedure, the good mechanical durability, the easy repairability, and the good corrosion protection.

Author Contributions: For research articles with several authors, a short paragraph specifying their individual contributions must be provided. The following statements should be used "conceptualization, J.O. and M.X.; methodology, W.Z.; validation, W.Z. and C.X.; formal analysis, W.Z.; investigation, W.Z.; resources, C.X.; writing—original draft preparation, J.O.; writing—review and editing, J.O.; visualization, C.X.; supervision, M.X.; project administration, J.O.; funding acquisition, J.O.

Funding: This research was funded by the National Natural Science Foundation of China (Grant No. 51563018).

Conflicts of Interest: The authors declare no conflict of interest.

References

1. Kendig, M.W.; Buchheit, R.G. Corrosion inhibition of aluminum and aluminum alloys by soluble chromates, chromate coatings, and chromate-free coatings. *Corrosion* **2003**, *59*, 379–400. [CrossRef]
2. Asadi, V.; Danaee, I.; Eskandari, H. The effect of immersion time and immersion temperature on the corrosion behavior of zinc phosphate conversion coatings on carbon steel. *Mater. Res.* **2015**, *18*, 706–713. [CrossRef]
3. Finšgar, M.; Jackson, J. Application of corrosion inhibitors for steels in acidic media for the oil and gas industry: A review. *Corros. Sci.* **2014**, *86*, 17–41. [CrossRef]
4. Marder, A.R. The metallurgy of zinc-coated steel. *Prog. Mater. Sci.* **2000**, *45*, 191–271. [CrossRef]

5. Liu, T.; Chen, S.; Cheng, S.; Tian, J.; Chang, X.; Yin, Y. Corrosion behavior of super-hydrophobic surface on copper in seawater. *Electrochim. Acta* **2007**, *52*, 8003–8007. [CrossRef]
6. Zhang, D.W.; Wang, L.T.; Qian, H.C.; Li, X.G. Superhydrophobic surfaces for corrosion protection: a review of recent progresses and future directions. *J. Coat. Technol. Res.* **2016**, *13*, 11–29. [CrossRef]
7. Boinovich, L.B.; Gnedenkov, S.V.; Alpysbaeva, D.A.; Egorkin, V.S.; Emelyanenko, A.M.; Sinebryukhov, S.L.; Zaretskaya, A.K. Corrosion resistance of composite coatings on low-carbon steel containing hydrophobic and superhydrophobic layers in combination with oxide sublayers. *Corros. Sci.* **2012**, *55*, 238–245. [CrossRef]
8. Boinovich, L.B.; Emelyanenko, A.M.; Modestov, A.D.; Domantovsky, A.G.; Emelyanenko, K.A. Not simply repel water: the diversified nature of corrosion protection by superhydrophobic coatings. *Mendeleev Commun.* **2017**, *27*, 254–256. [CrossRef]
9. Yin, K.; Yang, S.; Dong, X.R.; Chu, D.K.; Duan, J.A.; He, J. Robust laser-structured asymmetrical PTFE mesh for underwater directional transportation and continuous collection of gas bubbles. *Appl. Phys. Lett.* **2018**, *112*, 243701. [CrossRef]
10. Li, J.; Wu, R.; Jing, Z.; Yan, L.; Zha, F.; Lei, Z. One-step spray-coating process for the fabrication of colorful superhydrophobic coatings with excellent corrosion resistance. *Langmuir* **2015**, *31*, 10702–10707. [CrossRef] [PubMed]
11. Wang, N.; Xiong, D.S.; Deng, Y.L.; Shi, Y.; Wang, K. Mechanically robust superhydrophobic steel surface with anti-icing, UV-durability, and corrosion resistance properties. *ACS Appl. Mater. Interface* **2015**, *7*, 6260–6272. [CrossRef] [PubMed]
12. Zhu, X.T.; Zhang, Z.Z.; Men, X.H.; Yang, J.; Wang, K.; Xu, X.H.; Zhou, X.Y.; Xue, Q.J. Robust superhydrophobic surfaces with mechanical durability and easy repairability. *J. Mater. Chem.* **2011**, *21*, 15793–15797. [CrossRef]
13. Marmur, A.; Volpe, C.D.; Siboni, S.; Amirfazli, A.; Drelich, J.W. Contact angles and wettability: towards common and accurate terminology. *Surf. Innov.* **2017**, *5*, 3–8. [CrossRef]
14. Starostin, A.; Valtsifer, V.; Strelnikov, V.; Bormashenko, E.; Grynyov, R.; Bormashenko, Y.; Gladkikh, A. Robust technique allowing the manufacture of superoleophobic (omniphobic) metallic surfaces. *Adv. Eng. Mater.* **2014**, *16*, 1127–1132. [CrossRef]
15. Nosonovsky, M. Multiscale roughness and stability of superhydrophobic biomimetic interfaces. *Langmuir* **2007**, *23*, 3157–3161. [CrossRef] [PubMed]
16. Zhang, X.T.; Liang, J.; Liu, B.X.; Peng, Z.J. Preparation of superhydrophobic zinc coating for corrosion protection. *Colloids Surf. A* **2014**, *454*, 113–118. [CrossRef]
17. Brassard, J.D.; Sarkar, D.K.; Perron, J.; Audibert-Hayet, A.; Melot, D. Nano-micro structured superhydrophobic zinc coating on steel for prevention of corrosion and ice adhesion. *J. Colloid Interface Sci.* **2015**, *447*, 240–247. [CrossRef]
18. Qian, B.; Shen, Z. Fabrication of superhydrophobic surfaces by dislocation- selective chemical etching on aluminum, copper, and zinc substrates. *Langmuir* **2005**, *21*, 9007–9009. [CrossRef]
19. Yoshida, S. Quantitative evaluation of an epoxy resin dispersion by infrared spectroscopy. *Polym. J.* **2014**, *46*, 430–434. [CrossRef]
20. Kang, Y.S.; Lee, D.K.; Stroeve, P. FTIR and UV-vis spectroscopy studies of Langmuir–Blodgett films of stearic acid/γ-Fe_2O_3 nanoparticles. *Thin Solid Films* **1998**, *327*, 541–544. [CrossRef]
21. Raman, A.; Quiñones, R.; Barriger, L.; Eastman, R.; Parsi, A.; Gawalt, E. Understanding organic film behavior on alloy and metal oxides. *Langmuir* **2010**, *26*, 1747–1754. [CrossRef] [PubMed]
22. Moulder, J.F.; Stickle, W.F.; Sobol, P.E.; Bomben, K.D. *Handbook of X-ray Photoelectron Spectroscopy*; Perkin-Elmer Corporation: Eden Prairie, MN, USA, 1992.
23. Feng, L.B.; Zhang, H.X.; Mao, P.Z.; Wang, Y.P.; Ge, Y. Superhydrophobic alumina surface based on stearic acid modification. *Appl. Surf. Sci.* **2011**, *257*, 3959–3963. [CrossRef]
24. Peng, S.; Tian, D.; Yang, X.J.; Deng, W.L. Highly efficient and large-scale fabrication of superhydrophobic alumina surface with strong stability based on self-congregated alumina nanowires. *ACS Appl. Mater. Interfaces* **2014**, *6*, 4831–4841. [CrossRef]
25. Liu, L.J.; Xu, F.Y.; Ma, L. Facile fabrication of a superhydrophobic Cu surface via a selective etching of high-energy facets. *J. Phys. Chem. C* **2012**, *116*, 18722–18727. [CrossRef]
26. Huang, W.H.; Lin, C.S. Robust superhydrophobic transparent coatings fabricated by a low-temperature sol-gel process. *Appl. Surf. Sci.* **2014**, *305*, 702–709. [CrossRef]

27. Liu, L.J.; Feng, X.R.; Guo, M.X. Eco-friendly fabrication of superhydrophobic bayerite array on Al foil via an etching and growth process. *J. Phys. Chem. C* **2013**, *117*, 25519–25525. [CrossRef]
28. He, Z.; Ma, M.; Lan, X.; Chen, F.; Wang, K.; Deng, H.; Zhang, Q.; Fu, Q. Fabrication of a transparent superamphiphobic coating with improved stability. *Soft Matter* **2011**, *7*, 6435–6443. [CrossRef]
29. *Standard Test Methods for Measuring Adhesion by Tape Test, (Method B: Cross-Cut Tape Test)*; ASTM D 3359; American Society for Testing Materials: West Conshohocken, PA, USA, 2006.

Article

Lotus-Inspired Multiscale Superhydrophobic AA5083 Resisting Surface Contamination and Marine Corrosion Attack

Binbin Zhang [1,2,3,*], Weichen Xu [1,2,3], Qingjun Zhu [1,2,3], Shuai Yuan [1,2,4] and Yantao Li [1,2,3,*]

1 CAS Key Laboratory of Marine Environmental Corrosion and Bio-fouling, Institute of Oceanology, Chinese Academy of Sciences, No.7 Nanhai Road, Qingdao 266071, China; w.xu@qdio.ac.cn (W.X.); zhuqingjun@qdio.ac.cn (Q.Z.); ys520399@163.com (S.Y.)
2 Open Studio for Marine Corrosion and Protection, Pilot National Laboratory for Marine Science and Technology (Qingdao), No.1 Wenhai Road, Qingdao 266237, China
3 Center for Ocean Mega-Science, Chinese Academy of Sciences, No.7 Nanhai Road, Qingdao 266071, China
4 School of Civil Engineering, Qingdao University of Technology, No. 11 Fushun Road, Qingdao 266033, China
* Correspondence: zhangbinbin11@mails.ucas.ac.cn (B.Z.); ytli@qdio.ac.cn (Y.L.)

Received: 22 April 2019; Accepted: 11 May 2019; Published: 15 May 2019

Abstract: The massive and long-term service of 5083 aluminum alloy (AA5083) is restricted by several shortcomings in marine and industrial environments, such as proneness to localized corrosion attack, surface contamination, etc. Herein, we report a facile and cost-effective strategy to transform intrinsic hydrophilicity into water-repellent superhydrophobicity, combining fluorine-free chemisorption of a hydrophobic agent with etching texture. Dual-scale hierarchical structure, surface height relief and surface chemical elements were studied by field emission scanning electron microscopy (FE-SEM), atomic force microscopy (AFM), energy dispersive X-ray spectroscopy (EDS) and X-ray photoelectron spectroscopy (XPS), successively. Detailed investigations of the wetting property, self-cleaning effect, NaCl-particle self-propelling, corrosion and long-term behavior of the consequent superhydrophobic AA5083 surface were carried out, demonstrating extremely low adhesivity and outstanding water-repellent, self-cleaning and corrosion-resisting performance with long-term stability. We believe that the low cost, scalable and fluorine-free transforming of metallic surface wettability into waterproof superhydrophobicity is a possible strategy towards anti-contamination and marine anti-corrosion.

Keywords: fluorine free; silanization; superhydrophobic; corrosion protection; self-cleaning

1. Introduction

Metals and their alloys are central engineering materials in numerous industrial fields. As a typical representative, 5083 aluminum alloys (AA5083), featuring a high strength-to-weight ratio and good weldability, are widely employed in military equipment, marine constructions, automobile manufacture and aerospace applications. However, proneness to localized corrosion attacks [1–4] in corrosive environments restricts the large-scale and long-term application of AA5083 materials. Corrosion attacks in aggressive environments can produce a premature failure of AA5083 structural materials, resulting in environmental disruption, enormous economic loss, as well as catastrophic safety accident. Thus, how to endow AA5083 materials with superior corrosion resistance is an extensively concerning issue.

In recent years, inspired from the unique water-repellent property of natural organisms [5–9], the artificial fabrication of bionic superhydrophobic surfaces attracted intensive attention of scientists and engineers owing to their multi-functional applications, such as self-cleaning [10,11], oil–water separation [12,13], drag reduction [14,15], anti-icing/frosting [16,17], microdroplet transportation [18,19], water collection [20,21], marine anti-corrosion [22,23], anti-biofoulings [24,25], etc. It is believed and demonstrated that superhydrophobic surfaces could effectively reduce the solid/liquid interfacial contacts and provide a functional corrosion-resistant barrier. Thus, transforming surface wettability from intrinsic hydrophilicity to water-repellent superhydrophobicity is considered to be a possible strategy resisting marine corrosion attacks [26,27].

At present, much efforts [28–31] have been devoted to exploring the fabrication and investigating the consequent corrosion-resistant behavior of a superhydrophobic surface on aluminum/aluminum alloy substrates. For instance, Boinovich et al. [32,33] reported a combination of nanosecond laser texturing and fluorinated hydrophobic agent chemisorption, achieving a superhydrophobic aluminum-magnesium alloy with an extremely low corrosion current. Wang et al. [34] used a hydrothermal in situ growth method to fabricate superhydrophobic Mg-Al-layered double hydroxide films on 6061 aluminum alloy substrates, presenting a highly improved corrosion resistance. In our previous reports, we developed an ammonia etching approach [35] and an anodization method [36] followed by 1H,1H,2H,2H-Perfluorodecyltriethoxysilane chemisorption to achieve superhydrophobic surfaces on aluminum/aluminum alloy substrates with multiscale hierarchical topography, and greatly enhanced corrosion inhibition performance.

However, most of the fabrication methods above are hindered by several shortcomings, such as being time-consuming, high-cost and fluorine reagents employment, etc., restricting their large area usage and severely threatening ecological environments and human security. Until now, only limited attempts have been achieved to develop scalable and fluorine-free superhydrophobic aluminum surfaces for marine corrosion protection. So, these issues push us to explore a cost effective, facile and non-fluorinated approach, transforming intrinsically hydrophilic AA5083 with a Wenzel contact [37,38] to superhydrophobic AA5083 with a Cassie–Baxter contact [39,40].

Herein, we report a facile, cost-effective and non-fluorinated fabrication strategy to prepare a superhydrophobic surface on an AA5083 substrate by combining etching texture and a hexadecyltrimethoxysilane (HDTMS) hydrophobic molecules assembly. Detailed studies about the surface morphology and chemical composition were carried out successively. In addition, wetting property, self-cleaning ability, corrosion-resisting behavior and long-term stability were investigated to display the typical characteristics and promising functional applications.

2. Experimental Section

2.1. Materials and Reagents

A 5083 aluminum alloy (AA5083) plate with a 0.3 mm thickness was obtained from Dongguan Wanxing Metal Co., Ltd. (Dongguan, China), and tailored into 25 mm × 20 mm specimens. The main composition of the pristine AA5083 substrate was 4.0–4.9% Mg, 0.4–1.0% Mn, 0.25% Zn, 0.4% Si, 0.15% Ti, 0.1% Cu, 0.05–0.25% Cr, 0.1–0.4% Fe and the balance was Al. Sodium hydroxide (NaOH), sodium chloride (NaCl), methylene blue trihydrate ($C_{16}H_{18}ClN_3S{\cdot}3H_2O$), ethanol absolute and graphite powder were purchased from Sinopharm Chemical Reagent Co., Ltd. (Beijing, China). Manganese monoxide (MnO) was bought from Aladdin Industrial Corporation. Hexadecyltrimethoxysilane ($C_{19}H_{42}O_3Si$, HDTMS) was received from J&K Scientific Ltd. (Shanghai, China). All experimental reagents mentioned above were used as received without further purification.

2.2. Preparation of Superhydrophobic AA5083

The brief schematic illustration of the fabrication process is presented in Figure 1, involving an etching-texture process and HDTMS assembly. Firstly, the pristine AA5083 substrates were sanded through SiC sandpaper with different grades (400, 800, 1200, etc.) and cleaned by sonication in ethanol and deionized water for more than 5 min, respectively. The cleaned and pristine AA5083 specimens were rinsed with deionized water and dried under an air blower before etching. Subsequently, the pristine AA5083 samples were immersed in 15 g/L sodium hydroxide aqueous solution for 30 min to accomplish the etching texture. After deionized water cleaning, ethanol cleaning and an 80 °C oven drying treatment, the etching-textured AA5083 specimens were immersed in a 3 vol.% HDTMS/ethanol solution for 1 h to chemically assemble HDTMS molecules. After the modification process, the modified AA5083 specimens were heated at 120 °C in a drying oven for 20 min.

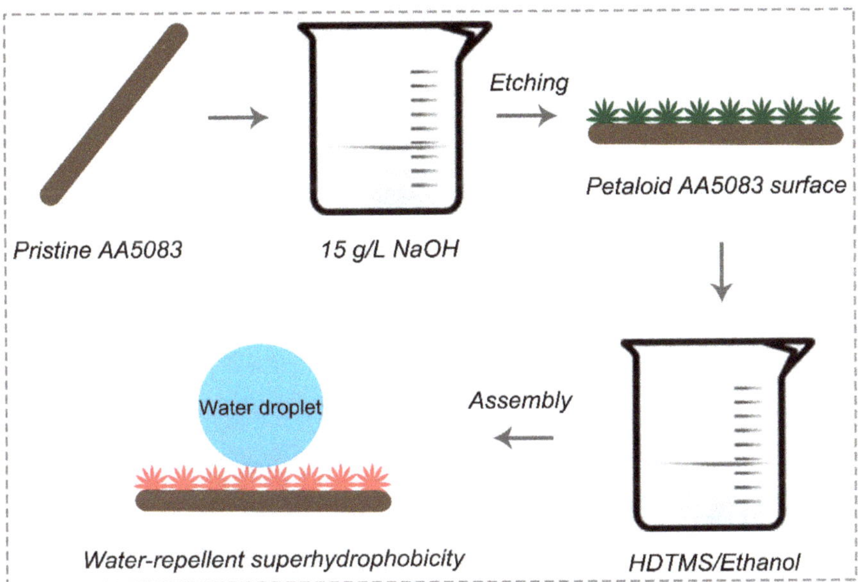

Figure 1. Schematic illustration of the fabrication process.

2.3. Characterization

The surface topography of pristine and as-fabricated superhydrophobic AA5083 surfaces were characterized by field emission scanning electron microscopy (FE-SEM, FEI Nova Nano SEM450, Hillsboro, OR, USA) and atomic force microscopy (AFM, Bruker Multimode 8, Karlsruhe, Germany). The AFM images were obtained under tapping mode. Energy dispersive X-ray spectroscopy (EDS, Oxford X-Max^{N}50, Hillsboro, OR, USA) and X-ray photoelectron spectroscopy (XPS, Thermo Scientific Escalab 250Xi, Massachusetts, MA, USA) were performed to determine the existence of the key elements upon the specimens. The XPS measurements were performed using a monochromated Al Kα irradiation and the chamber pressure was 3×10^{-8} Torr during the test. The binding energy of adventitious carbon C1s (284.8 eV) was used as a basic reference. A Dataphysics OCA25 instrument (Stuttgart, Germany) was used to measure the static water contact angles and sliding angles of the as-prepared superhydrophobic AA5083 samples. For each measurement, at least three different positions were performed to obtain average data.

2.4. Electrochemical Test

The electrochemical tests were all carried out in a 3.5 wt.% NaCl aqueous solution through an Ametek Parstat 4000+ electrochemical workstation. A typical three-electrode measure system, including counter electrode (Pt sheet), reference electrode (saturated silver/silver chloride) and working electrode (pristine/superhydrophobic AA5083), was employed to proceed with the electrochemical tests. Prior to the test, the working electrode was exposed to a 3.5 wt.% NaCl aqueous solution for more than 1 h, achieving a stable measuring system. Electrochemical impedance spectroscopy (EIS) was measured under OCP (open circuit potential) at a frequency range of 100 kHz–10 MHz. *ZsimpWin* software was utilized to analyze and fit the EIS data for anti-corrosion evaluation.

3. Results and Discussion

3.1. Surface Morphology and Wettability Behavior

The surface topography of pristine AA5083 and superhydrophobic AA5083 surfaces were characterized by FE-SEM. Figure 2 displays the SEM images of the pristine AA5083 and superhydrophobic AA5083 surface. As for pristine AA5083 shown in Figure 2a, some micro-grooves/scratches can be seen, which is attributed to the pre-treatment of the pristine AA5083 specimen. The surface of the pristine AA5083 is relatively smooth. For the etching-textured superhydrophobic AA5083 surface, displayed in Figure 2b, some micro-textured rough structure can be found. Figure 2c shows the higher magnification image of Figure 2b and presents an obviously nano-scale petaloid surface architecture of the as-prepared superhydrophobic AA5083 sample. These micro-nano hierarchical structures contribute to the increase of surface roughness and provide sufficient structural clearance for the formation of the trapped air cushion between the solid surface and the water droplet, which is beneficial for the final Cassie–Baxter contact state. Figure 2d shows the static water contact angle of pristine AA5083 and superhydrophobic AA5083 surface. The contact angle of the pristine AA5083 surface is about 81.6° ± 1°. After the etching texture and HDTMS assembly, however, the contact angle of the as-prepared superhydrophobic AA5083 surface is about 156.3° ± 1° with a sliding angle lower than 1°. The etching-textured, dual-scale surface rough structure and the low surface energy of the HDTMS molecules endowed the AA5083 substrate with a high static water contact angle and a low sliding angle.

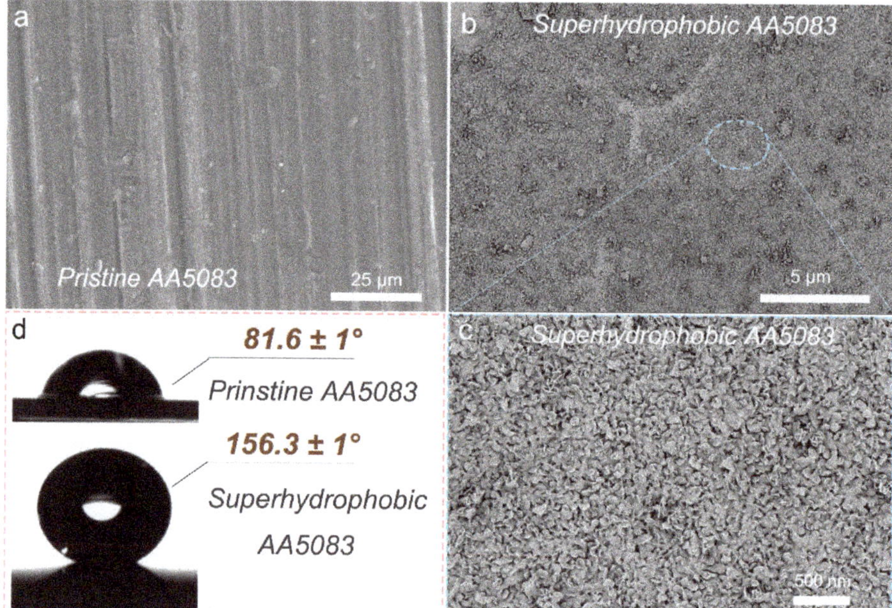

Figure 2. FE-SEM images (**a–c**) and static contact angles (**d**) of pristine 5083 aluminum alloy (AA5083) and as-fabricated superhydrophobic AA5083 surfaces.

As is well known, wetting behavior of a solid surface is mainly determined by roughness and surface energy. The microscale roughness and height relief of the pristine and as-fabricated superhydrophobic AA5083 surfaces were revealed by AFM, as shown in Figure 3. Figure 3a,b display the topographic fluctuation of pristine AA5083 and superhydrophobic AA5083 specimens. It was found that the height relief of the pristine AA5083 substrate was about 160 nm. On the contrary, the height relief of the as-prepared superhydrophobic AA5083 was approximately 970 nm, presenting a significant improvement. Figure 3c,d show the 3D AFM images of pristine AA5083 and superhydrophobic AA5083 surfaces. The surface roughness can be clearly observed and contrasted. Generally, R_a (average roughness), R_q (root mean square roughness) and R_{max} (maximum roughness) were utilized to present surface roughness. In this case, the surface of the pristine AA5083 was relatively smooth with R_a, R_q and R_{max} values (scanning area 10 μm × 10 μm) being 33.3 nm, 42.8 nm and 444 nm, respectively. While for the as-prepared superhydrophobic AA5083 surface, the R_a, R_q and R_{max} values (scanning area 20 μm × 20 μm) were apparently increased to 107 nm, 142 nm and 1229 nm, respectively. As discussed above, the resultant superhydrophobic AA5083 surface features an obviously improved surface roughness. The etching-textured process contributes to this roughness enhancement, which is in favor of the air cushion formation and Cassie–Baxter contact of the water/solid/air interface.

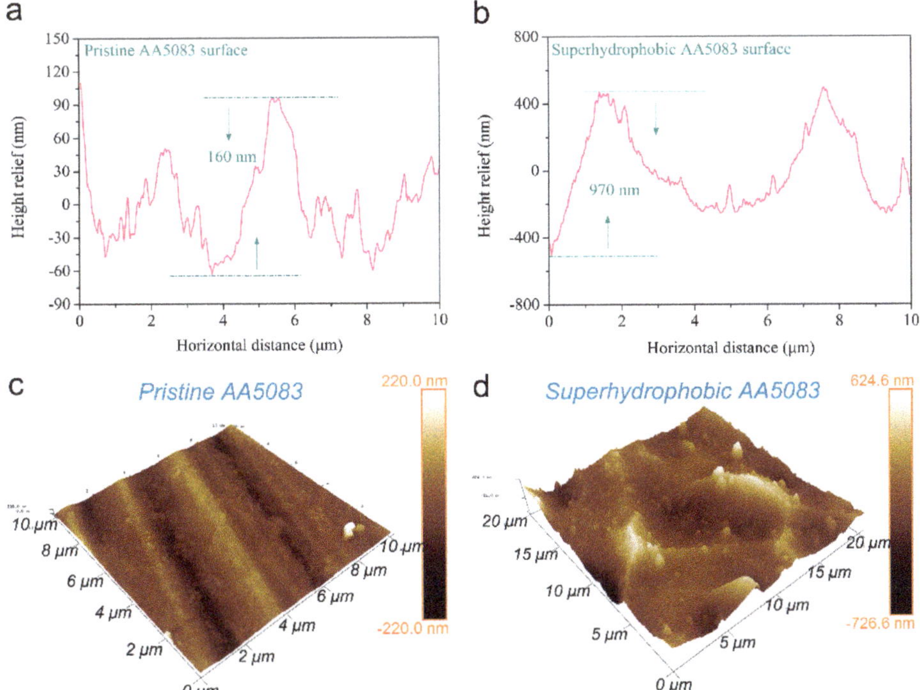

Figure 3. Surface relief and 3D atomic force microscopy (AFM) images of pristine AA5083 (**a**,**c**) and superhydrophobic AA5083 (**b**,**d**).

3.2. Low Surface Adhesivity

The larger etching-textured roughness and lower HDTMS surface energy played key roles in the resultant water-repellent superhydrophobicity. Figure 4a displays the optical image of spherical water droplets, illustrating a waterproof property. Figure 4b,c show the surface response with a dynamic jet of water flow and faucet water impact. It can be clearly seen that the water flow jet and faucet water cannot remain on the superhydrophobic AA5083 surface. The water flow reflects, bounces and finally reflecting/rolling away easily from the specimen. Figure 4d–f presents the optical images of platform movement to contact and departing of the water droplet using the Dataphysics OCA25 instrument. With the gradually approaching, contacting and departing, the as-prepared superhydrophobic AA5083 surface could completely depart from the water droplet after tight contact, suggesting an extremely low adhesivity.

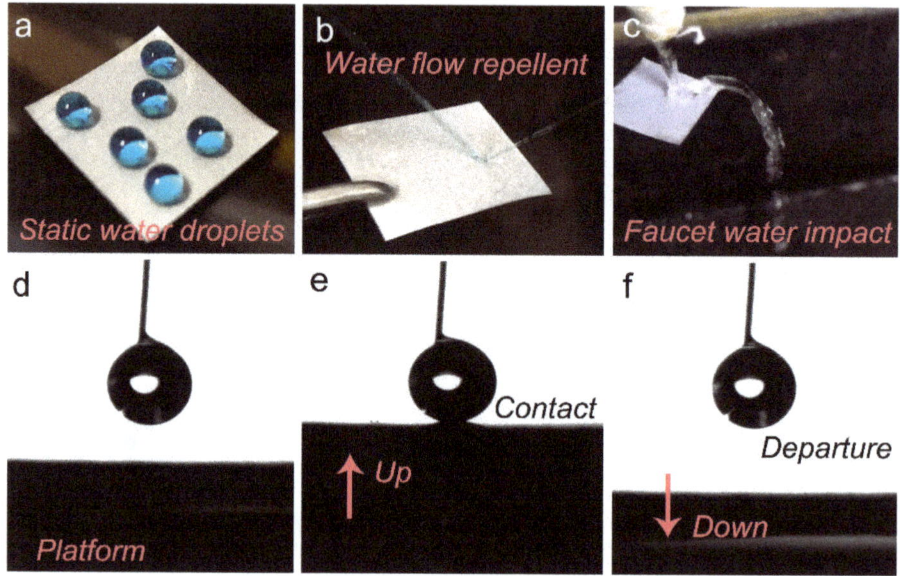

Figure 4. Optical images of the as-prepared superhydrophobic AA5083 surface (**a**) with spherical water droplets, (**b**) with a jet of water flow, (**c**) with faucet water impact, and (**d–f**) with platform movement to contact and departing of the water droplet.

The above wetting property and low adhesivity of the as-prepared superhydrophobic AA5083 can be explained by the combination action of the two-tier hierarchical rough structure and the HDTMS modification. Hierarchical petaloid rough structures benefit from the formation of the trapped air cushion, which significantly restrain the interfacial contact of the water–solid phase. Furthermore, the introduction of HDTMS molecules decrease the surface energy and further suppresses the penetration of water droplets into the surface structure. Thus, it can be concluded from the interaction between water and the superhydrophobic AA5083 surface that the water-repellent superhydrophobicity is attributed to the microscale roughness of the etching-textured surface and the low surface energy of HDTMS molecules.

3.3. Chemical Composition

The chemical composition of the pristine and HDTMS assembly superhydrophobic AA5083 surfaces were characterized by EDS (Figure 5a) and XPS (Figure 5b). The surface is rich in C, O, Mg, Al and Si elements as evidenced by the EDS spectrum, preliminarily verifying the assembly of the long carbon-chain tail on the as-prepared superhydrophobic AA5083 surface. As is well known, the XPS spectra present a surface chemical composition with a detecting depth of a few nanometers. The detailed elemental composition was further investigated through an XPS spectrum, as shown in Figure 5b. It is obvious that C, O, Si and Al elements were detected and demonstrated in the XPS spectrum of the as-prepared superhydrophobic AA5083 surface. Strong binding energy located at 284.7 eV, 532.4 eV, 102.3 eV and 74.6 eV were confirmed and ascribed to C 1s, O 1s, Si 2p and Al 2p, respectively, further confirming the existence of HDTMS species in the superhydrophobic AA5083 substrate. The C 1s, O 1s and Si 2p peaks in XPS spectra of the as-prepared superhydrophobic AA5083 sample were contributed to the HDTMS molecule assembly. This is in accordance with the EDS spectrum. It can be concluded, based on the analyses of surface topography, microscale roughness, wetting property and chemical composition, that the AA5083 substrates were endowed with excellent water repellent superhydrophobicity.

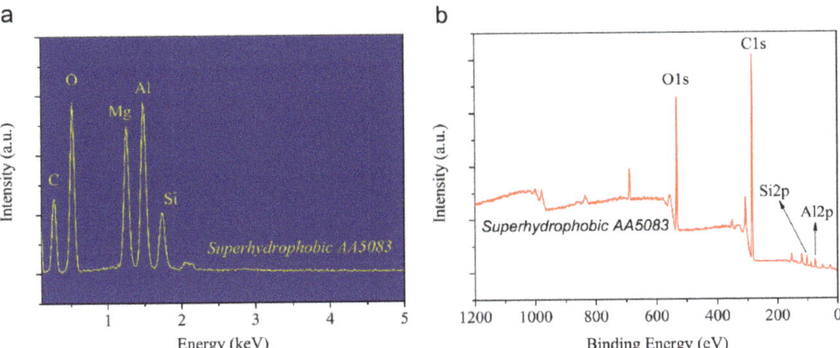

Figure 5. Energy dispersive X-ray spectroscopy (EDS) and X-ray photoelectron spectroscopy (XPS) spectra of the fabricated superhydrophobic AA5083 sample. (**a**) EDS spectrum, and (**b**) XPS spectrum.

3.4. Self-Cleaning Ability and NaCl Self-Propelling

As for real-world applications, the self-cleaning ability of superhydrophobic material is an essential and promising function. In this work, MnO powder and graphite powder were successively applied as surface contaminations of the as-prepared superhydrophobic AA5083 samples. The superhydrophobic AA5083 specimen was firstly inclined at an angle lower than 10°. Water droplets were subsequently dropped from above to evaluate the self-cleaning effect of the specimen. Figure 6a,b display the optical photos of the self-cleaning process. Given the excellent water repellence of the superhydrophobic AA5083 specimen, the water droplets can instantaneously roll off the sample surface, effectively picking up and taking away the MnO powder and graphite powder without difficulty. The water droplets rolled down the superhydrophobic AA5083 sample without moistening the solid surface, leaving several traces of rolling upon the surface. After washing, the contaminated AA5083 samples were totally clean and had no differences compared to the original uncontaminated specimen.

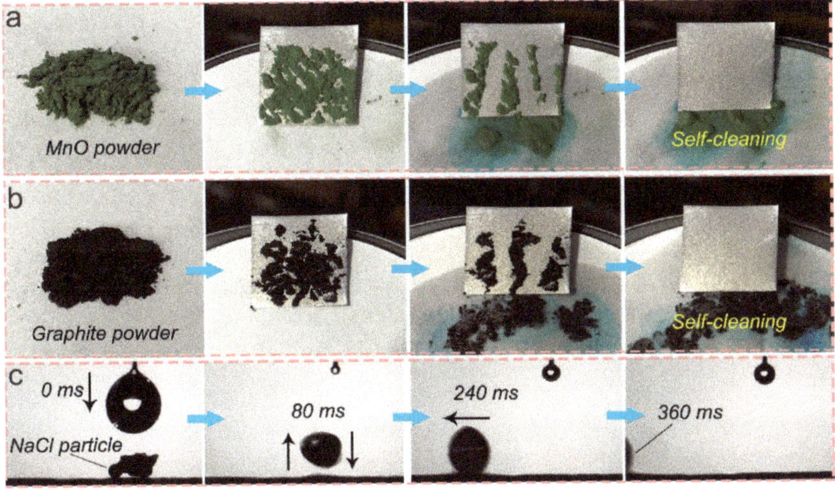

Figure 6. Self-cleaning ability with different surface contaminations: (**a**) MnO powder, (**b**) graphite powder, and (**c**) NaCl self-propelling property.

Figure 6c presents the NaCl self-propelling property of the superhydrophobic AA5083 surface. The NaCl particle was put on the surface of the sample. After the water droplet fell to the surface, the NaCl particle was fused with the water droplet. The outstanding water-repellence of the superhydrophobic AA5083 surface overcame the gravity of the NaCl particle, exhibiting a typical droplet/NaCl bouncing and deformation. An extremely low sliding angle propelled the droplet/NaCl coalition.The NaCl self-propelling process only takes 360 ms from the water droplet falling to the final sliding away. The self-cleaning and NaCl self-propelling ability mentioned above can be mainly ascribed to the extremely low surface energy, low adhesivity and water-repellence property of the surface. Therefore, it can be concluded that the fabricated superhydrophobic AA5083 substrate possesses an excellent self-cleaning ability resisting different surface contaminations.

3.5. Marine Corrosion Protection

In order to quantitatively characterize the corrosion behavior of the as-prepared superhydrophobic AA5083 film, we carried out and analyzed the electrochemical impedance spectroscopy (EIS) in an open circuit condition of a 3.5 wt.% NaCl aqueous corrosive solution. Figure 7a presents the EIS plots and fittings of pristine AA5083 and superhydrophobic AA5083 substrates. It can be seen from superhydrophobic AA5083 EIS plots that the diameter of the capacitive loop is much larger than that of the pristine AA5083 specimen. Figure 7b displays Bode plots of log |Z| vs. frequency and fittings of the pristine AA5083 and superhydrophobic AA5083 surfaces, presenting a three orders of magnitude higher impedance modulus value than that of the pristine AA5083 substrate. The anti-wetting and water-repellent property of the as-fabricated superhydrophobic AA5083 surface contributes to the remarkable improvement of impedance modulus.

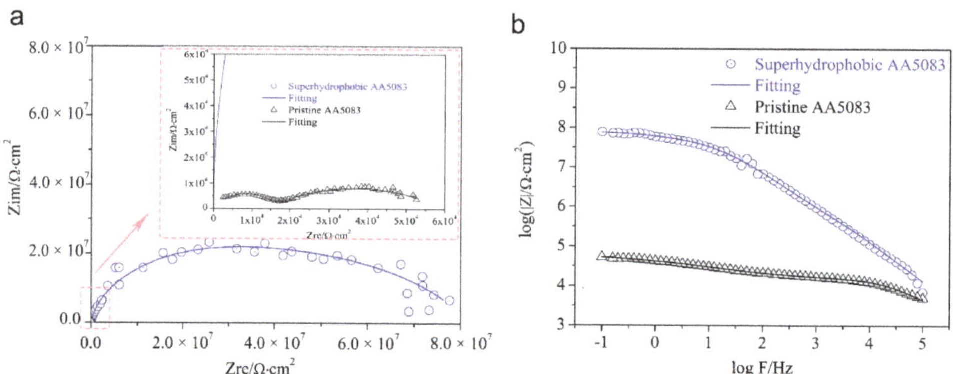

Figure 7. (**a**) Electrochemical impedance spectroscopy (EIS) plots, and (**b**) Bode plots of log |Z| vs. frequency and fittings of the pristine AA5083 and superhydrophobic AA5083 surfaces.

Different equivalent circuits were employed to analyze the EIS results through *ZsimpWin* software, as shown in Figure 8. Figure 8a,b display the equivalent circuit of pristine AA5083 and superhydrophobic AA5083 specimens. For the convenience of EIS data analyzing, R_s, R_{film}, R_{oxide} and R_{ct} in Figure 8 represent solution resistance, superhydrophobic film resistance, oxide resistance and charge transfer resistance, respectively. Q_{film}, Q_{oxide} and Q_{dl} represent the constant phase elements (CPE) modelling capacitance of the as-prepared superhydrophobic film, the oxide layer and the double-layer, respectively. Wherein, the impedance of the CPE could be defined as $1/Y_0(j\omega)^n$ [41,42], where Y_0, j, ω and n represent the modulus, imaginary number, angular frequency and the phase, respectively.

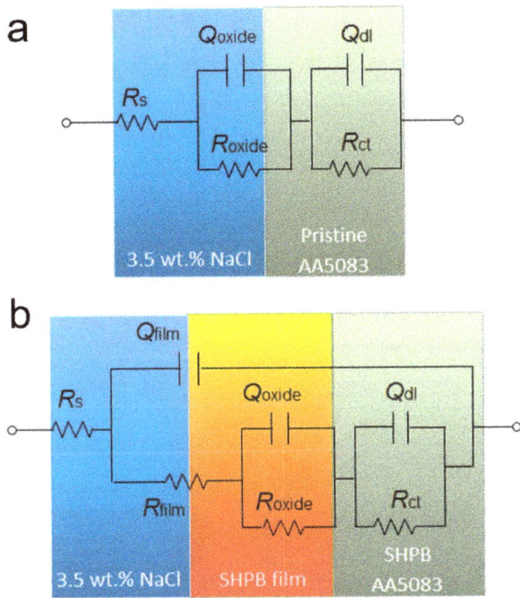

Figure 8. Equivalent circuit of (**a**) pristine AA5083 and (**b**) superhydrophobic AA5083.

The fitted electrochemical parameters are shown in Table 1. In General, R_{ct} is utilized to calculate and evaluate the corrosion resisting property of the fabricated protective layers. From Table 1, it can be clearly seen that the R_{ct} of the as-prepared superhydrophobic AA5083 specimen is 1.14×10^6 $\Omega \cdot cm^2$, while the R_{ct} of the pristine AA5083 sample is only 4.44×10^4 $\Omega \cdot cm^2$. The R_{ct} of superhydrophobic AA5083 is two orders of magnitude higher than that of pristine AA5083, demonstrating a remarkable enhanced corrosion-resisting performance. In addition, the Q_{dl} of pristine AA5083 and as-prepared superhydrophobic AA5083 was 6.77×10^{-6} $\Omega^{-1} \cdot cm^{-2} \cdot s^n$ and 3.54×10^{-10} $\Omega^{-1} \cdot cm^{-2} \cdot s^n$, respectively. The lower Q_{dl} value and higher R_{ct} value of the as-prepared superhydrophobic AA5083 substrate suggest that the charge transfer process of corrosive ions occurs with difficulty.

Table 1. The electrochemical parameters of simulated pristine AA5083 and superhydrophobic AA5083 surfaces in 3.5 wt.% NaCl aqueous solution.

Parameters	Specimens	
	Pristine AA5083	**Superhydrophobic AA5083**
R_s ($\Omega \cdot cm^2$)	2.43	4.09
Q_{film} ($\Omega^{-1} \cdot cm^{-2} \cdot s^n$)	/	1.19×10^{-10}
n_1	/	1
R_{film} ($\Omega \cdot cm^2$)	/	8.24×10^7
Q_{oxide} ($\Omega^{-1} \cdot cm^{-2} \cdot s^n$)	5.84×10^{-9}	1.66×10^{-9}
n_2	0.79	0.59
R_{oxide} ($\Omega \cdot cm^2$)	1.47×10^4	8.29×10^7
Q_{dl} ($\Omega^{-1} \cdot cm^{-2} \cdot s^n$)	6.77×10^{-6}	3.54×10^{-10}
n_3	0.44	0.80
R_{ct} ($\Omega \cdot cm^2$)	4.44×10^4	1.14×10^6
η (%)	/	96.11

In general, the R_{ct} value is utilized to calculate the inhibition efficiency (η) of the protective film using the formula $\eta = (R_{ct} - R_{ct}^0)/R_{ct}$ [43], in which R_{ct} and R_{ct}^0 represent the charge transfer resistance of the superhydrophobic AA5083 specimen and the pristine AA5083 specimen. It was calculated using the R_{ct} and R_{ct}^0 values discussed above that the inhibition efficiency in this case was approximately 96.11%, indicating an impressive performance resisting marine corrosion attack.

Figure 9 shows the anticorrosion mechanism of the as-fabricated superhydrophobic AA5083 surface. When the pristine AA5083 is exposed to environments containing aggressive corrosive ions, the natural oxide film breaks down at specific points leading to the formation of localized corrosion on AA5083 surface, viz. pitting corrosion or intergranular corrosion. On the contrary, the as-prepared superhydrophobic AA5083 surface could trap air within the micro-nano petaloid hierarchical structure, presenting a greatly decreasing fraction of the water/solid interface. The superior corrosion resistance is mainly attributed to the air cushion trapped in the petaloid rough structure, which suppresses the penetration and diffusion of aggressive corrosion species crossing the superhydrophobic protective film to the underlying AA5083 substrate. So, it can be implied that the as-fabricated superhydrophobic AA5083 specimen could largely improve the corrosion resistance of the substrate, displaying an outstanding protection ability towards marine corrosion attack.

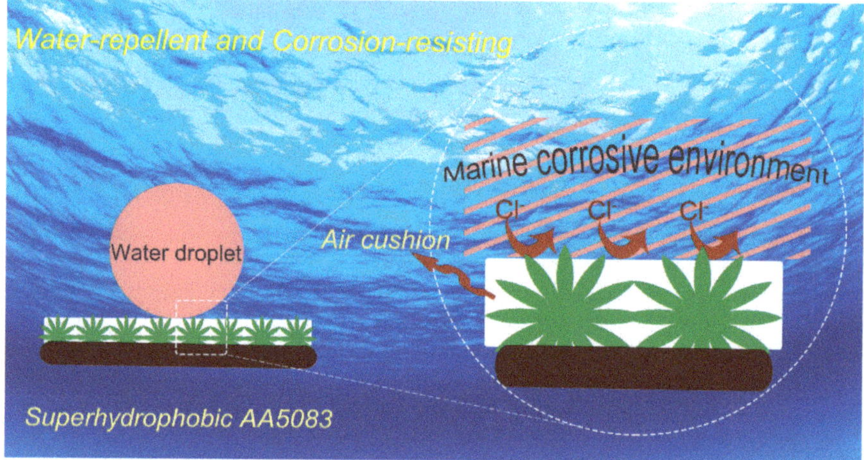

Figure 9. Anticorrosion mechanism of the as-prepared superhydrophobic AA5083.

3.6. Long-Term Stability

For practical applications, it is significant to characterize the stability of the superhydrophobic AA5083 specimen subjected to air exposure and a corrosive solution immersion for prolonged period. In this case, we carried out air exposure and a 3.5 wt.% NaCl aqueous solution immersion experiments to evaluate the durability of the as-produced superhydrophobic AA5083 substrate, as shown in Figure 10a,b. After 90 days of air exposure and 12 days of a 3.5 wt.% NaCl immersion/contact of the as-fabricated superhydrophobic AA5083 surface, it can be found from the contact angle and sliding angle variation that the as-prepared surface causes nearly no degradation of water-repellent superhydrophobicity with a contact angle higher than 155°, indicating outstanding long-term stability. The multiscale hierarchical structure and chemisorbed HDTMS hydrophobic molecules facilitate the formation of an air cushion, significantly contributing to the eventual water-repellence, air-exposure stability and corrosive-medium immersion stability.

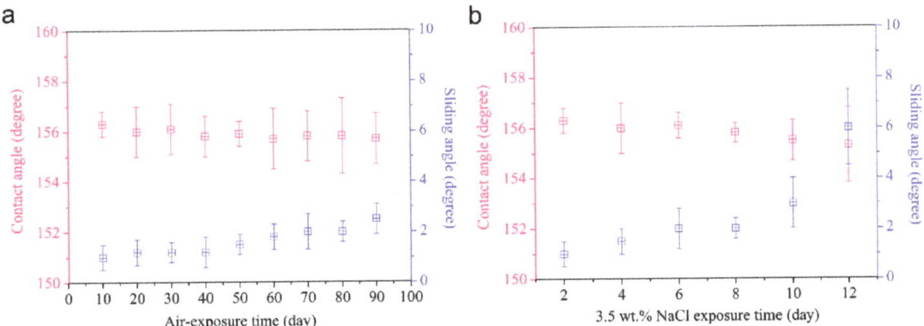

Figure 10. The variation of the contact angle and sliding angle under different exposure time in (**a**) air and (**b**) 3.5 wt.% NaCl aqueous solution.

4. Conclusions

In conclusion, to successfully design a lotus-inspired superhydrophobic AA5083 surface with a facile, cost-effective and fluorine-free strategy, we provide an approach combining etching texture followed by chemisorption of HDTMS hydrophobic molecules. It was shown that the surface topography of the as-prepared superhydrophobic AA5083 possesses a micro-nano hierarchical petaloid structure with a contact angle of 156.3° ± 1° and a sliding angle lower than 1°. The wetting behavior investigations presented water-repellence, extremely low adhesivity and a self-cleaning ability. The EIS and fitting results showed that the R_{ct} of the as-prepared superhydrophobic AA5083 specimen is two orders of magnitude higher than that of the pristine AA5083. In addition, the Q_{dl} of the as-prepared superhydrophobic AA5083 was four orders of magnitude lower than the pristine AA5083. The inhibition efficiency in this case was approximately 96.11%. In addition, after 90 days of air exposure and 12 days of a 3.5 wt.% NaCl immersion/contact, the as-fabricated superhydrophobic AA5083 surface sustained a durable and stable superhydrophobicity. Therefore, the resultant superhydrophobic AA5083 surface possesses superior corrosion-resisting performance and long-term stability. We greatly anticipate that this research work has important significance for the large-scale manufacturing of water-repellent superhydrophobic surfaces for long-term marine and industrial applications.

Author Contributions: B.Z. and Y.L. conceived the concept of this work; B.Z. conducted the fabrication and characterization and wrote the manuscript; and W.X., Q.Z. and S.Y. performed the electrochemical experiments.

Funding: This research work was financially supported by the National Natural Science Foundation of China (Grant No. 41806089), the Strategic Priority Program of Chinese Academy of Sciences (Grant No. XDA13040401).

Conflicts of Interest: The authors declare no conflicts of interest.

References

1. Yasakau, K.A.; Zheludkevich, M.L.; Lamaka, S.V.; Ferreira, M.G.S. Role of intermetallic phases in localized corrosion of AA5083. *Electrochim. Acta* **2007**, *52*, 7651–7659. [CrossRef]
2. Tan, L.; Allen, T.R. Effect of thermomechanical treatment on the corrosion of AA5083. *Corros. Sci.* **2010**, *52*, 548–554. [CrossRef]
3. Mills, R.J.; Lattimer, B.Y.; Case, S.W.; Mouritz, A.P. The influence of sensitization and corrosion on creep of 5083-H116. *Corros. Sci.* **2018**, *143*, 1–9. [CrossRef]
4. Liu, S.; Wang, X.; Tao, Y.; Han, X.; Cui, C. Enhanced corrosion resistance of 5083 aluminum alloy by refining with nano-CeB_6/Al inoculant. *Appl. Surf. Sci.* **2019**, *484*, 403–408. [CrossRef]
5. Su, B.; Tian, Y.; Jiang, L. Bioinspired interfaces with superwettability: From materials to chemistry. *J. Am. Chem. Soc.* **2016**, *138*, 1727–1748. [CrossRef] [PubMed]
6. Yu, Z.; Yun, F.F.; Wang, Y.; Yao, L.; Dou, S.; Liu, K.; Jiang, L.; Wang, X. Desert beetle-inspired superwettable patterned surfaces for water harvesting. *Small* **2017**, *13*, 1701403. [CrossRef] [PubMed]

7. Barthlott, W.; Neinhuis, C. Purity of the sacred lotus, or escape from contamination in biological surfaces. *Planta* **1997**, *202*, 1–8. [CrossRef]
8. Feng, L.; Li, S.; Li, Y.; Li, H.; Zhang, L.; Zhai, J.; Song, Y.; Liu, B.; Jiang, L.; Zhu, D. Super-hydrophobic surfaces: From natural to artificial. *Adv. Mater.* **2002**, *14*, 1857–1860. [CrossRef]
9. Gao, X.F.; Jiang, L. Biophysics: Water-repellent legs of water striders. *Nature* **2004**, *432*, 36. [CrossRef] [PubMed]
10. Zhang, B.; Li, Y.; Hou, B. One-step electrodeposition fabrication of a superhydrophobic surface on an aluminum substrate with enhanced self-cleaning and anticorrosion properties. *RSC Adv.* **2015**, *5*, 100000–100010. [CrossRef]
11. Zhang, B.; Zhao, X.; Li, Y.; Hou, B. Fabrication of durable anticorrosion superhydrophobic surfaces on aluminum substrates via a facile one-step electrodeposition approach. *RSC Adv.* **2016**, *6*, 35455–35465. [CrossRef]
12. Siddiqui, A.R.; Maurya, R.; Balani, K. Superhydrophobic self-floating carbon nanofiber coating for efficient gravity-directed oil/water separation. *J. Mater. Chem. A* **2017**, *5*, 2936–2946. [CrossRef]
13. Beshkar, F.; Khojasteh, H.; Salavati-Niasari, M. Recyclable magnetic superhydrophobic straw soot sponge for highly efficient oil/water separation. *J. Colloid Interface Sci.* **2017**, *497*, 57–65. [CrossRef] [PubMed]
14. Hwang, G.B.; Patir, A.; Page, K.; Lu, Y.; Allan, E.; Parkin, I.P. Buoyancy increase and drag-reduction through a simple superhydrophobic coating. *Nanoscale* **2017**, *9*, 7588–7594. [CrossRef]
15. Taghvaei, E.; Moosavi, A.; Nouri-Borujerdi, A.; Daeian, M.A.; Vafaeinejad, S. Superhydrophobic surfaces with a dual-layer micro- and nanoparticle coating for drag reduction. *Energy* **2017**, *125*, 1–10. [CrossRef]
16. Liu, Y.; Li, X.; Jin, J.; Liu, J.; Yan, Y.; Han, Z.; Ren, L. Anti-icing property of bio-inspired micro-structure superhydrophobic surfaces and heat transfer model. *Appl. Surf. Sci.* **2017**, *400*, 498–505. [CrossRef]
17. Zuo, Z.; Liao, R.; Zhao, X.; Song, X.; Qiao, Z.; Guo, C.; Zhuang, A.; Yuan, Y. Anti-frosting performance of superhydrophobic surface with ZnO nanorods. *Appl. Therm. Eng.* **2017**, *110*, 39–48. [CrossRef]
18. Yang, Y.; Li, X.; Zheng, X.; Chen, Z.; Zhou, Q.; Chen, Y. 3D-printed biomimetic super-hydrophobic structure for microdroplet manipulation and oil/water separation. *Adv. Mater.* **2018**, *30*, 1704912. [CrossRef] [PubMed]
19. Ding, G.; Jiao, W.; Wang, R.; Niu, Y.; Hao, L.; Yang, F.; Liu, W. A biomimetic, multifunctional, superhydrophobic graphene film with self-sensing and fast recovery properties for microdroplet transportation. *J. Mater. Chem. A* **2017**, *5*, 17325–17334. [CrossRef]
20. Wang, M.; Liu, Q.; Zhang, H.; Wang, C.; Wang, L.; Xiang, B.; Fan, Y.; Guo, C.F.; Ruan, S. Laser direct writing of tree-shaped hierarchical cones on a superhydrophobic film for high-efficiency water collection. *ACS Appl. Mater. Interfaces* **2017**, *9*, 29248–29254. [CrossRef] [PubMed]
21. Seo, D.; Lee, C.; Nam, Y. Influence of geometric patterns of microstructured superhydrophobic surfaces on water-harvesting performance via dewing. *Langmuir* **2014**, *30*, 15468–15476. [CrossRef] [PubMed]
22. Arukalam, I.O.; Oguzie, E.E.; Li, Y. Nanostructured superhydrophobic polysiloxane coating for high barrier and anticorrosion applications in marine environment. *J. Colloid Interface Sci.* **2018**, *512*, 674–685. [CrossRef] [PubMed]
23. Zhang, B.; Xu, W.; Zhu, Q.; Li, Y.; Hou, B. Ultrafast one step construction of non-fluorinated superhydrophobic aluminum surfaces with remarkable improvement of corrosion resistance and anti-contamination. *J. Colloid Interface Sci.* **2018**, *532*, 201–209. [CrossRef] [PubMed]
24. Selim, M.S.; El-Safty, S.A.; Fatthallah, N.A.; Shenashen, M.A. Silicone/graphene oxide sheet-alumina nanorob ternary composite for superhydrophobic antifouling coating. *Prog. Org. Coat.* **2018**, *121*, 160–172. [CrossRef]
25. Zhang, B.; Li, J.; Zhao, X.; Hu, X.; Yang, L.; Wang, N.; Li, Y.; Hou, B. Biomimetic one step fabrication of manganese stearate superhydrophobic surface as an efficient barrier against marine corrosion and *Chlorella vulgaris*-induced biofouling. *Chem. Eng. J.* **2016**, *306*, 441–451. [CrossRef]
26. Wang, J.; Liu, F.; Chen, H.; Chen, D. Superhydrophobic behavior achieved from hydrophilic surfaces. *Appl. Phys. Lett.* **2009**, *95*, 084104. [CrossRef]
27. Hoshian, S.; Jokinen, V.; Somerkivi, V.; Lokanathan, A.R.; Franssila, S. Robust superhydrophobic silicon without a low surface-energy hydrophobic coating. *ACS Appl. Mater. Interfaces* **2015**, *7*, 941–949. [CrossRef] [PubMed]
28. Kim, J.H.; Mirzaei, A.; Kim, H.W.; Kim, S.S. Realization of superhydrophobic aluminum surfaces with novel micro-terrace nano-leaf hierarchical structure. *Appl. Surf. Sci.* **2018**, *451*, 207–217. [CrossRef]

29. Yang, Z.; Liu, X.; Tian, Y. Hybrid laser ablation and chemical modification for fast fabrication of bio-inspired super-hydrophobic surface with excellent self-cleaning, stability and corrosion resistance. *J. Bionic Eng.* **2019**, *16*, 13–26. [CrossRef]
30. Zhang, X.; Zhao, J.; Mo, J.; Sun, R.; Li, Z.; Guo, Z. Fabrication of superhydrophobic aluminum surface by droplet etching and chemical modification. *Colloid Surf. A* **2019**, *567*, 205–212. [CrossRef]
31. Wang, G.; Shen, Y.; Tao, J.; Luo, X.; Jin, M.; Xie, Y.; Li, Z.; Guo, S. Facilely constructing micro-nanostructure superhydrophobic aluminum surface with robust ice-phobicity and corrosion resistance. *Surf. Coat. Technol.* **2017**, *329*, 224–231. [CrossRef]
32. Boinovich, L.B.; Modin, E.B.; Sayfutdinova, A.R.; Emelyanenko, K.A.; Vasiliev, A.L.; Emelyanenko, A.M. Combination of functional nanoengineering and nanosecond laser texturing for design of superhydrophobic aluminum alloy with exceptional mechanical and chemical properties. *ACS Nano* **2017**, *11*, 10113–10123. [CrossRef] [PubMed]
33. Boinovich, L.B.; Emelyanenko, K.A.; Domantovsky, A.G. Laser tailoring the surface chemistry and morphology for wear, scale and corrosion resistance superhydrophobic coatings. *Langmuir* **2018**, *34*, 7059–7066. [CrossRef] [PubMed]
34. Wang, F.; Guo, Z. Insitu growth of durable superhydrophobic Mg-Al layered double hydroxides nanoplatelets on aluminum alloys for corrosion resistance. *J. Alloys Comp.* **2018**, *767*, 382–391. [CrossRef]
35. Zhang, B.; Guan, F.; Zhao, X.; Zhang, Y.; Li, Y.; Duan, J.; Hou, B. Micro-nano textured superhydrophobic 5083 aluminum alloy as a barrier against marine corrosion and sulfate-reducing bacteria adhesion. *J. Taiwan Inst. Chem. Eng.* **2019**, *97*, 433–440. [CrossRef]
36. Zhang, B.; Hu, X.; Zhu, Q.; Wang, X.; Zhao, X.; Sun, C.; Li, Y.; Hou, B. Controllable *Dianthus caryophyllus*-like superhydrophilic/superhydrophobic hierarchical structure based on self-congregated nanowires for corrosion inhibition and biofouling mitigation. *Chem. Eng. J.* **2017**, *312*, 317–327. [CrossRef]
37. Wenzel, R.N. Resistance of solid surfaces to wetting by water. *Ind. Eng. Chem.* **1936**, *28*, 988–994. [CrossRef]
38. Wenzel, R.N. Surface roughness and contact angle. *J. Phys. Chem.* **1949**, *53*, 1466–1467. [CrossRef]
39. Cassie, A.B.D.; Baxter, S. Wettability of porous surfaces. *Trans. Faraday Soc.* **1944**, *40*, 546–551. [CrossRef]
40. Cassie, A.B.D. Contact angles. *Discuss. Faraday Soc.* **1948**, *3*, 11–16. [CrossRef]
41. Zhao, X.; Jin, Z.; Zhang, B.; Zhai, X.; Liu, S.; Sun, X.; Zhu, Q.; Hou, B. Effect of graphene oxide on anticorrosion performance of polyelectrolyte multilayer for 2A12 aluminum alloy substrates. *RSC Adv.* **2017**, *7*, 33764–33774. [CrossRef]
42. Zhao, X.; Zhang, B.; Jin, Z.; Chen, C.; Zhu, Q.; Hou, B. Epoxy coating modified by 2D MoS_2/SDBS: Fabrication, anticorrosion behavior and inhibition mechanism. *RSC Adv.* **2016**, *6*, 97512–97522. [CrossRef]
43. Zhang, B.; Zhu, Q.; Li, Y.; Hou, B. Facile fluorine-free one step fabrication of superhydrophobic aluminum surface towards self-cleaning and marine anticorrosion. *Chem. Eng. J.* **2018**, *352*, 625–633. [CrossRef]

Article

Optimization of Cathodic Protection Design for Pre-Insulated Pipeline in District Heating System Using Computational Simulation

Min-Sung Hong, Yoon-Sik So and Jung-Gu Kim *

School of Advanced Materials Engineering, Sungkyunkwan University, 300 Chunchun-Dong, Jangan-Gu, Suwon 440-746, Korea; smith803@skku.edu (M.-S.H.); soy2871@naver.com (Y.-S.S.)
* Correspondence: kimjg@skku.edu

Received: 13 May 2019; Accepted: 28 May 2019; Published: 30 May 2019

Abstract: Cathodic protection (CP) has been used as a primary method in the control of corrosion, therefore it is regarded as the most effective way for protecting buried pipelines. However, it is difficult to apply CP to a pipeline for district heating distribution systems, because the pipeline has thermally insulated coatings which could disturb the CP. Theoretical calculation and field tests alone are not enough for a reliable CP design, and therefore additional CP design methods such as computational analysis should be used. In this study, the CP design for pre-insulated pipelines is tested considering several environmental factors, such as temperature and coating defect ratio. Additionally, computational analysis is performed to verify and optimize the CP design. The simulation results based on theoretical methods alone failed to satisfy the CP criteria. Then, a re-design is conducted considering the IR drop. Consequently, all of the simulation results of defective pipelines satisfied the CP criteria after adding the proper CP current.

Keywords: cathodic protection; corrosion mitigation method; potentiodynamic polarization test; simulation; pre-insulated pipeline

1. Introduction

In district heating (DH) systems, heated water is distributed through a double-pipe network and transferred to buildings for use in space heating, hot water generation, and process heating [1]. DH systems have three main elements: the heat source, the distribution system, and the customer interface. The distribution system supplies hot water from the heat source to the heat consumer and returns with temperatures in the range of 40 °C to 120 °C [2]. Generally, pipelines in DH distribution systems use a thermally insulated coating to minimize heat loss during transfer. As shown in Figure 1, the coating consists of two layers: an inner layer of polyurethane foam (PUR) to reduce heat loss, and an outer layer of high-density polyethylene (HDPE) to protect the PUR [3]. The coatings effectively mitigate corrosion by blocking the outer environment, which contains corrosive elements such as water, oxygen, and chloride ions, when the coating is maintained perfectly. However, the HDPE is susceptible to unpredictable mechanical damage, and the PUR can be vanished by heat, humidity, and oxygen during its long operational life [4,5]. Several studies have reported that the main source of corrosion is groundwater introduced through failure of the HDPE and PUR [5–7].

Cathodic protection (CP) has been used as the primary method in the control of corrosion, in conjunction with protective coatings. CP can reduce the corrosion rate, and a properly maintained system will provide protection in accordance with the designed structural life [8]. The impressed current CP (ICCP) system has a power supply (rectifier) that generates larger potential differences between the anode and the structure [9]. For this reason, ICCP is applied to many industrial pipelines. However, despite the availability of CP, there are still several limitations in applying CP to pre-insulated pipelines [10]. The National Association of Corrosion Engineers (NACE) reported that the CP design for pre-insulated pipelines is ineffective because the protection current cannot reach the corroded area through the insulating layer [11]. Additionally, according to previous studies, the corrosion rate of PUR-insulated carbon steel is much lower than that of uninsulated (bare) steel, even when the PUR is fully immersed in groundwater [12]. Therefore, CP for pre-insulated pipelines may be unnecessary when the PUR layer is intact. However, the immersed PUR layer can deteriorate and vanish during long operating periods, causing exposure of the bare carbon steel to the corrosive environment. For this reason, it is important to apply CP to operating pipelines with external coating defects, as a precaution against sudden fracture. Nevertheless, it is difficult to design CP systems for operating pipelines using only theoretical methods and a limited number of standards. It is also difficult to verify the appropriate protecting current required to reach the external surface of the pipeline with proper CP potential. For this reason, additional CP design methods, such as computational analysis, should be applied to optimize the design [13–15]. To improve the reliability of simulation results, several essential factors should be considered, such as polarization data for real materials and appropriate environmental information.

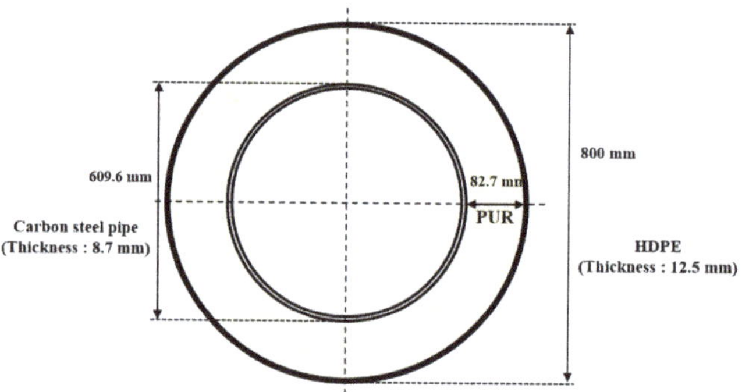

Figure 1. Schematic diagram of the pre-insulated coated pipe (600 A).

In this study, a CP system was designed for an existing pipeline with damaged insulation, taking into consideration environmental factors, such as corrosion properties of real materials, operating temperatures, and structural effects. Additionally, electrochemical tests were performed in synthetic groundwater to obtain input data for the computer simulation. Finally, a computational analysis was performed to verify and optimize the CP design of pre-insulated pipelines.

2. Materials and Methods

2.1. Materials and Test Conditions

The corrosion environment used was synthetic groundwater. Table 1 gives the chemical composition of the synthetic groundwater, and HNO_3 was used to control the pH of the solution. A welded carbon steel specimen consisting of a base metal, heat affected zone, and a weld metal was used during testing to calculate the required CP current. Table 2 shows the chemical composition of the SPW400 (carbon steel), and Table 3 shows the welding methods used in all experiments. The surface of the specimen was polished with 600-grit silicon carbide (SiC) paper, degreased with ethanol, and dried with N_2.

Table 1. Chemical composition of synthetic groundwater.

$CaCl_2$ (ppm)	$MgSO_4{\cdot}7H_2O$ (ppm)	$NaHCO_3$ (ppm)	H_2SO_4 (ppm)	HNO_3 (ppm)	pH	Resistivity (kΩ·cm)
133.2	59	208	85	22.2	6.8	1.736

Table 2. Chemical composition of SPW400 (wt.%).

Fe	C	P	S
Balance	0.25 Max.	0.04 Max.	0.04 Max.

Table 3. Welding procedure specification.

Welding Process	GTAW
Joint design	Single V joint with a 60° included angle and a 1.6 mm root face
Electrode	GTAW ER70S-G
Voltage	12–15 V
Current	100–180 A
Polarity	Direct Current Straight Polarity (DCSP)
Travel speed	20–30 cm/min
Welding atmosphere	Ar, 15–25 L/min

2.2. Electrochemical Test Methods

All electrochemical experiments were performed using a three-electrode system, in a 1000 mL Pyrex glass corrosion cell connected to an electrochemical apparatus. The test specimens were connected to a working electrode, a graphite rod was used as the counter electrode, and a saturated calomel electrode (SCE) was used as the reference electrode. The area of the test specimen exposed to the electrolyte was 2.25 cm^2 (1.5 cm × 1.5 cm). An open-circuit potential (OCP) was established within three hours to carry out the electrochemical test. Potentiodynamic polarization tests were carried out in accordance with ASTM G5-14 (Standard Reference Test Method for Making Potentiodynamic Anodic Polarization Measurements), using a VMP2 (Bio-Logic Science Instruments, Seyssinet-Pariset, France) with a potential sweep of 0.166 mV/sec, from an initial potential of −2000 mV versus the reference to a final potential of 200 mV versus the OCP. The electrochemical tests were performed at 80 °C, because a previous study found that the highest protection current for carbon steel was required at this temperature [7].

2.3. CP Design and Computational Analysis Method

The computational analysis tool BEASY S/W (BEASY Ltd., Southampton, England), which is based on the boundary element method (BEM), was used to conduct 3D modeling and computational analysis of the pre-insulated pipeline. The required CP current (I_{req}) for the pipeline was calculated, taking into consideration the current density of real material measured by electrochemical tests. The cathodic polarization curve, which was used as input data for the simulation, was obtained from the potentiodynamic polarization test, which incorporated the environmental information.

3. Results and Discussion

3.1. Potentiodynamic Polarization Tests

The applied current density (i_{app}) for the pre-insulated pipeline was calculated using the Evans diagram, as shown in Figure 2. According to the diagram, anodic current density is under activation control (activation polarization), and cathodic current density is limited at a higher current density (concentration polarization). As the applied current density for CP is increased, the potential and the corrosion current density are reduced simultaneously [16,17]. According to the previous study, since the pre-insulated pipeline has a high corrosion rate at 80 °C, the reasonable maximum CP potential is −1350 mV_{SCE} [7]. Figure 3 shows the results of the potentiodynamic polarization test in synthetic groundwater at 80 °C. The corrosion current density was determined using the Tafel extrapolation method. Table 4 shows the calculated CP current density, which will apply to the CP design. The applied current density was calculated as the difference between the anodic polarization curve and cathodic polarization curve at −1350 mV_{SCE}, as shown in Figure 2. The cathodic polarization curve, which contains the corrosion properties of real material, was used as the input data for computational analysis.

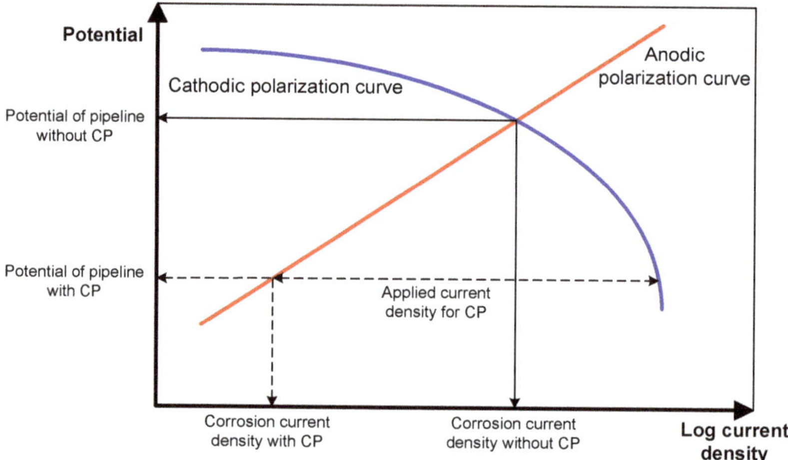

Figure 2. Evan's diagram, indicating the relationship between the applied current density and protection potential [16].

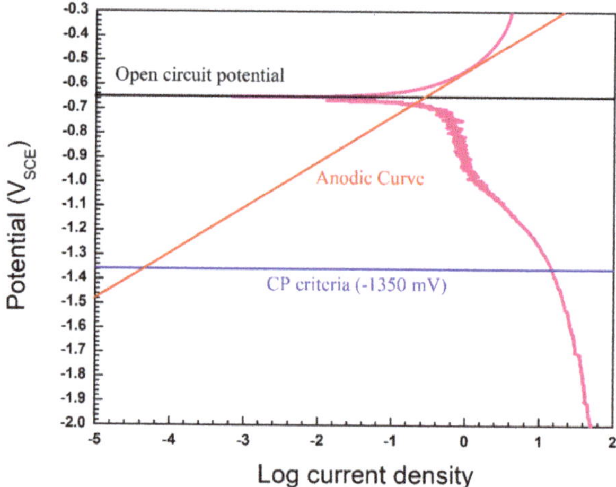

Figure 3. Potentiodynamic polarization curves in the synthetic groundwater at 80 °C.

Table 4. Results of Potentiodynamic Polarization Test at 80 °C in the Synthetic Groundwater.

Corrosion Potential (E_{corr}, mV_{SCE})	Corrosion Current Density (i_{corr}, A/m^2)	Bc (mV)	Ba (mV)	Applied Current Density (i_{app}, A/m^2)
−649	0.493	258.3	78.2	14.45

3.2. Cathodic Protection Design and Computational Analysis

The pre-insulated pipeline was connected every 6 m by welds, therefore, the CP design was preformed to 6 meters of 600 A pipe (Figure 1). In addition, ICCP anodes were installed at both edges of the pipeline, which are the parts most sensitive to corrosion because it will connect using welding. In this study, the CP design was applied to operating pipelines with slight defects. Therefore, the CP design was tested at a range of defect ratios (1, 5, 10, 20%), and it is assumed that the insulating part of the pipeline has no defect. Figure 4 shows the 3D modeling of the pipeline according to defect ratio. For modeling and calculations, the approach was based on the assumption that the crevice between the coating and pipeline was not effective as a CP [18]. Table 5 shows the basic design parameters related to the structural factors. The surface area of the pipe used in the CP design was 12.62 m^2, which included an additional 10% safety factor. The resistivity of soil was assumed to be 1000 Ω·cm, corresponding to a highly corrosive environment. The required current (I_{req}) for CP was calculated from the following equation [19,20]:

$$I_{req} = C_{defect} \cdot i_{app} \cdot A_{pipe}, \tag{1}$$

where C_{defect} is the defect ratio of the pipeline, i_{app} is the applied current density of the pipe material calculated from the electrochemical test, A_{pipe} is the surface area of the pipe. I_{req} is calculated with the defect ratio, as listed in Table 6.

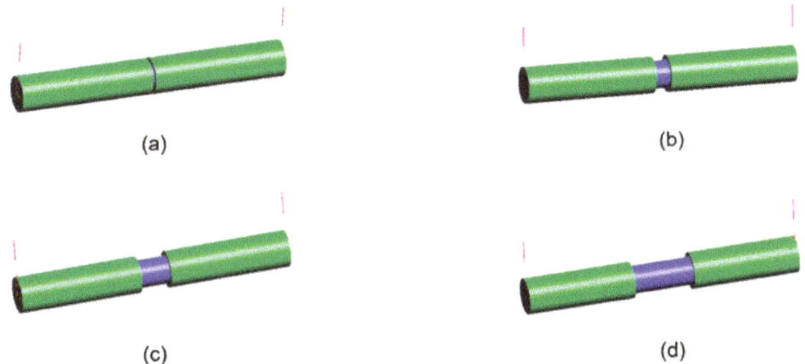

Figure 4. 3D modeling of pipeline according to the defect ratio: (**a**) 1%, (**b**) 5%, (**c**) 10%, (**d**) 20%.

Table 5. Basic design parameters related to the structural factors.

Pipeline (600 A)	**Diameter**	609.6 mm
	Length	6 m
	Surface Area	11.48 m^2
Resistivity of Soil		1000 Ω·cm
Temperature on the Pipeline		80 °C
CP Criteria		Under −1350 mV

Table 6. Required current calculation for cathodic protection (CP).

Applied Current Density (i_{app})	**Surface Area with 10% Safety Factor (A_{pipe})**	**Defect Ratio (C_{defect})**		**Required Current (I_{req})**
14.45 A/m^2	12.62 m^2	1%	0.01	1.824 A
		5%	0.05	9.120 A
		10%	0.1	18.241 A
		20%	0.2	36.483 A

The computational analysis was performed using the cathodic polarization curve data, obtained from the electrochemical tests. Figure 5 shows the simulation results for CP. All of the simulation results failed to satisfy the CP criteria for pre-insulated pipelines (under −1350 mV_{SCE}) because the IR drop caused by soil and structural factors was not considered in the CP design.

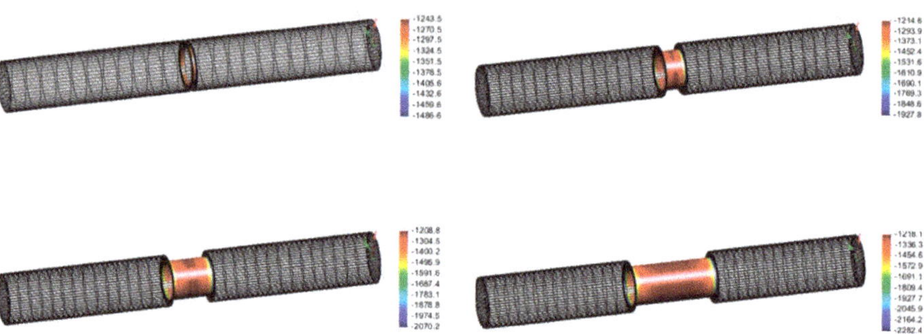

Figure 5. Simulation results (averaged protection potential, mV_{SCE}) according to the defect ratio: (**a**) 1%, (**b**) 5%, (**c**) 10%, (**d**) 20%.

The additional CP current required to satisfy the CP criteria should be calculated taking into consideration the polarization curve, as shown in Figure 6. The maximum CP potentials were defined based on the simulation results according to the defect ratio. Then, the applied current densities were calculated at the maximum CP potential from the simulation results, using the same method as above. To obtain the additional CP current densities, the difference was calculated between the calculated applied current densities, according to the defect ratio and applied current density at −1350 mV_{SCE}. The additional CP currents were then calculated using Equation (1). The calculated values are listed in Table 7.

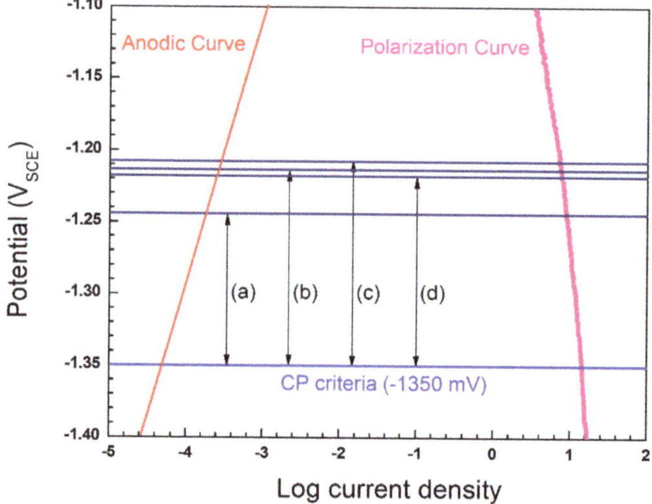

Figure 6. Calculation of additional current caused by IR drop in the polarization curve: potential difference of (**a**) 1%, (**b**) 5%, (**c**) 10%, (**d**) 20% defected pipelines.

Table 7. Results of calculated additional CP current and optimized current for CP.

Defect Ratio	Max. Potential in Previous Results	Additional Current Density	Additional Current	Optimized Current for CP
1%	−1243.5 mV	7.079 A/m^2	0.893 A	2.717 A
5%	−1214.6 mV	12.589 A/m^2	7.944 A	17.064 A
10%	−1208.8 mV	13.804 A/m^2	17.420 A	35.661 A
20%	−1218.1 mV	12.303 A/m^2	31.052 A	67.535 A

Then, the entire simulation was re-conducted. Figure 7 shows the optimized simulation results, and it was verified that all of the pipelines with different defect ratios satisfied the CP criteria. Another important point is over-protection due to the low CP criteria of district pipelines. The simulation results show that the minimum CP potentials have a range from −1.7 V_{SCE} to −2.6 V_{SCE}. This is quite a low potential value, which could cause hydrogen embrittlement risk. However, according to the international standards, such as NACE (RP0169-96), ARAMCO (SAES-X-400), and BSI (BS 7361-1), the over protection range of the steel pipeline ranges from −2.5 V_{SCE} to −5 V_{SCE}. Therefore, the simulation results can apply up to 10% of the defect ratio, which has a minimum potential of about −2.49 V_{SCE}. When the CP applies over 10% of the defect ratio, the site of defect should be previously investigated. Then, the anode should be installed as close as possible to the defect area, to avoid over protection and reduce CP current requirement. Therefore, the investigation of the defect area is one of the significant design parameters in practical CP installation.

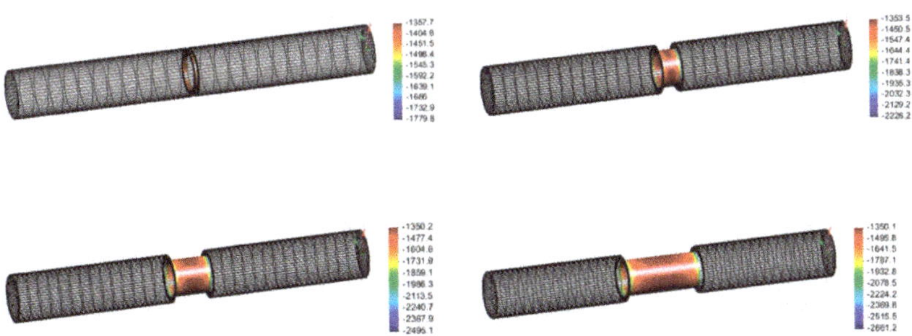

Figure 7. Optimized simulation results (averaged protection potential, mV_{SCE}) according to the defect ratio: (**a**) 1%, (**b**) 5%, (**c**) 10%, (**d**) 20%.

4. Conclusions

In this study, a credible CP design method for existing pre-insulated pipelines was conducted, taking into consideration the environmental factors, and computational analysis was performed to verify and optimize the CP design. According to the results, the following conclusions were drawn:

- The results of the simulations using the theoretical method failed to satisfy the CP criterion determined for heating pre-insulated pipeline. To solve the problem, a re-design was conducted, taking into consideration the IR drop caused by soil and structural factors. Consequently, after adding the proper CP current, all of the simulation results of defective pipelines satisfied the CP criteria.
- Incorporating practical corrosion properties of metal and environmental factors in the computational analysis improves the reliability of the CP design for a pipeline. For this reason, application of CP is recommended for pre-insulated pipelines, to mitigate external corrosion and reduce maintenance costs. The computational analysis is an essential step for credible CP design.

Author Contributions: Conceptualization, M.-S.H.; methodology, Y.-S.S.; software, M.-S.H.; validation, M.-S.H. and J.-G.K.; formal analysis, M.-S.H. and Y.-S.S.; investigation, M.-S.H.; resources, M.-S.H.; data curation, M.-S.H. and Y.-S.S.; writing—original draft preparation, M.-S.H.; writing—review and editing, M.-S.H. and J.-G.K.; visualization, M.-S.H.; supervision, J.-G.K.; project administration, M.-S.H. and J.-G.K.

Funding: This research was supported by the program for fostering next-generation researchers in engineering of National Research Foundation of Korea (NRF) funded by the Ministry of Science and ICT (2017H1D8A2031628).

Conflicts of Interest: The authors declare no conflict of interest.

References

1. Skagestad, B.; Mildenstein, P. *District Heating and Cooling Connection Handbook*; Netherlands Agency for Energy and the Environment: Amsterdam, The Netherlands, 2002; pp. 23–50.
2. Eliseev, K. District Heating System in Finland and Russia. Ph.D. Thesis, Mikkeli University of Applied Sciences, Mikkeli, Finland, 2011; pp. 4–20.
3. Choi, S.Y.; Yoo, K.Y.; Lee, J.B.; Shin, C.B.; Park, M.J. Mathematical modeling and control of thermal plant in the district heating system of Korea. *Appl. Therm. Eng.* **2010**, *30*, 2067–2072. [CrossRef]
4. Yarahmadi, N.; Sällström, J.H. Improved maintenance strategies for district heating pipe lines. In Proceedings of the 14th International Symposium on District Heating and Cooling, Stockholm, Sweden, 7–9 September 2014.
5. Choi, Y.-S.; Chung, M.-K.; Kim, J.-G. Effects of cyclic stress and insulation on the corrosion fatigue properties of thermally insulated pipeline. *Mater. Sci. Eng. A* **2004**, *384*, 47–56. [CrossRef]
6. Rassoul, E.-S.A.; Abdel-Samad, A.; El-Naqier, R. On the cathodic protection of thermally insulated pipelines. *Eng. Fail. Anal.* **2009**, *16*, 2047–2053. [CrossRef]
7. Kim, J.G.; Kim, Y.W. Cathodic protection criteria of thermally insulated pipeline buried in soil. *Corros. Sci.* **2001**, *43*, 2011–2021. [CrossRef]
8. Shreir, L. *Design and Operational Guidance on Cathodic Protection of Offshore Structures, Subsea Installations and Pipelines*; The Marine technology directorate limited publication: London, UK, 1990.
9. Von Baeckmann, W.; Schwenk, W.; Prinz, W. *Handbook of Cathodic Corrosion Protection*; Elsevier: Huston, TX, USA, 1997.
10. Gummow, R. Cathodic protection criteria—A critical review of NACE standard RP-01-69. *Mater. Perform.* **1986**, *25*, 9–16.
11. *Effectiveness of Cathodic Protection on Thermally Insulated Underground Metallic Surfaces*; NACE Technical Committee Report No. 24156; NACE Publication: Huston, TX, USA, 1992.
12. Kim, J.-G.; Kim, Y.-W.; Kang, M.-C. Corrosion Characteristics of Rigid Polyurethane Thermally Insulated Pipeline with Insulation Defects. *Corrosion* **2002**, *58*, 175–181. [CrossRef]
13. Parsa, M.; Allahkaram, S.; Ghobadi, A. Simulation of cathodic protection potential distributions on oil well casings. *J. Pet. Sci. Eng.* **2010**, *72*, 215–219. [CrossRef]
14. DeGiorgi, V.G.; Wimmer, S.A. Geometric details and modeling accuracy requirements for shipboard impressed current cathodic protection system modeling. *Eng. Anal. Bound. Elem.* **2005**, *29*, 15–28. [CrossRef]
15. Hong, M.-S.; Hwang, J.-H.; Kim, J.H. Optimization of the Cathodic Protection Design in Consideration of the Temperature Variation for Offshore Structures. *Corrosion* **2017**, *74*, 123–133. [CrossRef]
16. Roberge, P.R. *Handbook of Corrosion Engineering*; McGraw-Hill: New York, NY, USA, 2000.
17. Kim, Y.-S.; Lee, S.; Kim, J.-G. Influence of anode location and quantity for the reduction of underwater electric fields under cathodic protection. *Ocean Eng.* **2018**, *163*, 476–482. [CrossRef]
18. Fessler, R.R.; Markworth, A.J.; Parkins, R.N. Cathodic Protection Levels under Disbonded Coatings. *Corrosion* **1983**, *39*, 20–25. [CrossRef]
19. Veritas, D.N. *Cathodic Protection Design, Recommended Practice DNV-RP-B401*; DNV: Oslo, Norway, 2010.
20. Pedeferri, P. Cathodic protection and cathodic prevention. *Constr. Build. Mater.* **1996**, *10*, 391–402. [CrossRef]

Article

Microstructure and Corrosion Resistance of Zn-Al Diffusion Layer on 45 Steel Aided by Mechanical Energy

Jianbin Tong [1], Yi Liang [2,*], Shicheng Wei [2], Hongyi Su [2], Bo Wang [2], Yuzhong Ren [3], Yunlong Zhou [1] and Zhongqi Sheng [1]

1 College of Mechanical Engineering and Automation, Northeastern University, Shenyang 110819, China; evantjb@163.com (J.T.); zhouyunlong4317@163.com (Y.Z.); zhqsheng@mail.neu.edu.cn (Z.S.)

2 National Key Laboratory for Remanufacturing, Academy of Army Armored Forces, Beijing 100072, China; wsc33333@163.com (S.W.); zgysuhongyi993@163.com (H.S.); wangbobo421@163.com (B.W.)

3 Chongqing Dayou Surface Technology Co., Ltd., Chongqing 400020, China; renyuzhong666@126.com

* Correspondence: liangyi365@126.com; Tel.: +86-010-6671-8541

Received: 30 August 2019; Accepted: 17 September 2019; Published: 18 September 2019

Abstract: In harsh environments, the corrosion damage of steel structures and equipment is a serious threat to the operational safety of service. In this paper, a Zn-Al diffusion layer was fabricated on 45 steel by the Mechanical Energy Aided Diffusion Method (MEADM) at 450 °C. The microstructure and composition, the surface topography, and the electrochemical performance of the Zn-Al diffusion layer were analyzed before and after corrosion. The results show that the Zn-Al diffusion layer are composed of Al_2O_3 and Γ_1 phase (Fe11Zn40) and δ_1 phase ($FeZn_{6.67}$, $FeZn_{8.87}$, and $FeZn_{10.98}$) Zn-Fe alloy. There is a transition zone with the thickness of about 5 μm at the interface between the Zn-Al diffusion layer and the substrate, and a carbon-rich layer exists in this zone. The full immersion test and electrochemical test show that the compact corrosion products produced by the initial corrosion of the Zn-Al diffusion layer will firmly bond to the Zn-Al diffusion layer surface and fill the crack, which plays a role in preventing corrosion of the corrosive medium and reducing the corrosion rate of the Zn-Al diffusion layer. The salt spray test reveals that the initial corrosion products of the Zn-Al diffusion layer are mainly ZnO and $Zn_5(OH)_8Cl_2H_2O$. New corrosion products such as $ZnAl_2O_4$, FeOCl appear at the middle corrosion stage. The corrosion product $ZnAl_2O_4$ disappears, and the corrosion products $Zn(OH)_2$ and $Al(OH)_3$ appear at the later corrosion stage.

Keywords: Zn-Al diffusion layer; mechanical energy aided diffusion; microstructure; corrosion resistance; electrochemistry

1. Introduction

Metal materials have long been exposed to environments with high temperatures, high humidity, high salt spray, and intense sunlight, making their corrosion levels several times or even dozens of times higher than that in other environments at the same time [1–10]. In the process of metal corrosion, the mechanical properties and internal microstructure of the metal have changed. The corrosion harm includes not only the damage of internal metal structures but also the destruction of overall metal structures. Metal corrosion causes tremendous economic loss and becomes a severe threat to the development of various fields. Therefore, corrosion protection of metal materials is essential in many industrial applications [11–14]. In various environments, steel is the most used type of material in various facilities and equipment. Therefore, the corrosion protection of steel structures and equipment have flourished over recent decades [15–20].

Chemical heat treatment is usually used to improve the corrosion resistance, high-temperature oxidation resistance and hardness of metal parts [21–24]. However, due to the high temperature and long production time, it not only consumes a lot of energy, but also affects the mechanical properties of parts. Therefore, the Mechanical Energy Aided Diffusion Method (MEADM) that emerged in recent years has become an attractive metal materials anti-corrosion technology in the field of material surface strengthening. The MEADM, developed by the addition of mechanical energy (rotation, vibration, friction, etc.) in the traditional solid pack-cementation process, achieves the purpose of reducing the preparation temperature required and shortening the preparation time. As such, the MEADM develops rapidly. There are three steps in the MEADM as shown in Figure 1: (a) with the assistance of mechanical energy, the active powder particles in the diffusion agent rub and impact the surface of the heated substrate, resulting in micro vacancies and plastic deformation on the substrate surface; (b) the active particles enter the generated vacancies or adsorb into the deformed surface to form a surface solid solution or an intermetallic compound, forming an initial diffusion layer; (c) a dense protective diffusion layer is achieved by adsorbing the active particles in a continuous fashion.

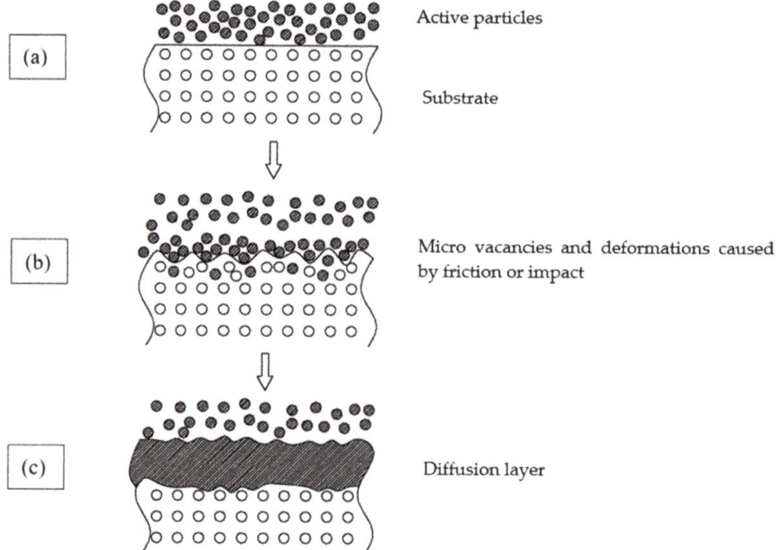

Figure 1. Schematic illustration of the mechanical energy aided diffusion processes.

Yuan et al. [25] have employed a pack-cementation process to produce aluminide coatings on both ferritic-martensitic (RAFM) and austenitic (316L) alloys at the temperature of 600–800 °C. Lee [26] has applied a pack-cementation process to form aluminide coatings at 850 °C for 2 h. However, Wang et al. [27,28] studied the MEADM and found that mechanical energy can improve the adhesion ability of Al powder on the surface of the substrate, increase the chemical activity and adhesion strength of Al, and thereby significantly reduce the aluminizing temperature, which can be reduced to 500 °C [27]. The MEADM attributes to the temperature reduction for the following reasons: (a) the powder, the workpiece, and the vessel wall collide with each other in the mechanical energy aided diffusion process to form flow and heat transfer, enhance the heat conduction and thereby accelerating the diffusion rate; (b) the moving powder particles hit the surface of the workpiece to cause activation and misalignment of the surface lattice atoms, forming supersaturated vacancies and thereby reducing the diffusion temperature; (c) the movement of the powder particles accelerates the chemical reaction of the diffusion and increases the concentration of the active Al atoms; (d) the movement of the powder particles purifies the surface of the workpiece.

Zhang et al. [29] have used the MEADM to prepare the Zn diffusion layer on Q235 steel, and found that the diffusion layer, composed of $FeZn_{15}$, $FeZn_{11}$, $FeZn_9$, and $FeZn_7$ phases, has excellent resistance to high temperature oxidation and corrosion. He et al. [30] obtained Al-Zn-Cr diffusion layer on 20 steel at 600 °C by the MEADM and concluded that the Al-Zn-Cr diffusion layer, a multi-layered structure, is resistant to high temperature oxidation and corrosion. In addition to excellent corrosion resistance and high temperature oxidation resistance, the Zn-Al diffusion layer (based on the MEADM) has the characteristics of uniform thickness, high hardness, good scratch resistance, high powder recyclability, and low environmental pollution. At present, Zn layers, Al layers, and Zn-Al layers are studied widely [27,28,31–33].

In this paper, in order to analyze the severe corrosion problem of steel structure and equipment in harsh corrosive environments, 45 steel was used as the substrate material, and the Zn-Al diffusion layer was prepared by the MEADM at 450 °C without changing the properties of the substrate. In this paper, we investigated the microstructures and element distribution of the Zn-Al diffusion layer. The 3D topography and corrosion mechanism of the Zn-Al diffusion layer surface were studied by full immersion test with 3.5 wt. % NaCl. The surface topography and corrosion product composition of the Zn-Al diffusion layer in different salt spray corrosion periods were analyzed. The results can provide a technical basis for long-term corrosion protection of steel structures and equipment in harsh environments.

2. Experimental

2.1. Experimental Materials and Sample Preparation Process

The 45 steel was used as the substrate material, and its specific components are shown in Table 1. The sample size was 10(L) × 10(W) × 4(H) mm.

The preparation process of the Zn-Al diffusion layer is mainly divided into three parts: pre-treatment, mechanical energy aided diffusion process, and post-treatment. The oil and rust on the surface of the substrate were removed by sanding, ultrasonic cleaning, and shot blasting (0.2 mm steel shot) during the pre-treatment. The mechanical energy aided diffusion processes are as follows:

(1) Stirring: The diffusion agent was weighed with a certain pre-calculated percentage (0.02% NH_4Cl, 0.05% Rare Earth, 49.93% Al_2O_3, 35% Zn, and 15% Al) and mixed uniformly with a mixer.

(2) Loading furnace: A part of the mixed agent was taken out into the rotary furnace (a special mechanical energy aided diffusion device), the workpiece was put into the furnace, then the remaining agent was put into the furnace, and the furnace lid was tightened finally.

(3) Heating and holding: The furnace was heated in the electric furnace and rotated at a constant rotational speed of 7 r/min. The timing started when the temperature reaches 450 °C. The heating was stopped when the set holding time (4 h) was reached.

(4) Cooling: The furnace was kept rotating at a constant speed (can be increased to 10 r/min). The furnace was cooled naturally to room temperature in the air.

(5) Separation: The workpiece and the powder in the furnace were separated by filtration. The post-treatment included cleaning, alcohol wiping, drying, and testing.

Table 1. Elemental composition of 45 steel (wt. %).

Element	C	Mn	Si	Cu	P	S	Cr	Ni	Fe
wt. %	0.42–0.50	0.5–0.8	0.17–0.37	≤0.25	≤0.035	≤0.035	≤0.25	≤0.3	rest

2.2. Experimental Equipment and Parameters

(1) The cross-sectional topography, composition and element distribution of the Zn-Al diffusion layer before and after corrosion were analyzed by Scanning Electron Microscope (SEM) (Nova Nano SEM50, FEI, Hillsboro, OR, US).

(2) The phase composition of the Zn-Al diffusion layer before and after corrosion was analyzed by X-ray Diffractometer (XRD) (Smartlab, Rigaku, Tokyo, Japan). The specific test conditions were Cu

target (9 KW), accelerating voltage 40 kV, tube current 40 mA, scanning rate 5°/min, test angle 10°–90°, and scanning step length 0.02°.

(3) The 3D surface topography of the sample in different immersion stages was measured by 3D Laser Scanning Microscope (LEXT OLS4100, OLYMPUS, Tokyo, Japan).

(4) The samples of different corrosion stages were obtained on Salt Spray Testing Chamber (YWX-010, SHUANGKE, Beijing, China). The specific test conditions were that the corrosive medium was 5 wt. % NaCl solution, the temperature was 35 °C, and the spray method was continuous.

(5) Electrochemical impedance spectroscopy (EIS) (amplitude 5 mV, scanning frequency range 10 mHz–100 kHz) and potentiodynamic polarization (scanning speed 1 mV/s, scanning range ±250 mV of electrode potential) were tested on an Electrochemical Workstation (IM6, Zahner, Kronach, Germany).

3. Results and Discussion

3.1. Cross-Sectional Topography and Composition Analysis of the Zn-Al Diffusion Layer

Figure 2 shows the SEM images and corresponding Energy Dispersive Spectroscopy (EDS) spectra of the Zn-Al diffusion layer in cross section. Figure 2b is an enlarged image of the rectangular area in Figure 2a,c,d are EDS spectra of the Zn-Al diffusion layer. It can be found that the main elements of the diffusion layer are Zn, Al, and Fe. After the Al powder is diffused by the MEADM at 450 °C, a high Al content is only detected near the diffusion layer surface, which indicates that the Al layer with a thickness of 2–4 μm is mainly present in the superficial layer. In the vertical direction of the Zn-Al diffusion layer, the content of Zn changes little, and the overall content shows a slight downward trend, whereas the content of Fe increases slowly.

On observation of Figure 2b, there is a transition zone with a thickness of 5 μm at the interface between the Zn-Al diffusion layer and the substrate. The content of Zn and Fe in the transition zone changes abruptly. The content of Zn reduces to almost 0. The Fe element content remains constant after a sharp increase. The transition zone is divided into two parts. The first part (Part A) is close to the Zn-Al diffusion layer and has a thickness of 3 μm. The other part (Part B) is close to the substrate and has a thickness of 2 μm. This part is formed by the diffusion of Zn into the substrate. A small number of pores (average diameter 0.6 μm) are found at the boundary of the transition zone by the SEM image. These pores are caused by a small amount of entrained air that cannot be discharged in time when the diffusion element permeates into the substrate. This is attributed to the fact that the ambient preparation temperature is not stable and the substrate temperature is low at the initial stage of the Zn-Al diffusion layer growth.

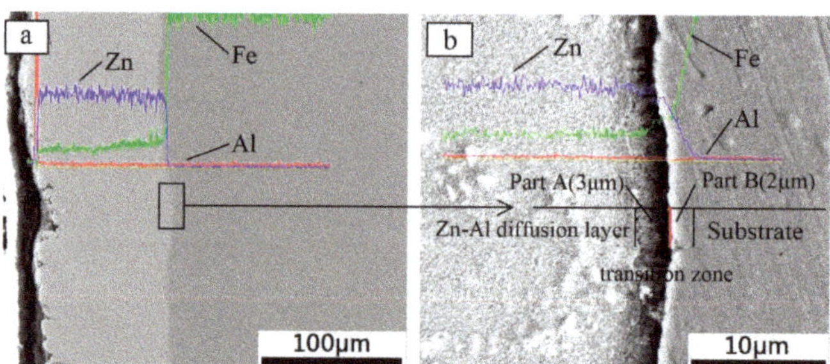

Figure 2. *Cont.*

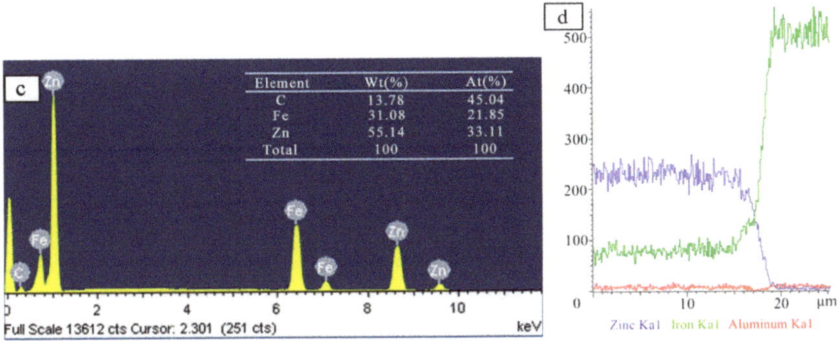

Figure 2. Scanning Electron Microscope (SEM) images and corresponding Energy Dispersive Spectroscopy (EDS) spectra of the Zn-Al diffusion layer in cross section: (**a**) sectional topography of the Zn-Al diffusion layer; (**b**) enlarged image of the rectangular area; (**c**,**d**) EDS spectra of the Zn-Al diffusion layer.

Figure 3 shows the XRD patterns of the Zn-Al diffusion layer. It is found that Al_2O_3 and Zn-Fe alloys of Γ_1 phase ($Fe_{11}Zn_{40}$) and δ_1 phase ($FeZn_{6.67}$, $FeZn8_{.87}$, $FeZn_{10.98}$) are mainly formed in the Zn-Al diffusion layer. Combined with SEM images and EDS spectra, the element ratios at positions 1, 2, 3, and 4 in Figure 4, are shown in Table 2. The Zn-Fe content ratios (W_{Zn}:W_{Fe}) are 10, 8.2, 8.4, and 6.4, respectively. According to XRD patterns, the Zn-Fe alloy near the surface of the Zn-Al diffusion layer is mainly $FeZn_{10.98}$, the Zn-Fe alloy near the boundary between the diffusion layer and the substrate is mainly $FeZn_{6.67}$, and the Zn-Fe alloy in the middle area of the diffusion layer is mainly $FeZn_{8.87}$.

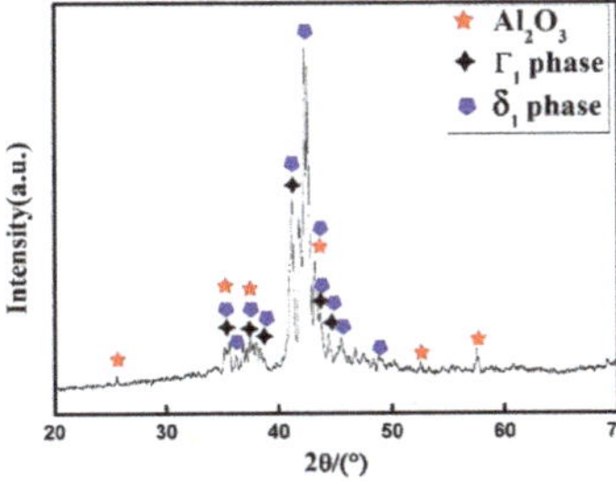

Figure 3. XRD patterns of the Zn-Al diffusion layer.

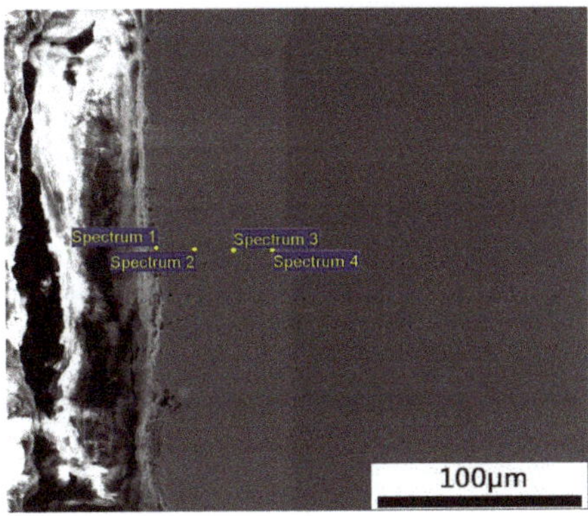

Figure 4. EDS spot scanning position distribution of the Zn-Al diffusion layer in cross section.

Table 2. Percentage of constituent elements of the Zn-Al diffusion layer in different positions.

Element	1	2	3	4
O	7.5	7.17	6.08	5.52
Zn	83.11	81.65	83.32	80.98
Al	1.07	1.26	0.66	0.88
Fe	8.32	9.92	9.94	12.62

The content and distribution of elements in the region near the interface of the Zn-Al layer were analyzed. The elements and ratios at the positions numbered by 1, 2, 3, 4, and 5 in Figure 5 are shown in Table 3. The Zn-Fe content ratios (W_{Zn}:W_{Fe}) in the positions numbered by 1, 2, 3, 4, and 5 are 6.8, 6.2, 5, 5.87, and 3, respectively, further indicating that the Zn-Fe alloy near the substrate is $FeZn_{6.67}$. Analyses of SEM images and EDS spectra reveal that the closer to the substrate the Zn-Al diffusion layer is, the higher the C content is. At the positions numbered by 3, 4, and 5, the C content reaches 51.85%, 47.12%, and 47.78%, respectively, indicating that the region of the Zn-Al diffusion layer forms a carbon-rich layer near the substrate. The C content of the carbon-rich layer is much higher than that of the 45 steel substrate. This is because Zn and Al are non-carbide forming elements, which cause the C atom crowding-out effect during the formation of the diffusion layer [34]. Therefore, the diffusion layer will have a carbon-rich layer near the substrate.

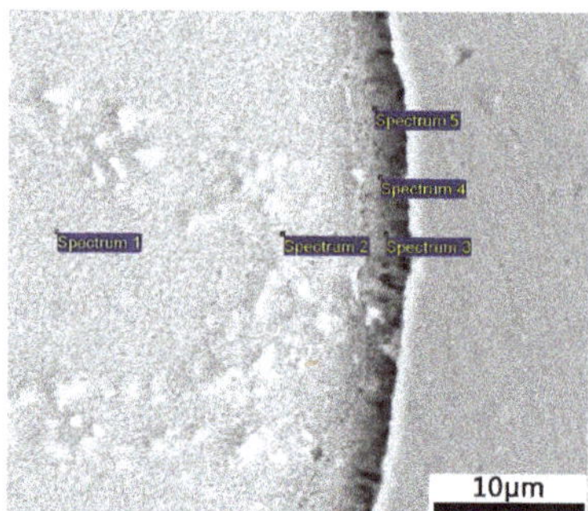

Figure 5. Position distribution of EDS spot scanning in the region near the interface of the Zn-Al layer.

Table 3. Percentage of constituent elements in the region near the interface of the Zn-Al layer.

Element	1	2	3	4	5
C	—	—	51.85	47.12	47.78
Fe	12.80	13.86	7.94	7.70	13.12
Zn	87.20	86.14	40.22	45.18	39.10

3.2. Corrosion Resistance

3.2.1. 3D Surface Topography Analysis of the Zn-Al Diffusion Layer during Full Immersion

In order to study the corrosion resistance of the Zn-Al diffusion layer, the samples were immersed fully in 3.5 wt. % NaCl solution, and the 3D topography of the samples in different immersion stages was characterized at room temperature. The topographical images shown in Figure 6 are composed of a real graph and a color graph, and the color graph is the color mark of the height of the Zn-Al diffusion layer in different regions of the real one.

Observing the surface topography, it is found that the surface of the Zn-Al diffusion layer is uniform and has a low roughness before the full immersion test, and the drop between the high and low points is within 54 μm. The reason is that the substrate was subjected to shot blasting during pre-treatment, which caused a certain roughness on the surface of the substrate. Therefore, the Zn-Al diffusion layer is uniformly distributed on the surface of the substrate, which causes a certain degree of surface drop.

After 240 h full immersion, the surface color of the sample changed. Compared with the surface before the full immersion, more pronounced peak and pit features are exhibited, and the maximum drop between the high and low points increase to 74 μm. After full immersion, some corrosion products accumulate on the Zn-Al diffusion layer surface and form some corrosion pits, which aggravates the surface roughness. With the immersion time extended to 600 h, a large number of white corrosion products appear on the Zn-Al diffusion layer surface. As the accumulated corrosion products further increase, the area and depth of corrosion pits further increase, and the maximum drop between the high and low points reaches 110 μm. When the sample is immersed for 1000 h, the corrosion products on the Zn-Al diffusion layer surface further increase, and the surface corrosion pits are obvious. The maximum drop on the surface is increased to 131 μm and some red rust spots are observed in the real image. At this moment, a small number of corrosion pits have penetrated the entire Zn-Al diffusion layer to the substrate.

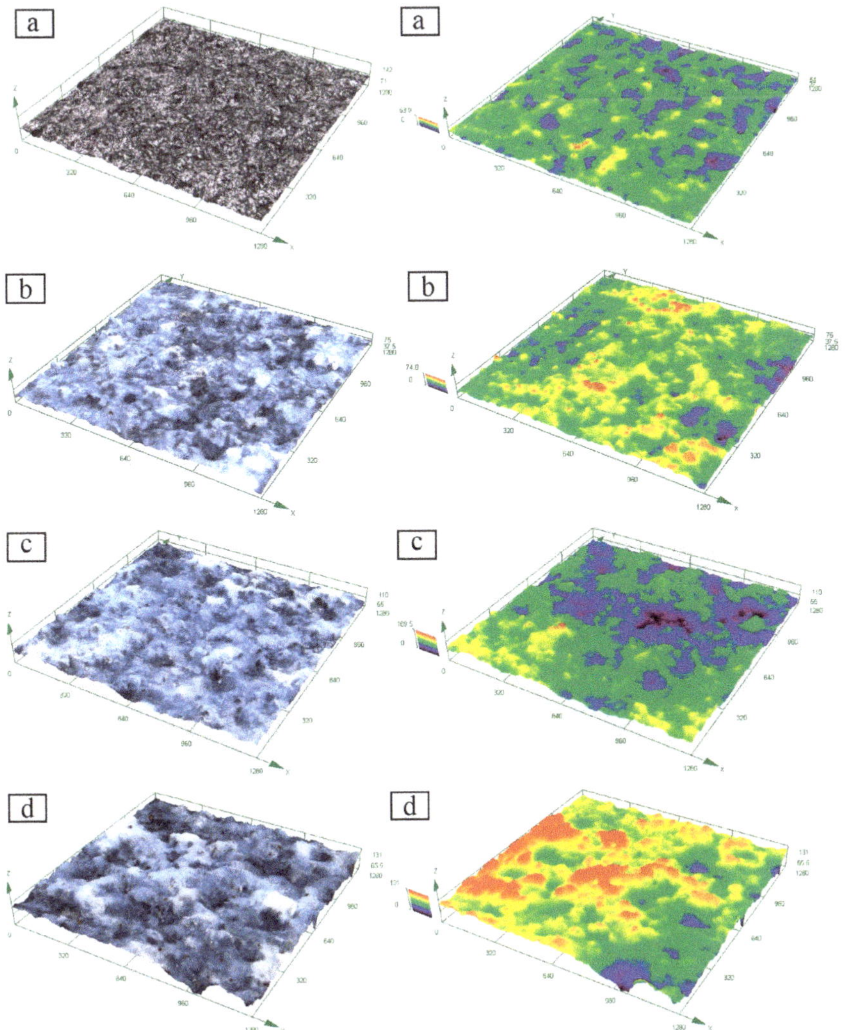

Figure 6. 3D topography of the Zn-Al diffusion layer during full immersion: (**a**) before immersion; (**b**) immersed for 240 h; (**c**) immersed for 600 h; (**d**) immersed for 1000 h.

Figure 7 depicts the microscopic corrosion topography and EDS spectra of the Zn-Al diffusion layer after immersing for 360 h and ultrasonic cleaning for 10 min. The corrosion topography show that the metal powder on the Zn-Al diffusion layer surface is actively dissolved, and the flocculent corrosion products deposit on the surface, covering the entire surface. The energy spectrum analysis of the filler in the surface crack of the diffusion layer is shown in Table 4. In addition to the elements of O, Al, Fe, and Zn, the Cl element which is the main element causing corrosion is detected. It indicates that the filler in the surface crack is corrosion products produced by Cl^- corroding in solution. After ultrasonic cleaning, the corrosion products are still present in the crack, indicating that the corrosion products are firmly bonded to the Zn-Al diffusion layer. The firmly combined corrosion products fill the crack to help block the intrusion tunnel of the corrosive medium, which can slow down the corrosion rate of the Zn-Al diffusion layer and improve the protection ability for the substrate [35,36].

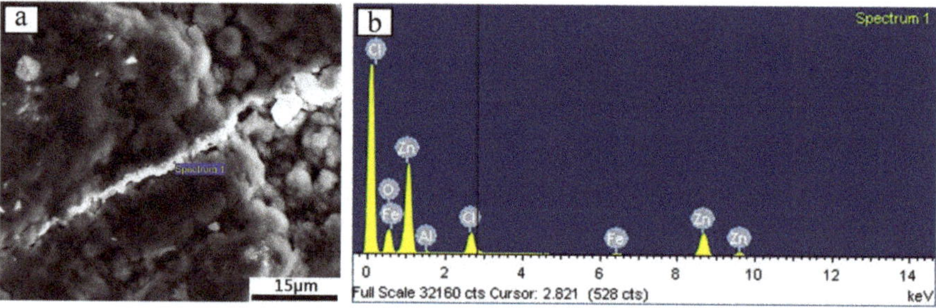

Figure 7. (**a**) Microstructure of the Zn-Al layer for 360 h immersion; (**b**) EDS spectra of the Zn-Al layer for 360 h immersion.

Table 4. Main elemental weight ratio and atomic ratio of the Zn-Al diffusion layer for 360 h immersion.

Element	Weight Ratio (%)	Atomic Ratio (%)
O	29.15	59.85
Al	0.67	0.81
Cl	9.29	8.60
Fe	1.55	0.91
Zn	59.35	29.82

3.2.2. Corrosion Behavior Analysis of the Zn-Al Diffusion Layer

In order to further study the corrosion resistance of the Zn-Al diffusion layer, a neutral salt spray test was developed. The sample was placed in a salt spray test chamber. The surface corrosion topography and the corrosion product changes of the Zn-Al diffusion layer in different corrosion stages were studied.

The surface corrosion topography and corrosion product XRD results of the Zn-Al diffusion layer in different salt spray corrosion stages are shown in Figures 8 and 9. At the initial stage of salt spray corrosion (within 168 h), a layer of flocculent corrosion products uniformly forms on the Zn-Al diffusion layer surface. The corrosion products cover the surface, so that the tunnels (the corrosion solution can invade the substrate through these tunnels) are reduced, thereby the corrosion resistance of the Zn-Al diffusion layer is improved. XRD results show that the corrosion products on the Zn-Al diffusion layer surface are mainly composed of ZnO, Al_2O_3, and $Zn_5(OH)_8Cl_2H_2O$. From the corrosion products formed, with the electrochemical reaction proceeding, Na^+ moves toward the cathodic region, and Cl^- moves toward the anodic region. Zinc hydroxychloride ($Zn_5(OH)_8Cl_2H_2O$) and Zinc oxide (ZnO) gradually form in the anodic dissolution region.

When corroding to the middle stage of corrosion (480 h), the flocculent corrosion products on the surface have become the needle-like corrosion products that are shown by a network-like structure on the surface. According to the XRD results, it is obvious that the main corrosion products are comprised by ZnO, Al_2O_3, $Zn_5(OH)_8Cl_2H_2O$, $ZnAl_2O_4$, and FeOCl. Compared with the initial corrosion stage, the number of $Zn_5(OH)_8Cl_2H_2O$ on the surface increases, and the density of corrosion product layer increases, which helps slow down the corrosion from corrosive medium and reduce the corrosion rate of the Zn-Al diffusion layer. In addition, the newly formed corrosion product, iron oxychloride (FeOCl), is a structurally unstable corrosion intermediate that can accelerate corrosion. However, when FeOCl releases Cl^-, it can form FeO(OH), and the migrated OH^- can react with the metallic ions in the corrosive medium to develop new products. These products cover the Zn-Al diffusion layer surface, which further suppress the corrosion of the Zn-Al diffusion layer to some extent.

In the later stage of corrosion (1000 h), more agglomerated products appear on the surface of the diffusion layer. Compared with the initial stage and the middle stage of corrosion, a small number of flocculent and needle-like corrosion products distribute on the surface. The agglomerated corrosion product layer is easy to fall off. Therefore, the corrosion solution easily passes through the pores between the corrosion products and penetrates the diffusion layer. At this time, the corrosion resistance of the Zn-Al diffusion layer is weakened. XRD analyses show that $ZnAl_2O_4$ disappear on the surface and $Zn(OH)_2$ and $Al(OH)_3$ appear in comparison with the middle corrosion stage. The disappearance of $ZnAl_2O_4$ is due to the decrease of Al content in the diffusion layer and $ZnAl_2O_4$ formed and the shedding of $ZnAl_2O_4$ of the surface with the prolongation of corrosion time.

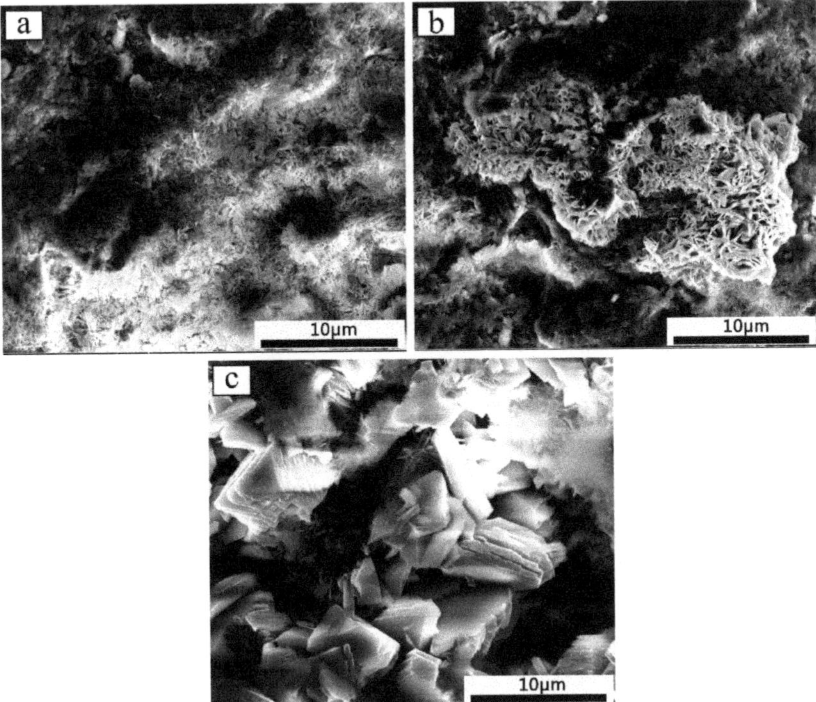

Figure 8. Surface topography of the Zn-Al diffusion layer during salt spray corrosion: (**a**) 168 h; (**b**) 480 h; (**c**) 1000 h.

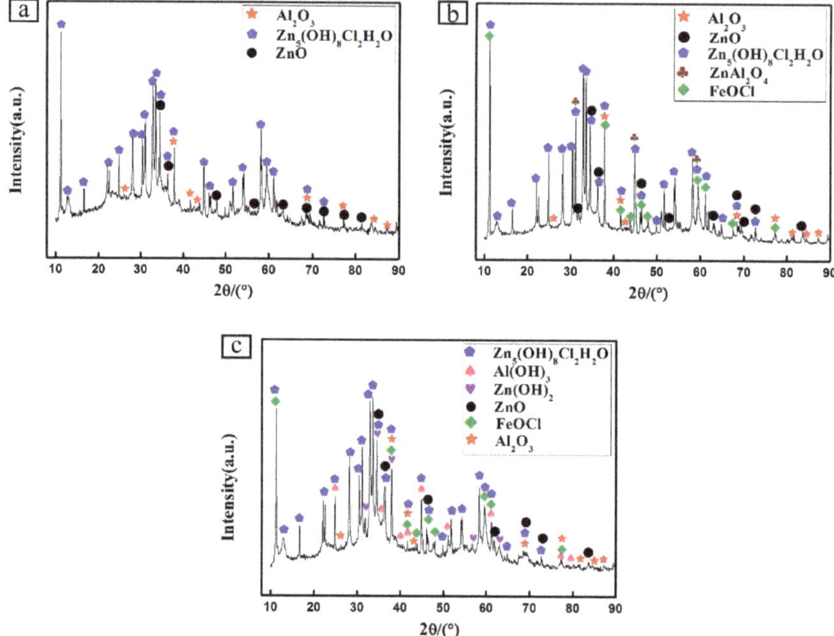

Figure 9. XRD patterns of the Zn-Al diffusion layer corrosion products during salt spray corrosion: (**a**) 168 h; (**b**) 480 h; (**c**) 1000 h.

3.3. Electrochemical Performance Analysis of the Zn-Al Diffusion Layer

Figure 10 depicts the potentiodynamic polarization curves of the Zn-Al diffusion layer in 3.5 wt. % NaCl solution for different immersion times, and Table 5 shows the corresponding polarization curve fitting data. Through the polarization curve analysis, it is found that as the immersion time is prolonged, the self-corrosion potential of the Zn-Al diffusion layer increases significantly, but the self-corrosion current density decreases by an order of magnitude. In other words, the corrosion rate of the Zn-Al diffusion layer decreases with the prolongation of immersion time in a certain time range.

In the whole process of full immersion corrosion, the anode Tafel slope βa of the Zn-Al diffusion layer is less than the cathode Tafel slope βc, indicating that the corrosion reaction of the Zn-Al diffusion layer is mainly controlled by the cathodic reaction. The specific reaction is as follows: the anodic reaction $Zn - 2e^- \rightarrow Zn^{2+}$, $Al - 3e^- \rightarrow Al^{3+}$; the cathodic reaction $O_2 + 2H_2O + 4e^- \rightarrow 4OH^-$; the total reaction $2Zn + O_2 + 2H_2O \rightarrow 2Zn(OH)_2$, $4Al + 3O_2 + 6H_2O \rightarrow 4Al(OH)_3$ [37]. It is found that the cathode Tafel slopes βc and the anode Tafel slopes βa are not much different during the initial immersion stage. At this time, the anodic reaction is that Zn and Al in the Zn-Al diffusion layer dissolve to produce Zn^{2+} and Al^{3+} in the corrosive medium, and the cathode absorbs oxygen to form OH^-. With the prolongation of immersion time, the anode Tafel slope βa remains unchanged, but the cathode Tafel slope βc increases gradually. It indicates that the cathodic oxygen-absorbing reaction (the formation of the corrosion products layer) controls the corrosion reaction of the Zn-Al diffusion layer with the βc value increases. The accumulating rate of electrons in the cathode region accelerates, resulting in a decrease of the self-corrosion potential difference and the corrosion current density between anode and cathode. In addition, the polarization resistance Rp also increases greatly with the prolongation of immersion time, indicating that the corrosion products formed on the Zn-Al diffusion layer surface accumulate gradually and the corrosion rate of the Zn-Al diffusion layer decreases. The compact corrosion products formed adhere to the surface, which acts as a protective layer and slows the corrosion rate. The results of the polarization potential test are also consistent with the surface analysis results of the full immersion test.

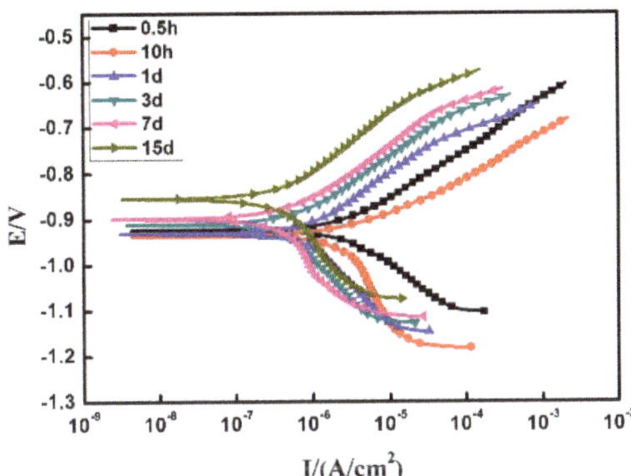

Figure 10. Potentiodynamic polarization curve of the Zn-Al diffusion layer.

Table 5. Polarization curve fitting data of the Zn-Al diffusion layer.

Time	E (V)	Icorr (A/cm^2)	βa (mV)	βc (mV)	Rp (Ω/cm^2)
0.5 h	−0.9440	2.77×10^{-6}	113.41	130.35	9124.1
10 h	−0.9369	2.84×10^{-6}	79.651	366.16	9779.4
1 d	−0.9310	2.70×10^{-6}	162.71	326.42	15,522
3 d	−0.9110	1.03×10^{-6}	148.74	453.58	46,121
7 d	−0.8990	8.09×10^{-7}	136.29	607.70	60,865
15 d	−0.8563	8.89×10^{-7}	152.31	547.2	68,548

Figure 11 presents the EIS of the Zn-Al diffusion layer in different immersion time, and Table 6 shows the impedance modulus in different immersion times. The impedance modulus diagram (Figure 11a) shows that the impedance modulus of the low-frequency region decreases within 24 h of immersion. When the immersion time is 72–360 h, the impedance modulus of the low-frequency region increases sharply, indicating that the corrosion rate of the Zn-Al diffusion layer decreases at this time. The phase angle diagram (Figure 11b) shows that only a time constant characteristic appears in the Zn-Al diffusion layer, and the peaks in the phase angle diagram gradually become higher as the immersion time is prolonged. It is concluded that the prolongation of immersion time (more than 24 h) makes the Zn-Al diffusion layer to be an isolating layer with high resistance and low capacitance, which plays a protective role for the Zn-Al diffusion layer. In the Nyquist diagram (Figure 11c), there is only one capacitive reactance arc in the Zn-Al diffusion layer, and the radius of the capacitive reactance arc decreases firstly and then increases with the prolongation of the immersion time, which is consistent with the change law of the corrosion resistance of the Zn-Al diffusion layer in the impedance modulus diagram and phase angle diagram.

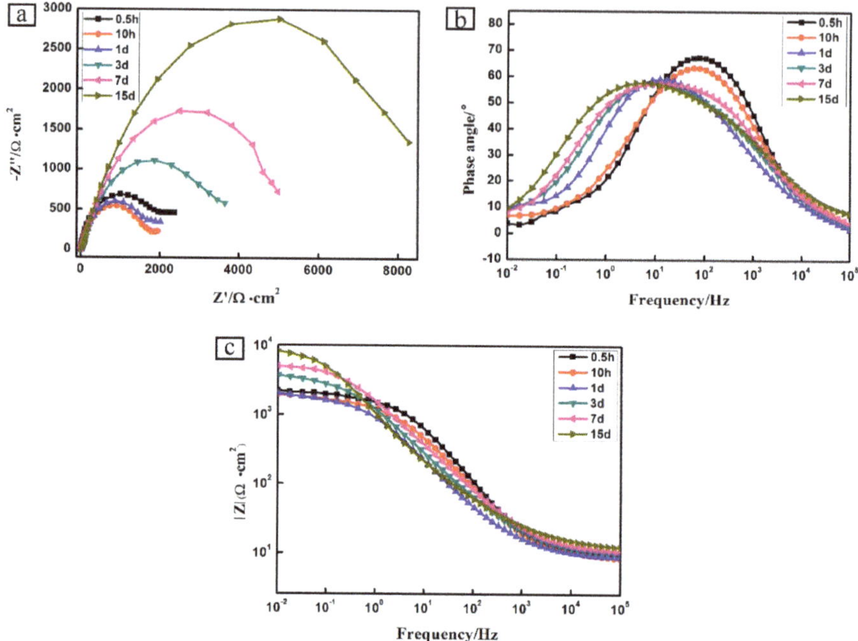

Figure 11. EIS spectra plots of different immersion time of the Zn-Al diffusion layer: (**a**) Bode impedance modulus; (**b**) Bode phase angle; (**c**) Nyquist.

Table 6. The impedance modulus and corrosion rate of different immersion time of the Zn-Al diffusion layer.

Time (h)	0.5	10	24	72	168	360
Impedance modulus ($\Omega\cdot cm^2$)	2378.6	1954.2	2024.8	3649.6	4976.9	8267.7
Corrosion rate ($A/(cm^2\cdot h)$)	2.74×10^{-6}	2.76×10^{-6}	2.89×10^{-6}	1.01×10^{-6}	7.55×10^{-7}	7.18×10^{-7}

The corrosion of the Zn-Al diffusion layer is a controlled process in which the electrochemical reaction gradually changes to the diffusion of corrosive medium or corrosion products. During the initial immersion stage, the impedance modulus of the low-frequency region in the 3.5 wt. % NaCl solution is low. The reactions on the Zn-Al diffusion layer surface are mainly zinc oxide, aluminum oxide, and zinc aluminum active dissolution. When the immersion time is 10 and 24 h, the impedance modulus of the low-frequency region decreases. The prolongation of the immersion time makes the electrolyte solution continuously penetrate the Zn-Al diffusion layer, causing continuous corrosion damage of the Zn-Al diffusion layer and the decreases of the impedance modulus. Therefore, as the immersion time is prolonged, the impedance modulus of the low-frequency region decreases. However, as the immersion time continues to increase, the corrosion products continuously deposit on the surface and fill into the crack. The tunnel from the corrosive medium to the substrate reduces, slowing the penetration rate of the corrosive electrolyte. Therefore, the impedance modulus of the Zn-Al diffusion layer rises rapidly in the 3.5 wt. % NaCl solution, and the corrosion rate decreases remarkably.

Two equivalent electrical circuits shown in Figure 12 were utilized to fit the EIS data and account for the corrosion behavior of the Zn-Al diffusion layer. The first equivalent circuit (Rs(Qc(Rc(QctRct)))) was used to fit the EIS data displaying the Zn-Al diffusion layer within 24 h of immersion, whereas the second one (Rs(QcRc)(QctRct)) was used for the EIS data displaying the impedance after immersing for 24 h. In Figure 12, Rs is the solution resistance, Rc is the Zn-Al diffusion layer resistance, Rct is the charge transfer resistance, Qc is the Zn-Al diffusion layer capacitance, and Qct is the electric double layer equivalent capacitance between the Zn-Al diffusion layer and the substrate.

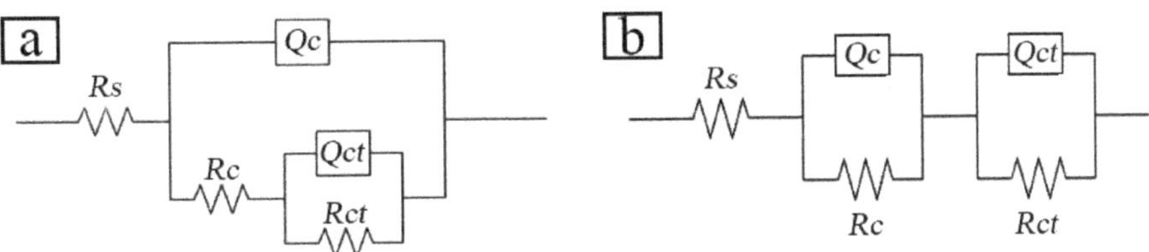

Figure 12. EIS equivalent circuit models of the Zn-Al diffusion layer in different immersion times: (**a**) for fitting the diffusion layer impedance data within 24 h of immersion(Rs(Qc(Rc(Qct Rct)))); (**b**) for fitting the diffusion layer impedance data after immersing for 24 h (Rs(QcRc)(Qct Rct)).

The equivalent circuit fitting data of the Zn-Al diffusion layer is shown in Table 7. When the immersion time is 0–360 h, the Zn-Al diffusion layer capacitance value Qc in the corrosion solution increases gradually. At this stage, the Zn-Al diffusion layer is invaded by a corrosive medium, increasing the value of Qc. The microscopic fluctuation of the Zn-Al diffusion layer surface leads to a large or small deviation of the electric double layer capacitance Qct, so that the Qct value changes irregularly, that is, the dispersion phenomenon. When immersed for 0–24 h, the Rc value decreases gradually, and the Rct value decreases slightly, which is consistent with the radius variation of the capacitive reactance arc in the Nyquist diagram. Analyses of the diagram data reveal that the diffusion layer has a much larger resistance Rc than the solution resistance Rs. In other words, the penetration of the corrosion solution makes the Zn-Al diffusion layer resistance Rc decrease. As the immersion time continues to increase, the Rc and Rct values increase gradually. The Zn-Al diffusion layer reacts with the corrosive solution, which generates a large number of corrosion products on the surface or inside and accumulates continuously. The phenomenon of the above reaction reduces the porosity of the Zn-Al diffusion layer, and strengthens the self-sealing effect, so that it slows down the progress of the electrochemical reaction.

Table 7. Fitting data of the equivalent circuit of the Zn-Al diffusion layer.

Time (h)	0.5	10	24	72	168	360
Rs ($\Omega \cdot cm^2$)	9.738	8.945	8.715	9.728	10.2	10.73
Qc (10^{-4} $F \cdot cm^{-2}$)	0.463	0.786	1.913	3.691	6.799	12.12
Rc ($k\Omega \cdot cm^2$)	1.582	1.553	1.338	2.431	2.156	2.901
Qct (10^{-4} $F \cdot cm^{-2}$)	2.166	4.582	2.934	4.332	2.165	13.53
Rct ($k\Omega \cdot cm^2$)	1.929	1.637	1.799	3.652	5.542	7.239

4. Conclusions

(1) The Zn-Al diffusion layer aided by mechanical energy is composed of mainly Al_2O_3 and Γ_1 phase ($Fe_{11}Zn_{40}$) and δ_1 phase ($FeZn_{6.67}$, $FeZn_{8.87}$, $FeZn_{10.98}$) Zn-Fe alloy. The $FeZn_{10.98}$ alloy is mainly located in the region near the Zn-Al diffusion layer surface, and the $FeZn_{6.67}$ alloy is mainly located in the interface between the Zn-Al diffusion layer and the substrate. There is a transition zone between the substrate and the Zn-Al diffusion layer with a thickness of about 5 μm, and a carbon-rich layer exists in this zone. In the direction of perpendicular to the surface, the content of Zn reduces gradually, the content of Fc increases slowly, and abrupt changes occur in this transition zone.

(2) As the immersion time is prolonged, the corrosion products accumulate on the surface and form corrosion pits, which increase the surface roughness. At the same time, the corrosion products of the Zn-Al diffusion layer accumulate on the surface to fill the surface crack and prevent the erosion of the corrosive medium, thereby reducing the corrosion rate of the Zn-Al diffusion layer and exerting the effect of enhancing the corrosion resistance.

(3) The results of salt spray corrosion test show that the corrosion products of the Zn-Al diffusion layer surface are flocculent in the initial corrosion stage, and the corrosion products are mainly ZnO and $Zn_5(OH)_8Cl_2H_2O$. In the middle corrosion stage, the needle-like corrosion products appear on the Zn-Al diffusion layer surface, and new corrosion products are $ZnAl_2O_4$ and FeOCl. In the later corrosion stage, the agglomerated corrosion products appear, the $ZnAl_2O_4$ disappears in the corrosion products, and $Zn(OH)_2$ and $Al(OH)_3$ appear. The appearance of $Zn_5(OH)_8Cl_2H_2O$ increases the compactness of the corrosion products layer and slows down the corrosion rate of the Zn-Al diffusion layer. FeOCl is unstable and reacts with OH^- to form FeO(OH) that accumulates on the surface, which further slows down the corrosion rate.

(4) The results of the Zn-Al diffusion layer electrochemical test show that as the immersion time is prolonged, the self-corrosion potential of the Zn-Al diffusion layer shifts gradually to the positive electrode after moving to the negative electrode. The self-corrosion current density first rises and then falls, the polarization resistance Rp increases, and the corrosion resistance of the Zn-Al diffusion layer increases gradually.

(5) The impedance modulus of low-frequency region decreases, and the radius of the capacitive reactance arc becomes smaller within 0–24 h. At this time, the corrosion solution invades the Zn-Al diffusion layer. The corrosion diffusion stage is 72–360 h, and the impedance modulus and the capacitive reactance arc radius of the low-frequency region increases gradually. In this stage, the corrosion rate of the Zn-Al diffusion layer gradually decreases. After immersing for 360 h, the Zn-Al diffusion layer still has excellent corrosion resistance.

Author Contributions: Y.L., S.W. and Z.S. designed the experiments, J.T., Y.Z. and Y.R. performed the experiments, J.T., H.S. and B.W. analyzed the data, J.T., Y.L. and B.W. wrote the paper.

Funding: This research was funded by The National Natural Science Foundation of China (51675533 and 51701238), and Equipment Pre-research Sharing Technology Project of '13th five-year' (41404010205).

Conflicts of Interest: The authors declare no conflict of interest.

References

1. Pan, C.; Lv, W.; Wang, Z.; Su, W.; Wang, C.; Liu, S. Atmospheric Corrosion of Copper Exposed in a Simulated Coastal-Industrial Atmosphere. *J. Mater. Sci. Technol.* **2017**, *33*, 587–595. [CrossRef]
2. Li, D.-L.; Fu, G.-Q.; Zhu, M.-Y.; Li, Q.; Yin, C.-X. Effect of Ni on the corrosion resistance of bridge steel in a simulated hot and humid coastal-industrial atmosphere. *Int. J. Min. Met. Mater.* **2018**, *25*, 325–338. [CrossRef]
3. Hofmeister, M.; Klein, L.; Miran, H.; Rettig, R.; Virtanen, S.; Singer, R. Corrosion behaviour of stainless steels and a single crystal superalloy in a ternary LiCl–KCl–CsCl molten salt. *Corros. Sci.* **2015**, *90*, 46–53. [CrossRef]
4. Cao, M.; Liu, L.; Fan, L.; Yu, Z.; Li, Y.; Oguzie, E.E.; Wang, F. Influence of Temperature on Corrosion Behavior of 2A02 Al Alloy in Marine Atmospheric Environments. *Materials* **2018**, *11*, 235. [CrossRef] [PubMed]
5. Arrabal, R.; Pardo, A.; Merino, M.C.; Merino, S.; Mohedano, M.; Casajús, P. Corrosion behaviour of Mg/Al alloys in high humidity atmospheres. *Mater. Corros.* **2015**, *62*, 326–334. [CrossRef]
6. Li, S.X.; Hihara, L.H. Atmospheric corrosion initiation on steel from predeposited NaCl salt particles in high humidity atmospheres. *Br. Corros. J.* **2013**, *45*, 49–56. [CrossRef]
7. Arwati, I.A.; Majlan, E.H.; Daud, W.R.W.; Shyuan, L.K.; Arifin, K.B.; Husaini, T.; Alfa, S.; Ashidiq, F. Temperature Effects on Stainless Steel 316L Corrosion in the Environment of Sulphuric Acid (H_2SO_4). *Iop Conf. Ser. Mater. Sci. Eng.* **2018**, *343*, 012016. [CrossRef]
8. Liu, Y.; Zhang, B.; Zhang, Y.; Ma, L.; Yang, P. Electrochemical polarization study on crude oil pipeline corrosion by the produced water with high salinity. *Eng. Fail. Anal.* **2016**, *60*, 307–315. [CrossRef]
9. Liu, Y.; Zhang, Y.; Yuan, J. Influence of produced water with high salinity and corrosion inhibitors on the corrosion of water injection pipe in Tuha oil field. *Eng. Fail. Anal.* **2014**, *45*, 225–233. [CrossRef]
10. Eashwar, M.; Subramanian, G.; Palanichamy, S.; Rajagopal, G. The influence of sunlight on the localized corrosion of UNS S31600 in natural seawater. *Biofouling* **2011**, *27*, 837–849. [CrossRef]

11. Hou, B.; Li, X.; Ma, X.; Du, C.; Zhang, D.; Zheng, M.; Xu, W.; Lu, D.; Ma, F. The cost of corrosion in China. *npj Mater. Degrad.* **2017**, *1*, 4. [CrossRef]
12. Liu, X.; Hou, P.; Zhao, X.; Ma, X.; Hou, B. The polyaniline-modified TiO2 composites in water-based epoxy coating for corrosion protection of Q235 steel. *J. Coat. Technol. Res.* **2018**, *16*, 71–80. [CrossRef]
13. Tiong, U.H.; Clark, G. Impact of Mechanical Strain Environment on Aircraft Protective Coatings and Corrosion Protection. *J. Aircr.* **2011**, *48*, 1315–1330. [CrossRef]
14. Ammar, A.U.; Shahid, M.; Ahmed, M.K.; Khan, M.; Khalid, A.; Khan, Z.A. Electrochemical Study of Polymer and Ceramic-Based Nanocomposite Coatings for Corrosion Protection of Cast Iron Pipeline. *Materials* **2018**, *11*, 332. [CrossRef] [PubMed]
15. Li, J.; Syed, J.A.; Gao, Y.; Lu, H.; Meng, X. Electrodeposition of Ni(OH)2 reinforced polyaniline coating for corrosion protection of 304 stainless steel. *Appl. Surf. Sci.* **2018**, *440*, 1011–1012.
16. Qiu, S.; Chen, C.; Zheng, W.; Li, W.; Zhao, H.; Wang, L. Long-term corrosion protection of mild steel by epoxy coating containing self-doped polyaniline nanofiber. *Synth. Met.* **2017**, *229*, 39–46. [CrossRef]
17. Heakal, F.E.-T.; Elkholy, A.E. Gemini surfactants as corrosion inhibitors for carbon steel. *J. Mol. Liq.* **2017**, *230*, 395–407. [CrossRef]
18. Standish, T.E.; Zagidulin, D.; Ramamurthy, S.; Keech, P.G.; Noël, J.J.; Shoesmith, D.W. Galvanic corrosion of copper-coated carbon steel for used nuclear fuel containers. *Corros. Eng. Sci. Technol.* **2017**, *52*, 65–69. [CrossRef]
19. Usman, B.J.; Umoren, S.A.; Gasem, Z.M. Inhibition of API 5L X60 steel corrosion in CO_2 -saturated 3.5% NaCl solution by tannic acid and synergistic effect of KI additive. *J. Mol. Liq.* **2017**, *237*, 146–156. [CrossRef]
20. Kraljić, M.; Mandic, Z.; Duić, L. Inhibition of steel corrosion by polyaniline coatings. *Corros. Sci.* **2003**, *45*, 181–198. [CrossRef]
21. Pournazari, S.; Deen, K.M.; Maijer, D.M.; Asselin, E. Effect of retrogression and re-aging (RRA) heat treatment on the corrosion behavior of B206 aluminum-copper casting alloy. *Mater. Corros.* **2018**, *69*, 998–1015. [CrossRef]
22. Li, S.; Dong, H.; Li, P.; Chen, S. Effect of repetitious non-isothermal heat treatment on corrosion behavior of Al-Zn-Mg alloy. *Corros. Sci.* **2018**, *131*, 278–289. [CrossRef]
23. Adnan, M.A.; Kee, K.-E.; Raja, P.B.; Ismail, M.C.; Kakooei, S. Influence of Heat Treatment on the Corrosion of Carbon Steel in Environment Containing Carbon Dioxide and Acetic Acid. *Iop Conf. Ser. Mater. Sci. Eng.* **2018**, *370*, 012039. [CrossRef]
24. Zakay, A.; Aghion, E. Effect of Post-heat Treatment on the Corrosion Behavior of AlSi10Mg Alloy Produced by Additive Manufacturing. *JOM* **2019**, *71*, 1150–1157. [CrossRef]
25. Yuan, X.M.; Yang, H.G.; Zhao, W.W.; Zhan, Q.; Hu, Y. The Pack-Cementation Process of Iron-Aluminide Coating on China Low Activation Martensitic and 316L Austenitic Stainless Steel. *Fusion Sci. Technol.* **2011**, *60*, 1065–1068. [CrossRef]
26. Lee, B.W. Effect of diffusion coatings on the high temperature properties of nickel-chromium-superalloys. *Int. J. Mod. Phys. B* **2018**, *32*, 1840056. [CrossRef]
27. Wang, X.C.; Li, M.S.; Li, Q.G.; Zhang, J.; Ma, J. Microstructure Analysis of Mechanical Energy Aided Aluminized Layer on Stainless Steel. *Adv. Mater. Res.* **2011**, *306*, 505–508. [CrossRef]
28. Wang, X.C.; Wei, J.; Zhang, J.; Ma, J.; Liu, S.; Bin Yi, X. Mechanism of Mechanical Energy Aided Aluminizing. *Adv. Mater. Res.* **2013**, *669*, 267–272. [CrossRef]
29. Zhang, Y.; Yao, C.W.; Zhang, L.Y.; Xiang, H.; Li, H.T. Microstructure and Properties of Zincizing Layer Prepared by Mechanical Energy Aided Diffusing Method. *Mater. Test.* **2014**, *38*, 60–65.
30. He, Z.X.; Su, X.P.; Peng, H.P.; Liu, Y.; Wu, C.J.; Tu, H. Al-Zn-Cr Diffusion Process Aided by Mechanical Energy and Microstructure of Alloying Layer. *China Surf. Eng.* **2016**, *29*, 44–51.
31. Wang, X.C.; Wei, J.; Zhang, L.L.; Bin Yi, X.; Zhang, J.; Li, M.S. Microstructure Analysis of Ti-Al Intermetallic Compound Aluminized Layer Generated in Low Temperature. *Appl. Mech. Mater.* **2014**, *638*, 1508–1511. [CrossRef]
32. Ma, Q.H.; Fu, D.H.; Li, Z.B. The Influence of Strcture and Mechanical Performance in Application of Zincing by Aid of Mechanical-energy on the High-tensile Steel. *China Surf. Eng.* **2010**, *23*, 78–81.
33. Zhu, W.W.; Wang, X.D.; Chen, X.P.; Mi, F.Y. Study on Process of Mechanical Energy Aided Aluminizing for Precipitation Hardening Stainless Steel. *Hot Work. Technol.* **2011**, *24*, 192–194. [CrossRef]

34. Liu, Q.D.; Zhao, S.J.; Liu, W.Q. Combination of TEM and 3D atom probe study of microstructure and alloy carbide in a Nb-V micro-alloyed steel tempered at 650 °C. *Met. Mater. Int.* **2013**, *19*, 777–782. [CrossRef]
35. Asami, K.; Kikuchi, M. In-depth distribution of rusts on a plain carbon steel and weathering steels exposed to coastal–industrial atmosphere for 17 years. *Corros. Sci.* **2003**, *45*, 2671–2688. [CrossRef]
36. De La Fuente, D.; Díaz, I.; Simancas, J.; Chico, B.; Morcillo, M. Long-term atmospheric corrosion of mild steel. *Corros. Sci.* **2011**, *53*, 604–617. [CrossRef]
37. Hoar, T.P. Electrochemical principles of the corrosion and protection of metals. *J. Chem. Technol. Biotechnol.* **2010**, *11*, 121–130. [CrossRef]

Article

Mechanical and Corrosion Resistance Enhancement of Closed-Cell Aluminum Foams through Nano-Electrodeposited Composite Coatings

Yiku Xu [1,*], Shuang Ma [1], Mingyuan Fan [1], Hongbang Zheng [1], Yongnan Chen [1,*], Xuding Song [2] and Jianmin Hao [1]

1 School of Materials Science and Engineering, Chang'an University, Xi'an 710064, China; 2017131019@chd.edu.cn (S.M.); 2018231004@chd.edu.cn (M.F.); 2017902285@chd.edu.cn (H.Z.); jmhao@chd.edu.cn (J.H.)
2 Key Laboratory of Road Construction Technology and Equipment, MOE, Chang'an University, Xi'an 710064, China; songxd@chd.edu.cn
* Correspondence: xuyiku23@hotmail.com (Y.X.); frank_cyn@163.com (Y.C.); Tel.: +86-1502-918-9267 (Y.X.); +86-1338-494-8620 (Y.C.)

Received: 26 July 2019; Accepted: 26 September 2019; Published: 29 September 2019

Abstract: This work aims to improve the properties of aluminum foams including the mechanical properties and corrosion resistance by electrodepositing a SiC/TiN nanoparticles reinforced Ni–Mo coating on the substrate. The coatings were electrodeposited at different voltages, and the morphologies of the coating were detected by SEM (scanning electron microscope) to determine the most suitable voltage. We used XRD (x-ray diffraction) and TEM (transmission electron microscope) to analyze the structure of the coatings. The aluminum foams and the substrates on which the coatings were electrodeposited at a voltage of 6.0 V for different electrodeposition times were compressed on an MTS (an Electro-mechanical Universal Testing Machine) to detect the mechanical properties. The corrosion resistance before and after the electrodeposition experiment was also examined. The results showed that the coating effectively improved the mechanical properties. When the electrodeposition time was changed from 10 min to 40 min, the W_v of the aluminum foams increased from 0.852 J to 2.520 J and the σ_s increased from 1.06 MPa to 2.99 MPa. The corrosion resistance of the aluminum foams was significantly improved after being coated with the Ni–Mo–SiC–TiN nanocomposite coating. The self-corrosion potential, pitting potential, and potential for primary passivation were positively shifted by 294 mV, 99 mV, and 301 mV, respectively. The effect of nanoparticles on the corrosion resistance of the coatings is significant.

Keywords: aluminum foam; electrodeposition; compression test; corrosion resistance

---

1. Introduction

Aluminum foams have the characteristics of both metal materials and foam materials due to their special structure. They are functional materials with the properties of both structural materials and functional materials [1–4]. Aluminum foams have a unique stress–strain curve including a linear elastic region, a plastic collapse region, and a densification region, which makes aluminum foam materials suitable for use as an energy absorber [5]. Due to its excellent properties including light weight, high sound absorption and insulation performance, heat resistance, and high cushioning performance, it is widely used in sound absorption and sound insulation structures such as sound barriers and sound insulation boards, and for energy absorption and collision protection in automobiles [6–9]. However, the high porosity of aluminum foam significantly lowers its mechanical strength. When applied in an engineering field, it often fails prematurely, which greatly limits its potential range of applications.

For example, when an aluminum foam is used for sound insulation and heat transfer in an environment that requires a certain load, the aluminum foam will fail and be crushed after the load acts on it for a long period of time [10]. The seawater, micro-organisms, and salt spray in a marine environment can corrode aluminum foams when they are used in marine transportation applications. In order to expand the range of uses of aluminum foams and make them better for practical applications, surface modification methods are used to simultaneously improve their mechanical properties and corrosion resistance. An aluminum foam with a high energy absorption capacity should have a longer and higher stress platform. At present, using the same material to thicken the foam pillar is a common method for improving the energy absorption capacity [11–14]. However, this method has certain limitations. When the pillar of the aluminum foam is thickened, the platform stress will be improved, but the aluminum foam will be dense [15–17]. Densification may limit the ability of aluminum foams to absorb energy. At present, the commonly used surface modification methods for aluminum foams include micro-arc oxidation, anodization, electro-less plating, sol–gel deposition, and electrodeposition [18–20]. Of these methods, electrodeposition is widely used because it is simple, low cost, and easy to control.

At present, there are a number of reports on the use of deposited layers to enhance the properties of aluminum foams. Yuttanant Boonyongmaneerat et al. [21] electrodeposited a nanocrystalline Ni–W coating on open-cell aluminum foams to improve their properties including compressive strength and energy abortion. Zhendong Li et al. [22] confirmed that a thermally evaporating Zn film could significantly enhance open-cell aluminum foams and increase their yield strength. Liu Huan et al. [10] studied the enhancements that a Ni coating could provide to closed-cell aluminum foams. It demonstrated that a Ni coating could improve the properties of aluminum foams including both the mechanical and corrosion resistance properties. Jiaan Liu et al. [23] showed that the corrosion resistance of closed-cell aluminum foams could be improved by an electro-less Ni–P coating. Due to the excellent mechanical properties, corrosion resistance, and wear resistance of the Ni–Mo coating, it often used as an protective coating [24–26]. SiC nanoparticles have a high degree of hardness, wear resistance, and thermal stability, and TiN nanoparticles have a high degree of hardness, high strength, and corrosion resistance [27–29]. As far as we know, no research has been done on the use of a Ni–Mo coating and a duplex nanoparticles reinforced Ni–Mo coating to enhance the properties of closed-cell aluminum foams.

In this work, the influence of a Ni–Mo coating and a duplex nanoparticles reinforced Ni–Mo coating on the mechanical properties and corrosion resistance of closed-cell aluminum foams was studied. The effects of electrodeposition voltage and electrodeposition time on the morphology, mechanical properties, and corrosion resistance of the closed-cell aluminum foams were investigated. The deposition mechanism of the duplex nanoparticles reinforced Ni–Mo coating is also discussed.

2. Materials and Methods

2.1. Samples and Solution

We used the method of melt foaming to prepare the closed-cell aluminum foams in this experiment. The density of the samples was 0.2 g/cm^3. The pore diameter was 4 mm. We used an electrical discharge machine to reduce the sample's dimensions to 20 mm × 20 mm × 9 mm.

To create a good bond between the substrate and the coating, the aluminum foam was pretreated before the electrodeposition experiment. The aluminum foam sample was immersed in a 10~15% H_2SO_4 solution at 60 °C for 1–3 min. After immersion, the oil was removed. Then, a 5% NaOH solution was used to remove the Al_2O_3 film from the surface of the samples. The sample was immersed for 2 min. Finally, the aluminum foam was immersed for 5 min in a 10% HNO_3 solution. The corrosion products were removed and activated. After each of the steps was completed, the sample was washed with distilled water to prevent the pretreatment liquid from being contaminated. After the pretreatment steps were completed, the aluminum foam sample was placed in the electrolyte immediately to prevent it from being oxidized in the air.

Table 1 shows the electrolyte components that were used in this experiment. The electrolyte was composed of analytically pure reagent and distilled water. The added SiC and TiN nanoparticles (Shanghai Chaowei Nanotechnology Co. Ltd., Nanxiang Hi-Tech Industrial Park, Jiading District, Shanghai) both had a mean particle diameter of 20 nm and purity of 99 wt.%. Since nanoparticles tend to agglomerate in the electrolyte, SDS was chosen as the dispersing agent. The electrolyte was subjected to ultrasonic treatment for 2 h and the electrolyte was stirred using a magnetic stirrer with a speed of 300 rpm during the electrodeposition experiment. Electrolyte (200 mL) was placed in a bath, pure nickel plate (99.99 wt.%) was used as the anode, and the aluminum foam sample was used as the cathode. The anode and the cathode had a distance of 30 mm between them.

Table 1. Components of the electrolyte for the electrodeposition of the Ni–Mo–SiC–TiN coating.

Bath Composition	Concentration	Purpose
$NiSO_4{\cdot}6H_2O$	0.27 mol·dm^3	Ni source
$Na_2MoO_4{\cdot}2H_2O$	0.032 mol·dm^{-3}	Mo source
$Na_3C_6H_5O_7{\cdot}2H_2O$	0.52 mol·dm^{-3}	Complexing agent
NH_4Cl	0.65 mol·dm^{-3}	Buffer
SDS	0.1 g·dm^{-3}	Surfactant
SiC	5 g·dm^{-3}	Composite phase
TiN	5 g·dm^{-3}	Composite phase

2.2. Morphology Investigation

The surface morphology and a cross-section of the coating were observed using a Hitachi S4800 field scanning electron microscope (SEM, Hitachi, Ltd., Tokyo, Japan). The elements were analyzed by energy dispersive x-ray spectroscopy. The structure of the coating was examined by D8 ADVANCE x-ray diffraction (XRD, Bruker, Karlsruhe, Germany). Cu-k$_\alpha$ radiation was selected, and the 2θ range was 20~80°. In order to further analyze the specific structure of the nanocomposite coating and the distribution of nanoparticles, the coating was examined by an FEI Talos F200X transmission electron microscope (TEM, FEI™, Hillsboro, OR, USA) including high-resolution TEM (HR-TEM) and selected area electron diffraction (SAED).

2.3. Properties Investigation

An electrodeposited aluminum foam with the dimensions of 20 mm × 30 mm × 40 mm was subjected to a quasi-static compression test on an MTS (an Electro-mechanical Universal Testing Machine, American MTS Corporation, MN, USA) with a selected load of 10 KN, a compression speed of 5 mm/min, and a compression amount greater than 70%.

The corrosion resistance of the sample at room temperature was measured by the three-electrode working system. In this experiment, a 3.5 wt% NaCl solution was used as the etching solution. The working electrode was the aluminum foam sample, the reference electrode was the saturated calomel electrode, and the counter electrode was the platinum plate electrode. The selected voltage range was −2 to 1 V and the scan rate was 2 mV/s.

The samples were placed in an immersion test for 120 h to measure the corrosion rate at 25 °C. The immersion solution was a 3.5 wt% NaCl. The samples were weighed to calculate the mass loss every 24 h. Distilled water was used to rinse the samples, and they were dried thoroughly before each weighing. The weight of a sample was expressed as the average of three measurements. The analytical balance that was used to weigh the samples had an accuracy of 0.01 mg.

3. Theoretical Models

Electrodeposition of metals and alloys refers to the reduction of metal ions from an electrolyte, where electrons (*e*) are provided by an external power supply. The reaction time and the current can optimize the thickness of a coating. Molybdenum cannot be electrodeposited from the electrolyte solution, but the co-deposition of nickel and molybdenum can be achieved using sodium citrate as an inducer. During the electrodeposition of Ni–Mo composite coatings on an aluminum foam, the following chemical reactions occur at the cathode and anode [30]:

Anode:

$$Ni - 2e \rightarrow Ni^{2+} \tag{1}$$

Cathode:

$$Ni^{2+} + 2\,e \rightarrow Ni \tag{2}$$

$$MoO_4{}^{2-} + 2H_2O + 2e \rightarrow MoO_2 + 4OH^- \tag{3}$$

$$NiCit^- + MoO_2 \rightarrow [NiCitMoO_2]^-{}_{ads} \tag{4}$$

$$[NiCitMoO_2]^-{}_{ads} + 2H_2O + 4e \rightarrow Mo + NiCit^- + 4OH^- \tag{5}$$

With respect to the co-deposition of nanoparticles with a Ni–Mo matrix, the processes include three main steps, as illustrated in Figure 1. According to Gugliemi's absorption model, Ni ions and Mo ions in the electrolyte solution are first adsorbed on the nanoparticles to form Ni/Mo ionic clouds. Under the electric field force, metal ions and ionic clouds move toward the cathode and are tightly adsorbed on the aluminum foams. Then, the Ni and Mo ions adsorbed on the surface of the nanoparticles are reduced partially at the surface of the foam. Simultaneously, nanoparticles are trapped by the metal matrix and embedded in the Ni–Mo plating layer.

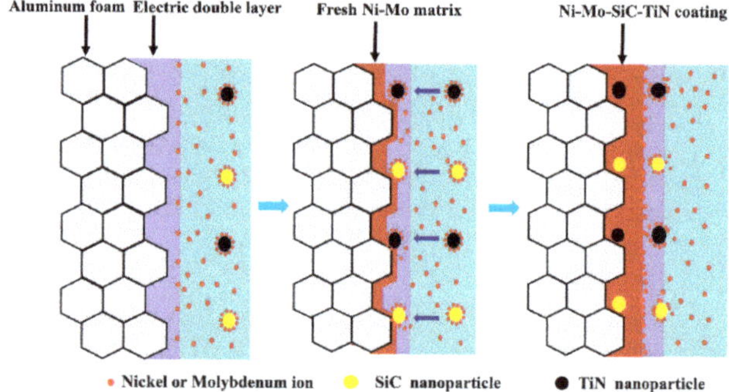

Figure 1. A schematic diagram representing the electrodeposition process of a duplex nanoparticles reinforced Ni–Mo coating.

Based on a theoretical model of Cu electrodeposition, the model of the electrodeposited Ni–Mo alloy coating in this experiment is now described [31].

The plating deposition rate is expressed by P%, and its expression is:

$$P\% = [(P_2)_i - P_{1i}]/P_{1i} \tag{6}$$

where P_1 indicates the mass of the substrate before the electrodeposition experiment; P_2 indicates the mass of the aluminum foam covered with a coating; and i indicates the sample number. P% is the ratio of the mass of the aluminum foam covered with a Ni–Mo coating to the mass of the aluminum foam before electrodeposition.

$$P\% = \frac{M_{NiMo}}{\rho * V_i} = (MM_{NiMo} {*} n_{NiMo})/(\rho {*} V_i) \tag{7}$$

where M_{NiMo} is the mass of the deposited Ni–Mo alloy coating; MM_{Ni-Mo} is the molar mass of Ni–Mo alloy; and ρ and V_i indicate the density and volume of the aluminum foam before the electrodeposition experiment, respectively.

The Ni–Mo alloy that was formed in this experiment is a Ni–Mo solid solution. When 1 mole of Ni–Mo alloy coating is deposited, 14 moles of electron are required. Then, P% also can be expressed as:

$$P\% = (MM_{NiMo} * n_e) / \ (14 * \rho * V_i) \tag{8}$$

where e is the electric charge of an electron.

It is known that $n_e = q/(N_a * e)$, $q = i * t$. Then,

$$P\% = MM_{NiMo} / (14 * N_a * e] * [(i * t) / (\rho * V_i) \tag{9}$$

where t is the electrodeposition time (in minutes); N_a is Avogadro's number with a value of 6.02×10^{23}; MM_{NiMo} is 331 g/mol; and e is 1.6×10^{-19} C. Then,

$$P\% = 2.45 \times 10^{-4} * [(i * t) / (\rho * V_i)]. \tag{10}$$

As this experiment was carried out under a certain voltage, the expression is written as

$$P\% = 2.45 \times 10^{-4} * [(u * t) / (\rho * V_i * r)] \tag{11}$$

where u is the electrodeposition voltage and r is the total resistance.

The relationship between the deposition rate of a Ni–Mo coating, the electrodeposition voltage u, and the time t can be obtained by Equation (11), and the P% that is obtained by experiments can be verified using theoretical calculations.

4. Results and Discussion

4.1. Coating Characterization

Figure 2b show the SEM images of the two kinds of nanoparticles with an original size of approximately 20 nm. As can be seen, both kinds of nanoparticles were agglomerated due to the surface effect.

Figure 2d–f show the morphologies of the electrodeposited duplex nanoparticles reinforced Ni–Mo coatings, applying electrodeposition voltages ranging from 2.5 V to 6.0 V, respectively. It has been reported that nanoparticles can make a coating have a finer grain and a higher microhardness. In accordance with this, SiC- and TiN-reinforced coatings have structures with finer grain sizes than Ni–Mo composite coatings. As the voltage increased, coating particles were gradually formed and completely covered the substrate. When the voltage was increased to 6.0 V, a uniform coating was prepared on the aluminum foam. The SEM image shown in Figure 2i revealed that the surface of the coating had nanoparticles dispersed upon it. At the voltage of 6.0 V, a nodular and homogenous Ni–Mo coating was also obtained. It is known that a larger electrodeposition voltage can increase the nucleation driving force, so plating particles are formed. The metal ion deposition rate was sufficiently high to form a uniform and dense coating on the substrate at the voltage of 6.0 V. Figure 2h shows the morphology of a cross-section of aluminum foam that was subject to electrodeposition for 10 min at 6.0 V. The coating had a thickness of about 25 μm. The bond between the plating layer and the substrate was good, the thickness of the plating layer was relatively uniform, and there were no cracks or discontinuities.

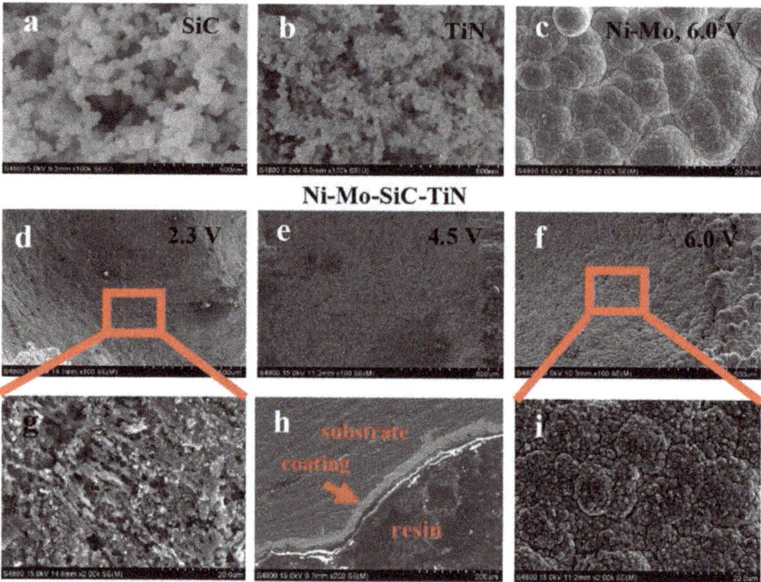

Figure 2. SEM (scanning electron microscope) images of (**a**) SiC nanoparticles, (**b**) TiN nanoparticles, (**c**) the Ni–Mo coating, and the duplex nanoparticles reinforced Ni–Mo coating electrodeposited (**d**) at 2.3 V, (**e**) 4.5 V, and (**f**) 6.0 V. The enlarged SEM images of the duplex nanoparticles reinforced Ni–Mo coating (**g**) at 2.3 V, (**i**) 6.0 V; and (**h**) morphology of a cross-section of the duplex nanoparticles reinforced Ni–Mo coating.

Figure 3 shows the XRD patterns of the coatings. The body-centered cubic structure that corresponds to nickel's (111), (200), and (220) diffraction peaks. No diffraction peak of molybdenum was detected, indicating that the nickel atom and the molybdenum atom existed in the form of a Ni–Mo solid solution. The nanoparticles did not change the structure of the Ni–Mo coating. In the XRD patterns, there were no diffraction peaks related to nanoparticles. This is mainly because the size of the nanoparticles was too small, their content too low, and the distribution was uniform [32]. The intensity of the peaks of the XRD patterns of the coatings electrodeposited at 6.0 V for different times were different. We used the Scherrer formula to calculate the crystallite size:

$$D = K\lambda/(\beta COS\theta) \tag{12}$$

where λ represents the wavelength of the x-ray (0.15406 nm); K is the Scherrer constant (0.9); β is the full width of the reflection line at half maxima; and θ is a Bragg diffraction angle.

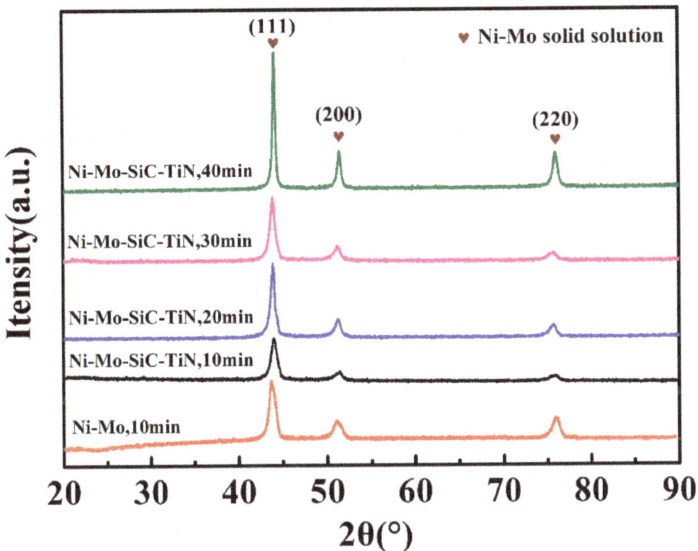

Figure 3. XRD (x-ray diffraction) patterns of coatings electrodeposited by different times.

Table 2 shows the results. As the electrodeposition time increased, the grains of the coatings accumulated and the crystallite size increased. Comparing the crystallite size of the Ni–Mo coating to that of the duplex nanoparticles reinforced Ni–Mo coating, it can be seen that the nanoparticles decreased the crystallite size of the coating. Nanoparticles can inhibit the grain growth because they provide nucleation dots.

Table 2. Crystallite size of different coatings electrodeposited at 6.0 V.

Coatings (Electrodeposition Time)	Crystallite Size (nm)
Ni–Mo (10 min)	13.31
Ni–Mo–SiC–TiN (10 min)	12.14
Ni–Mo–SiC–TiN (20 min)	17.01
Ni–Mo–SiC–TiN (30 min)	20.36
Ni–Mo–SiC–TiN (40 min)	24.96

Figure 4 shows the EDS (energy dispersive spectrometer) elements mapping of the duplex nanoparticles reinforced Ni–Mo coating. Ni, Mo, Si, C, Ti, and N elements were detected. The existence of Si, C, Ti, and N elements indicates that duplex nanoparticles were successfully electrodeposited in the Ni–Mo composite coating. The EDS element mapping demonstrates the specific distribution of duplex nanoparticles. Nanoparticles were uniformly dispersed in the coating, but partial agglomeration occurred. Since the coating used for EDS (energy dispersive spectrometer) detection was a 100 nm thin layer, the distribution of nanoparticles inside the coating can be known.

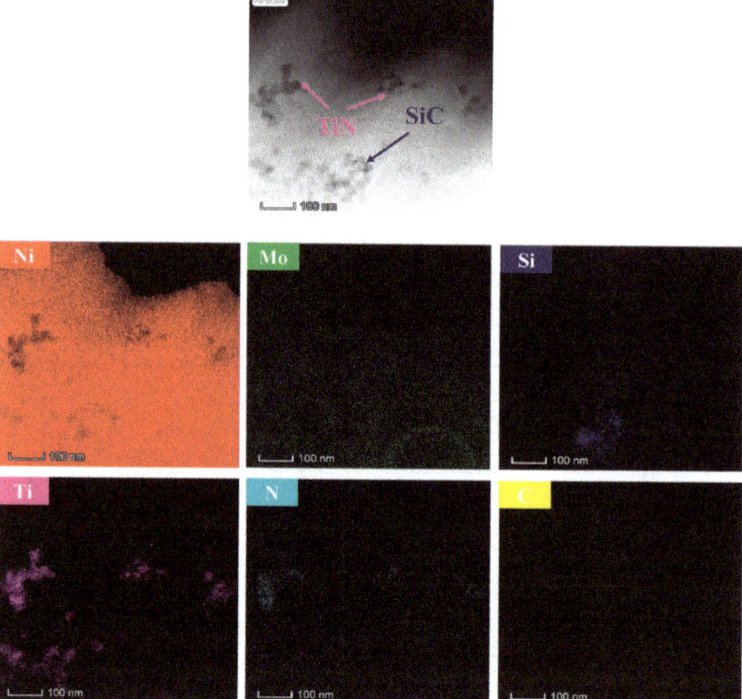

Figure 4. EDS (energy dispersive spectrometer) element mapping of the Ni–Mo–SiC–TiN nanocomposite coating.

Figure 5a presents the TEM images under a bright field. From the images, it was found that the nanoparticles were tightly embedded in the Ni–Mo metal matrix and there were no voids between them. The interface between the nanoparticles and the Ni–Mo metal matrix was clear and there were no harmful interfacial reaction products. The selected area electron diffraction rings in Figure 5b correspond to the (111), (200), (220), and (311) crystal faces of the nickel–molybdenum solid solution, respectively. The fast Fourier transform (FFT) and inverse FFT of the nanoparticles in Figure 5a are shown in Figure 5c. The nanoparticles were proven to be 6H–SiC, which have a hexagonal closed-packed (HCP) structure with a Lattice constant of 3.08 Å.

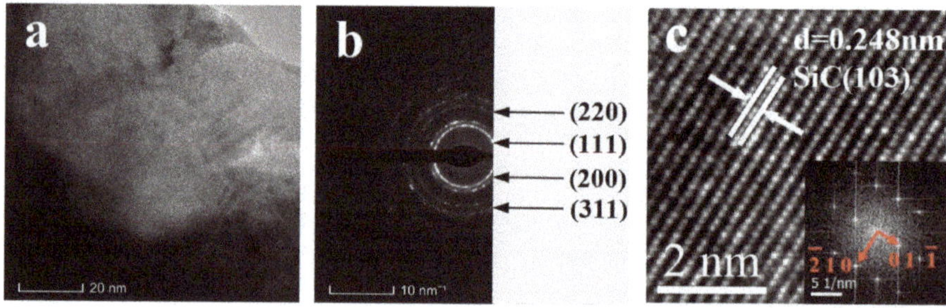

Figure 5. (**a**) Bright field image of the duplex nanoparticles reinforced Ni–Mo coating; (**b**) diffraction ring of the metal matrix; and (**c**) HR-TEM image of the nanoparticles.

4.2. Mechanical Behavior

Figure 6a shows the stress–strain curves of the aluminum foam and the aluminum foams subjected to electrodeposition for different times. The enlarged elastic region of the curve is shown in Figure 6b. The stress–strain curve of the aluminum foam includes three parts: an elastic deformation stage, a yield stage, and a densification stage. In the initial stage of the compression experiment, the stress increased as the strain increased. The relationship between stress and strain was linear. When the curve entered the yield stage, the stress appeared to be small or substantially constant as the strain increased. In the densification stage, since the pores inside the aluminum foam burst and collapsed, the stress increased sharply at this stage and the strain remained substantially unchanged.

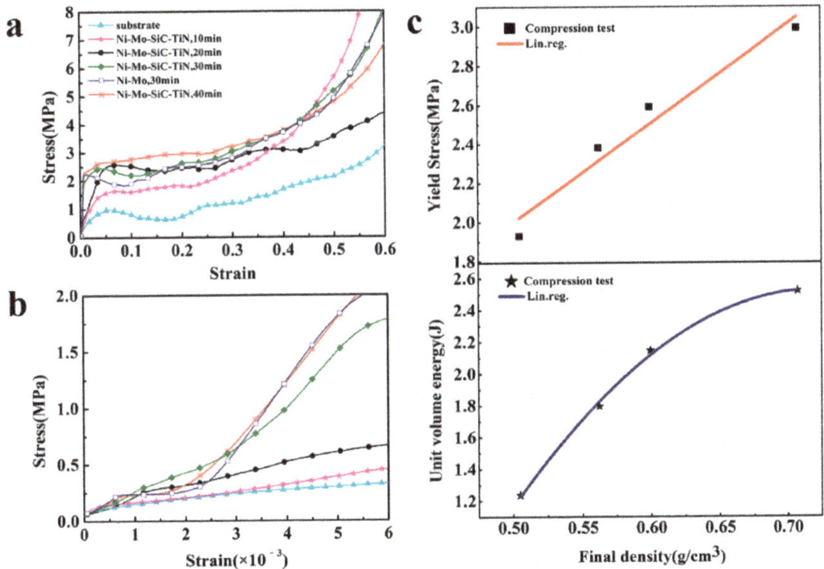

Figure 6. (**a**) Stress–strain curves of aluminum foam and aluminum foam subject to electrodeposition at the voltage of 6.0 V; (**b**) the enlarged elastic region of (**a**). (**c**) the yield strength and unit volume energy in function of the final density of samples.

Compared with the aluminum foam matrix, the elastic modulus and the platform stress of the aluminum foams after the coating was deposited were improved. When the strain remained the same, the stress of the aluminum foam after electrodeposition was larger than that of the aluminum foam substrate. There are two main reasons for the increase in strength and elastic modulus of aluminum foams with electrodeposited coatings. The first reason is that, due to the particularity of the structure of the aluminum foam, the deformation of the aluminum alloy during the compression experiment was not synchronized, resulting in separation of the coating from the substrate. The second reason is the friction and extrusion between the coating and the substrate. The stress–strain curve also showed that the coating increased the energy absorption of the aluminum foam.

The density of an aluminum foam affects its mechanical strength. An increase in density will increase the compressive properties. The density is related to the electrodeposition time. The electrodeposition time determines the quality of the coating that is deposited on the substrate, so the quality affects the mechanical strength of the substrate. Table 3 lists the coefficient of variation P% of different electrodeposition times. The resistance, which includes the external contact resistance r_1, the solution resistance r_2, and the resistance of the cathode film r_3 during the electrodeposition process, are all uncertain. The resistance at 10 min of electrodeposition was used as the resistance in this experiment.

Table 3. Characteristics of different samples.

Samples	Deposition Time t (min)	Initial Density ρ_0 (g/cm^3)	Final Density ρ (g/cm^3)	Coefficient of Variation P% (%)	Yield Strength σ_s (MPa)	Elastic Modulus (MPa)	W_v (J)
Substrate	0	/	/	/	1.06	43.67	0.852
Ni–Mo–SiC–TiN	10	0.4562	0.5051	10.7	1.93	54.39	1.237
Ni–Mo–SiC–TiN	20	0.4553	0.5623	23.5	2.38	106.27	1.795
Ni–Mo–SiC–TiN	30	0.4320	0.5996	38.8	2.59	248.63	2.146
Ni–Mo	30	0.4326	0.5976	38.2	2.52	240.46	2.069
Ni–Mo–SiC–TiN	40	0.4612	0.7075	53.4	2.99	344.75	2.520

The coefficient of variation of 20 min, 30 min, and 40 min of electrodeposition, as calculated by the established electrodeposition theoretical model, was 21.35%, 33.76%, and 42.16%, respectively, while the P% obtained from the experiment was 23.5%, 38.8%, and 53.4%, respectively. Density of Ni–Mo–SiC–TiN coatings does not obviously change with deposition time. As the deposition time increased, the error between the theoretical model and the results obtained from deposition rate increased. The main reason for this is that an increase in the electrodeposition time will cause a large change in resistance.

Table 3 also lists the density, yield strength, densification strain, and W_v of the aluminum foam after the electrodeposition experiments. W_v represents the energy absorbed per unit volume when the aluminum foam is deformed. When comparing the compression properties of the samples after different electrodeposition times, it was found that the stress–strain curves of aluminum foams moved upward with the increase of electrodeposition time. This is mainly because an increase in electrodeposition time will increase the quality of the coating on the aluminum foam. The quality of the coating on the surface increases the strength and stiffness of the aluminum foam. From the stress–strain curve, it can be seen that the curve appeared to fluctuate in the stress platform stage, which is due to the instability of the aluminum foam. This can be attributed to the non-uniformity of the aluminum foam's cell structure and its rough surface. When the stress–strain curve passes the linear elastic phase, the stress tends to decrease; the reasons for this are discussed in the literature [33,34].

The stress remained almost constant as the strain increased in the stress platform stage, which allowed the sample to absorb a large amount of energy during the compression process. Figure 6 shows the absorbed energy per unit volume of the aluminum foam during the quasi-static compression experiment in the stress–strain curve. Its calculation expression is [35]

$$W_v = \int_0^{\varepsilon_D} \sigma(\varepsilon)\,d\varepsilon \tag{13}$$

where ε_D represents the densification strain, which corresponds to a sharp rise in stress during compression because the aluminum foam is crushed and deformed and the cell structure completely collapses, and $\sigma(\varepsilon)$ represents the stress.

Gibson et al. proposed the following relationship between the densification strain of closed-cell aluminum foam, ε_D [36], and the relative density $\overline{\rho}$:

$$\varepsilon_D = 1 - 1.4\,\overline{\rho} \tag{14}$$

where $\overline{\rho}$ is the ratio of the apparent density ρ of the aluminum foam to the density ρ_s (2.70 g/cm^3) of the aluminum foam matrix.

The relationship between the W_v and the apparent density ρ of aluminum foams is obtained from the above two formulas:

$$W_v = \int_0^{1-0.518\rho} \sigma(\varepsilon)d\varepsilon. \tag{15}$$

The relationship indicates that W_v is related to the density of aluminum foams.

The specific relationship between the density and mechanical properties of samples was explored after the electrodeposition experiments, the density of aluminum foams after the electrodeposition of a coating between the W_v, and yield strength σ_s were respectively fitted. The fitting results are shown in Figure 6c, d. The relationship between the unit volume energy absorption W_v and the density ρ is

$$W_v = a + b_1\rho + b_2\rho^2. \tag{16}$$

From the fitting results, the value of a, b_1, and b_2 is −12.05942 ± 1.72093, 40.51736 ± 5.70059, and −28.13555 ± 4.65988, respectively. The value of the correlation coefficient R^2 is 0.99376.

The relationship between the yield strength σ_s and the density ρ is

$$\sigma_s = d + c\rho. \tag{17}$$

From the fitting results, the value of d, c, and the correlation coefficient R^2 is −0.5350 ± 0.44692, 5.0664 ± 0.74711, and 0.93748, respectively.

Due to the particular structure of each cell of the aluminum foams and the differences in the deposition rate, the data shown in Figure 6c are relatively discrete.

When the aluminum foams were subjected to electrodeposition for 10 min, 20 min, 30 min, and 40 min, W_v was quadratic with ρ, and σ_s increased linearly with ρ. Comparing the mechanical properties of the substrates, which were coated with a Ni–Mo coating and a duplex nanoparticles reinforced Ni–Mo coating, the addition of nanoparticles only slightly increased W_v and σ_s. This limited enhancement of the compressive properties is due to the small amount of nanoparticles in the Ni–Mo coatings.

4.3. Corrosion Resistance

The corrosion resistance of the aluminum foam and the aluminum foams with an electrodeposited Ni–Mo coating and a duplex nanoparticles reinforced Ni–Mo coating was detected. The obtained polarization curves are shown in Figure 7a. Table 4 lists the corrosion parameters extracted from the polarization curves. After the Ni–Mo coating was electrodeposited on the aluminum foam, the corrosion potential of the aluminum foam was positively shifted from −1160 mV to −937 mV and the corrosion current density decreased from 4.48×10^{-5} A/cm^2 to 3.90×10^{-5} A/cm^2. The pitting potential and the potential for primary passivation were positively shifted by 48 mV and 225 mV, respectively. The positive shift of the Zero current potential was due to changes in the hydrogen evolution reduction process. Both the aluminum foam and the aluminum foam with an electrodeposited Ni–Mo coating formed passive films. The aluminum foam formed a passive film because of the oxide layer. The oxygen-rich surface reacted with the etching solution to form an adsorption layer. The adsorption layer prevented contact of the etching solution with the surface of the plating layer to prevent the hydration of nickel, which is the first step in the formation of a passive nickel film on the surface of the aluminum foam covered with the Ni–Mo coating.

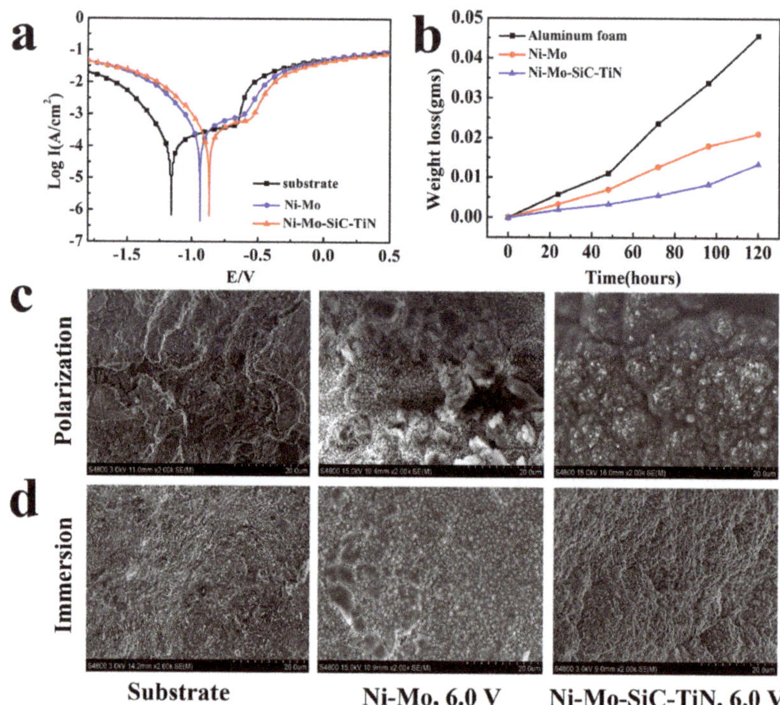

Figure 7. The polarization curves of the samples (**a**) and weight loss versus time curves after the immersion test (**b**); SEM images of the substrate and coatings after the polarization test (**c**) and immersion test (**d**).

Compared with the substrate, the corrosion resistance of the samples after electrodeposition was greatly improved. The Ni–Mo coating was found to effectively protect the aluminum foam from corrosion. In order for a corrosive liquid to have a corrosive effect on the aluminum foam's substrate, the passivation film on the surface of the aluminum foam must first be destroyed. The Cl^- in the etching solution was found to easily pass through the passivation film due to the small radius and adsorb on the samples to hinder the adsorption of oxygen. The cations in the passivation film combined with the Cl^- to form a soluble chloride. The substrate was partially exposed due to the local corrosion. The aluminum foam had a galvanic effect with the oxide film to form a corrosive micro-battery. The matrix and the impurity elements Ca, Ti, and Si, which were contained in the aluminum foam, also formed a corrosive micro-battery. This resulted in an uneven accumulation and distribution of Cl^-, which exacerbated the local corrosion.

The coating was able to effectively protect the aluminum foam matrix mainly because the electrodeposited coating could separate the aluminum foam matrix from the etching solution so the Ni-Mo coating had an initial corrosion. The amorphous Ni–Mo alloy coating had good corrosion resistance, and the Mo element could easily form an inert oxide with oxygen in solution to prevent further corrosion of the coating [37,38]. The plating layer was uniform and compact. The coating had a thickness of about 25 μm. The pinholes and cracks in the plating layer were reduced. These all made the distribution of Cl^- become uniform. The corrosion on the surface of the coating was relatively uniform. Then, there was a better corrosion resistance.

From the polarization curves of the samples, corrosion parameters can be obtained. The corrosion potential was shifted from –0.937 V for the Ni–Mo coating to –0.866 V for the duplex nanoparticles reinforced Ni–Mo coating, and the corrosion current density was reduced from 3.90×10^{-5} A/cm^2 to

2.72×10^{-5} A/cm^2. This change illustrates that the SiC and TiN nanoparticles can have an improvement on the corrosion resistance of the Ni–Mo coating. The reason for this is that these two kinds of nanoparticles are inert nanoparticles that have a certain degree of corrosion resistance. Dispersed nanoparticles can enhance the corrosion resistance of the coating because the nanoparticles can block the etching solution and the coating from coming into contact.

Figure 7c shows the SEM of the sample after the polarization experiment in corrosive solution. The aluminum foam was severely corroded, and there were many corrosion products and corrosion pits on the aluminum foam. An EDS spectrum analysis was performed on the corrosion surface, and the oxygen content in the corrosion product was found to be 17.04 wt.%. Compared with the aluminum foam, the Ni–Mo coating provided better protection to the substrate. Corrosion occurred on the surface of the coating when corrosion occurred. Local corrosion cracks could be observed, and the oxygen content in the corrosion products decreased to 8.03 wt.%. After adding duplex nanoparticles, the number of corrosion products was significantly reduced. This is because the two inert types of nanoparticles protected the matrix coating. Nanoparticles filled the voids in the matrix coating and improved the compactness of the coating. The smaller contact area effectively reduced the corrosion rate. The low content (4.69 wt.%) of oxygen also indicated an improvement in corrosion resistance.

Figure 7d shows the corrosion morphologies of the aluminum foam, the Ni–Mo coating, and the duplex nanoparticles reinforced Ni–Mo coating deposited at the voltage of 6.0 V after the immersion test. There are many corrosion pits on the surface of the substrate. The aluminum foam was obviously corroded because there was no protection of the coatings. The Ni–Mo coating was slightly corroded, and only a few corrosion products existed. Almost no corrosion was observed on the surface of the duplex nanoparticles reinforced Ni–Mo coating due to the protection of the nanoparticles.

Table 4 lists the corrosion rates of different samples. Compared with the aluminum alloy matrix, the corrosion rates of the Ni–Mo coating and the duplex nanoparticles reinforced Ni–Mo coating were improved by 51.9% and 72.5%, respectively.

Table 4. The corrosion parameters extracted from the polarization curves and weight loss vs. time curves.

Passivation Parameters	Substrate	Ni–Mo	Ni–Mo–SiC–TiN
Potential for primary passivation (E_{pp}, mV)	−1130	−905	−829
Breakdown potential (E_b, mV)	−653	−605	−554
Corrosion potential (E_{corr}, mV)	−1160	−937	−866
Corrosion current density (I_{corr}, A/cm^2)	4.48×10^{-5}	3.90×10^{-5}	2.72×10^{-5}
β_a (mV/decade)	69.08	47.14	33.32
β_c (mV/decade)	25.47	29.88	40.45
Corrosion rate (g/cm^2·h)	3.8643×10^{-4}	1.8583×10^{-4}	1.0643×10^{-4}

5. Conclusions

1. A uniform and dense duplex nanoparticles-reinforced Ni–Mo coating with a thickness of 25 μm was obtained by electroplating on the aluminum foam surface for 10 min at 6.0 V. The bond between the substrate and the coating was good.
2. The duplex nanoparticles reinforced Ni–Mo coating had a structure of FCC. The crystallite size of the Ni–Mo coatings was decreased from 13.31 nm to 12.14 nm after adding the duplex nanoparticles. The results indicate that increasing the electrodeposition time can effectively enlarge the crystallite size.
3. After the aluminum foams were coated with a duplex nanoparticles-reinforced Ni–Mo coating, there was a significant improve in the mechanical properties of the aluminum foams. When the electrodeposition time was 40 min, the W_v of the aluminum foam increased from 0.852 J to 2.520 J,

and the σ_s increased from 1.06 MPa to 2.99 MPa. The addition of nanoparticles made a limited improvement to the mechanical properties.

4. The duplex nanoparticles-reinforced Ni–Mo coating was found to have better corrosion resistance. Compared to the aluminum foams, the self-corrosion potential, the pitting potential, and the potential for primary passivation were positively shifted by 294 mV, 99 mV, and 301 mV, respectively. The corrosion rate of the aluminum foam covered with a Ni–Mo coating was reduced by 51.9%. After adding nanoparticles, the corrosion rate was reduced by 72.5%. The nanoparticles obviously improved the corrosion resistance.

Author Contributions: Conceptualization, S.M., M.F., and H.Z.; Funding Acquisition, Y.X., Y.C., and J.H.; Investigation, Y.X., S.M., X.S., and M.F.; Methodology, Y.X. and S.M.; Project Administration, Y.X., Y.C., and J.H.; Data Curation, M.F. and H.Z.; Writing—Original Draft, S.M.; Writing—Review & Editing, Y.X., Y.C., J.H., and X.S.

Funding: This work was financially supported by the National Natural Science Foundation of China (No. 51,301,021), the China Postdoctoral Science Foundation (No. 2016M592730), the Fundamental Research Funds for the Central Universities (Nos. 300,102,318,205; 310,831,161,020; 310,831,163,401; 300,102,319,304), the Innovation and Entrepreneurship Training Program of Chang' an University (No. 201,910,710,144), the Key projects of Shaanxi Natural Science Foundation (2019JZ-27), and the Shaanxi Natural Science Basic Research Program-Shaanxi Coal (2019JLM-47).

Conflicts of Interest: The authors declare no conflicts of interest.

References

1. Banhart, J.; Baumeister, J.; Weber, M. Damping properties of aluminum foams. *Mater. Sci. Eng. A* **1996**, *205*, 221–228. [CrossRef]
2. Uzun, A.; Karakoc, H.; Gokmen, U.; Cinici, H. Investigation of mechanical properties of tubular aluminum foams. *Int. J. Mater. Res.* **2017**, *107*, 996. [CrossRef]
3. Gong, L.; Kyriakides, S.; Jang, W.Y. Compressive response of open-cell foams. Part I: Morphology and elastic properties. *Int. J. Solids Struct.* **2005**, *42*, 1355–1379. [CrossRef]
4. Song, H.W.; Fan, Z.J.; Yu, G.; Wang, Q.C.; Tobota, A. Partition energy absorption of axially crushed aluminum foam-filled hat sections. *Int. J. Solids Struct.* **2005**, *42*, 2575–2600. [CrossRef]
5. Wang, W.; Burgueño, R.; Hong, J.W.; Lee, I. Nano-deposition on 3-d open-cell aluminum foam materials for improved energy absorption capacity. *Mater. Sci. Eng.* **2013**, *572*, 75–82. [CrossRef]
6. Kim, A.; Hasan, M.A.; Nahm, S.H. Evaluation of compressive mechanical properties of Al-foams using electrical conductivity. *Compos. Struct.* **2005**, *71*, 191–198. [CrossRef]
7. Rajendran, R.; Sai, K.P.; Chandrasekar, B.; Gokhale, A.; Basu, S. Preliminary investigation of aluminium foam as an energy absorber for nuclear transportation cask. *Mater. Des.* **2008**, *29*, 1732–1739. [CrossRef]
8. Liang, L.S.; Yao, G.C.; Wang, L.; Ma, J.; Hua, Z.S. Sound absorption of perforated closed-cell aluminum foam. *Chin. J. Nonferrous Met.* **2010**, *20*, 2372–2376.
9. Zhang, C.J.; Feng, Y.; Zhang, X.B. Mechanical properties and energy absorption properties of aluminum foamfilled square tubes. *Trans. Nonferrous Met. Soc. China* **2010**, *20*, 1380–1386. [CrossRef]
10. Liu, H.; Yao, G.C.; Cao, Z.K.; Hua, Z.K.; Shi, J.C. Properties of aluminum foams with electrodeposited Ni coatings. *Chin. J. Nonferrous Met.* **2012**, *22*, 2572–2577.
11. Marchi, C.S.; Mortensen, A. Deformation of open-cell aluminum foam. *Acta Mater.* **2001**, *49*, 3959–3969. [CrossRef]
12. Jung, A.; Natter, H.; Diebels, S.; Lach, E.; Hempelmann, R. Nanonickel Coated Aluminum Foam for Enhanced Impact Energy Absorption. *Adv. Eng. Mater.* **2011**, *13*, 23–28. [CrossRef]
13. Jung, A.; Lach, E.; Diebels, S. New hybrid foam materials for impact protection. *Int. J. Impact Eng.* **2014**, *64*, 30–38. [CrossRef]
14. Yi, F.; Zhu, Z.; Zu, F.; Hu, S.; Yi, P. Strain rate effects on the compressive property and the energy-absorbing capacity of aluminum alloy foams. *Mater. Charact.* **2001**, *47*, 417–422. [CrossRef]
15. Evans, A.G.; Hutchinson, J.W.; Ashby, M.F. Multifunctionality of cellular metal systems. *Prog. Mater. Sci.* **1998**, *43*, 171–221. [CrossRef]
16. Gibson, L. Mechanical behavior of metallic foams. *Annu. Rev. Mater. Sci.* **2000**, *30*, 191–227. [CrossRef]

17. Baumeister, J.; Banhart, J. Weber M. Aluminium foams for transport industry. *Mater. Des.* **1997**, *18*, 217–220. [CrossRef]
18. Barchi, L.; Bardi, U.; Caporali, S.; Fantini, M.; Scrivani, A.; Scrivani, A. Electroplated bright aluminium coatings for anticorrosion and decorative purposes. *Prog. Org. Coat.* **2010**, *68*, 120–125. [CrossRef]
19. Ma, J.; He, Y.D.; Wang, J.; Sun, B.D. High temperature corrosion behavior of microcrystalline aluminide coatings by electro-pulse deposition. *Trans. Nonferrous Met. Soc. China* **2008**, *18*, 13–17.
20. Lu, J.; Do, I.; Drzal, L.T.; Worden, R.M.; Lee, I. Nanometal-decorated exfoliated graphite nanoplatelet based glucose biosensors with high sensitivity and fast response. *ACS Nano* **2008**, *2*, 1825–1832. [CrossRef]
21. Boonyongmaneerat, Y.; Schuh, C.A.; Dunand, D.C. Mechanical properties of reticulated aluminum foams with electrodeposited Ni–W coatings. *Scr. Mater.* **2008**, *59*, 336–339. [CrossRef]
22. Li, Z.D.; Huang, Y.J.; Wang, X.F.; Wang, X.F.; Wang, D.; Han, F.S. Enhancement of open cell aluminum foams through thermal evaporating Zn film. *Mater. Lett.* **2016**, *172*, 120–124. [CrossRef]
23. Liu, J.; Zhu, X.Y.; Jothi, S.; Diao, W.; Yu, S. Increased Corrosion Resistance of Closed-Cell Aluminum Foams by Electroless Ni-P Coatings. *Mater. Trans.* **2011**, *52*, 2282–2284. [CrossRef]
24. Leszczyńska, A.; Winiarski, J.; Szczygiel, B.; Szczygiel, I. Electrodeposition and characterization of Ni–Mo–ZrO_2 composite coatings. *Appl. Surf. Sci.* **2016**, *369*, 224–231. [CrossRef]
25. Alizadeh, M.; Cheshmpish, A. Electrodeposition of Ni-Mo/Al_2O_3 nano-composite coatings at various deposition current densities. *Appl. Surf. Sci.* **2019**, *466*, 433–440. [CrossRef]
26. Beltowska-Lehman, E.; Indyka, P. Kinetics of Ni–Mo electrodeposition from Ni-rich citrate baths. *Thin Solid Film* **2012**, *520*, 2046–2051. [CrossRef]
27. Chang, C.S.; Hou, K.H.; Ger, M.D.; Chung, C.K.; Lin, J.F. Effects of annealing temperature on microstructure, surface roughness, mechanical and tribological properties of Ni–P and Ni–P/SiC films. *Surf. Coat. Technol.* **2016**, *288*, 135–143. [CrossRef]
28. Zhou, Y.; Xie, F.Q.; Wu, X.Q.; Zhao, W.D.; Chen, X. A novel plating apparatus for electrodeposition of Ni-SiC composite coatings using circulating-solution co-deposition technique. *J. Alloy Compd.* **2017**, *699*, 366–377. [CrossRef]
29. Kumar, K.A.; Kalaignan, G.P.; Muralidharan, V.S. Pulse and Pulse Reverse Current Electrodeposition and Characterization of Ni–W–TiN Composites. *Sci. Adv. Mater.* **2012**, *4*, 1039–1046. [CrossRef]
30. Chen, F.C.; Liu, X.J. Study on Electrodeposition of Ni-Mo Alloy. *J. Hunan Univ. Nat. Sci.* **2014**, *41*, 44–67.
31. Antenucci, A.; Guarino, S.; Tagliaferri, V.; Ucciardello, N. Improvement of the mechanical and thermal characteristics of open cell aluminum foams by the electrodeposition of Cu. *Mater. Des.* **2014**, *59*, 124–129. [CrossRef]
32. Li, B.S.; Li, X.; Huan, Y.X.; Xia, W.Z.; Zhang, W.W. Influence of alumina nanoparticles on microstructure and properties of Ni-B composite coating. *J. Alloy Compd.* **2018**, *762*, 132–142. [CrossRef]
33. Smith, B.H.; Szyniszewski, S.; Hajjar, J.F.; Schafer, B.W.; Arwade, S.R. Steel Foam for Structures: A Review of Applications, Manufacturing and Material Properties. *J. Constr. Steel Res.* **2012**, *71*, 1–10. [CrossRef]
34. Jang, W.Y.; Kyriakides, S.; Kraynik, A.M. On the compressive strength of open-cell metal foams with Kelvin and random cell structures. *Int. J. Solids Struct.* **2010**, *47*, 2872–2883. [CrossRef]
35. Lan, F.C.; Zeng, F.B.; Zhou, Y.J.; Chen, J.Q. Progress on Research of Mechanical Properties of Closed-cell Aluminum Foams and Its Applications in Automobile Crashworthiness. *Chin. J. Mech. Eng.* **2014**, *50*, 97–112. [CrossRef]
36. Gibson, L.J.; Ashby, M.F. *Cellular Solids Structures and Properties-Second Edition*; Cambridge University Press: Cambridge, UK, 1999; p. 528.
37. Li, N.; Gao, C.H. Microstructure and electrochemical properties of the electrodeposited Ni-Mo/ZrO_2 Alloy coating. *Mater. Sci. Technol. Lond.* **2011**, *19*, 104–109.
38. Xu, Y.; Ma, S.; Fan, M.Y.; Chen, Y.; Song, X.; Hao, J. Design and properties investigation of Ni-Mo composite coating reinforced with duplex nanoparticles. *J. Surf. Coat. Technol.* **2019**, *363*, 51–60. [CrossRef]

Article

Electrochemical Corrosion Behavior of Ni-Fe-Co-P Alloy Coating Containing Nano-CeO_2 Particles in NaCl Solution

Xiuqing Fu [1,2,*], Wenke Ma [1], Shuanglu Duan [1], Qingqing Wang [1] and Jinran Lin [1,2]

1 College of Engineering, Nanjing Agricultural University, Nanjing 210095, China
2 Key laboratory of Intelligence Agricultural Equipment of Jiangsu Province, Nanjing 210031, China
* Correspondence: fuxiuqing@njau.edu.cn; Tel.: +86-139-1387-8179

Received: 29 July 2019; Accepted: 14 August 2019; Published: 16 August 2019

Abstract: In order to study the effect of nano-CeO_2 particles doping on the electrochemical corrosion behavior of pure Ni-Fe-Co-P alloy coating, Ni-Fe-Co-P-CeO_2 composite coating is prepared on the surface of 45 steel by scanning electrodeposition. The morphology, composition, and phase structure of the composite coating are analyzed by means of scanning electron microscope (SEM), energy dispersive spectroscopy (EDS), and X-ray diffraction (XRD). The corrosion behavior of the coatings with different concentrations of nano-CeO_2 particles in 50 g/L NaCl solution is studied by Tafel polarization curve and electrochemical impedance spectroscopy. The corrosion mechanism is discussed. The experimental results show that the obtained Ni-Fe-Co-P-CeO_2 composite coating is amorphous, and the addition of nano-CeO_2 particles increases the mass fraction of P. With the increase of the concentration of nano-CeO_2 particles in the plating solution, the surface flatness of the coating increases. The surface of Ni-Fe-Co-P-1 g/L CeO_2 composite coating is uniform and dense, and its self-corrosion potential is the most positive; the corrosion current and corrosion rate are the smallest, and the charge transfer resistance is the largest, showing the best corrosion resistance.

Keywords: scanning electrodeposition; Ni-Fe-Co-P-CeO_2 composite coating; electrochemical corrosion behavior; corrosion mechanism

---

1. Introduction

Corrosive environments are one of the most common service environments for metal components in engineering. Due to the high chemical activity of Fe in such environments, the engineering application of steel components therein is facing severe challenges due to insufficient corrosion resistance [1,2]. Surface modification is one of the most effective ways to solve this problem. The electroplating process for the preparation of nanocomposites is a process for the co-deposition of nanoparticles and metal ions on the surface of a cathode workpiece via the electrochemical principle, and a process to obtain nanocomposites that demonstrate superior performance [3,4]. Scanning electrodeposition technology, as an extension of electroplating technology, is widely used in machinery, aerospace, electronics industry, etc., due to its controllability, high efficiency, selectivity, and superior coating performance [5,6]. In recent years, many scholars have devoted to improving the performance of traditional nickel-based alloy coatings. Usually, tungsten [7], copper [8], iron [9], cobalt [10], and other metal ions [11] are introduced into the electrolyte, thereby processing a multi-component alloy. The multi-component overcomes the shortcomings of unary and binary alloy coatings, and has good wear resistance and corrosion resistance, which meet the varying performance requirements of composite materials [12,13].

It has been found in research that the application properties and functions of alloy coating can be further improved by co-depositing second phase nano-oxide particles in a nickel-based alloy coating [14,15]. The rare earth element cerium is the only stable tetravalent element. It has a unique oxidizing property and a large effective nuclear charge number, which can catalyze many reactions, and is widely used in various applications. Cerium oxide is a typical rare earth oxide with good wear resistance and corrosion resistance, and can be used as a nanoparticle reinforcement phase in various applications [16–18]. In order to further improve the corrosion resistance of traditional nickel-based coatings, Ni-Fe-Co-P-CeO_2 composite coatings are prepared by scanning electrodeposition technology. The concentration of nano-CeO_2 particles in the plating solution is applied to the coating of Ni-Fe-Co-P alloys. The influence of appearance and structure, and the electrochemical corrosion behavior of composite coatings, provide a reference for the development of new composite materials.

2. Materials and Methods

2.1. Experimental Principle

The scanning electrodeposition test apparatus is shown in Figure 1, wherein the anode nozzle is mounted on the machine tool spindle; the workpiece is mounted on the workpiece mounting platform by tightening the fixing screws. During the scanning electrodeposition process, the anode bed of the anode nozzle reciprocates in the Y direction, and the water pump presses the plating solution from the reservoir into the anode nozzle through the inlet tube and sprays it on the surface of the workpiece at high speed to spray the plating solution in the electrodeposition chamber. The liquid return tube flows back to the reservoir to realize the circulation of the plating liquid. After the power is turned on, the plating solution sprayed on the surface of the workpiece through the anode nozzle forms a closed loop, and under the action of the external electric field, a redox reaction occurs to realize deposition of metal ions. The scanning length during the test is 20 mm, and the scanning speed is 13.5 mm/s. The height between the bottom of the anode nozzle and the workpiece processing surface is 1.5 mm.

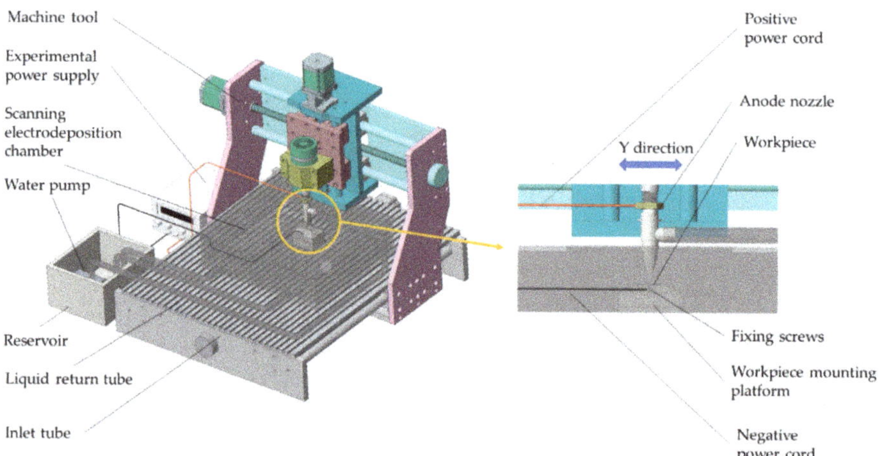

Figure 1. Scanning electrodeposition test device.

2.2. Materials and Methods

Fourty five steel with dimensions of 25 mm × 10 mm × 8 mm was used as substrate material, and its chemical composition is listed in Table 1. Table 2 shows the formulation of the plating solution used. The drugs used are of analytical grade and are prepared with deionized water. The particle size of the nano-CeO_2 particles in the test was 100 nm, and the concentration of nano-CeO_2 particles in the plating solution was 0, 0.5, 1, and 1.5 g/L, respectively. The cathode workpieces are polished with 800# and 1500# water sandpaper, respectively, and the workpiece is pretreated before scanning electrodeposition process; the specific process is shown in the Table 3, after each step is rinsed with deionized water. The workpiece that has been subjected to the pre-treatment is placed in a spray electrodeposition test apparatus for a sputtering test. The current during the spray electrodeposition process is 0.6 A, the pH of the plating solution is 1.0–1.5, the bath temperature is 60 °C, and the plating time is 20 min. After the end of the scanning electrodeposition process, the workpiece was subjected to ultrasonic cleaning and drying treatment, and performance studies were performed.

Table 1. Chemical composition of 45 steel (mass fraction).

C	Si	Mn	Cr	Ni	Cu
0.42~0.50%	0.17~0.37%	0.50~0.80%	≤0.25%	≤0.30%	≤0.25%

Table 2. Composition of plating solution.

Plating Solution Composition	Content (g/L)
Nickel sulfate hexahydrate ($NiSO_4{\cdot}6H_2O$)	120
Nickel chloride hexahydrate ($NiCl_2{\cdot}6H_2O$)	40
Ferrous sulfate ($FeSO_4{\cdot}7H_2O$)	20
Cobalt chloride ($CoCl_2{\cdot}6H_2O$)	10
Phosphoric acid (H_3PO_3)	30
Orthoboric acid (H_3BO_3)	30
Citric acid ($C_6H_8O_7$)	10
Thiourea (CH_4N_2S)	0.01
Sodium dodecyl sulfate ($C_{12}H_{25}SO_4Na$)	0.08

Table 3. The process of workpiece pretreatment.

Step	Solution Formula	Content (g/L)	Process Parameters
Electric net degreasing	Sodium hydroxide (NaOH)	25	Current = 1 A Power-on time = 20 s pH = 13
	Sodium carbonate (Na_2CO_3)	21	
	Trisodium phosphate anhydrous (Na_3PO_4)	50	
	Sodium chloride (NaCl)	2	
Weak activation	Hydrochloric Acid (HCl)	25	Current = 1 A Power-on time = 30 s pH = 0.3
	Sodium chloride (NaCl)	140	
Strong activation	Trisodium citrate dihydrate ($Na_3C_6H_5O_7{\cdot}2H_2O$)	140	Current = 1 A Power-on time = 20 s pH = 4
	Citric acid ($C_6H_8O_7$)	94	
	Nickel chloride hexahydrate ($NiCl_2{\cdot}6H_2O$)	3	

2.3. *Characterization*

The morphology of the coating was observed by scanning electron microscopy (FEI-SEM, Quanta FEG250; FEI Instruments, Hillsboro, OR, USA), with an accelerating voltage of 15 kV and image type of secondary electron image (SEI); the chemical composition of the coating was determined by energy dispersive spectroscopy (EDS, XFlash 5030 Bruker AXS, Inc., Berlin, Germany), with an accelerating voltage of 16 kV and the working distance of 11 mm; the phase structure of the coating was analyzed by X-ray diffraction (XRD, PANalytical X'pert; PANalytical Inc., Almelo, The Netherlands), with a radiation source of Cu Kα (λ = 0.15405 nm), operating voltage of 40 kV, scan rate of 5 °/min, and scanning range (2θ) of 10° ~ 80°, using HighScore Plus 3.0 to analyze the results.

The corrosion resistance of the coating was detected by electrochemical test of the three-electrode system (Figure 2). The working electrode is the workpiece, and the auxiliary electrode is Pt piece; the reference electrode is saturated calomel electrode (SCE), and the Tafel polarization curve measurement and electrochemical impedance spectroscopy (EIS) are completed by electrochemical workstation CS350 (Wuhan Corrtest Instruments Corp., Ltd., Wuhan, China). In the test, the workpiece to be tested was encapsulated with epoxy resin and immersed in a 50 g/L NaCl solution, and the Tafel polarization curve of the coating was obtained by a potentiodynamic scanning method and then obtained by polarization curve epitaxy. Corrosion potential, corrosion current, and other parameters were used to explore the corrosion resistance of the coating and the substrate. Under the open circuit potential, the impedance spectrum of the coating in NaCl solution was tested by the alternating current impedance method (EIS). The test frequency was 0.01–10^5 Hz, and the scanning direction was from high frequency to low frequency. The impedance fitting of different coatings was performed by Zview 2 software analysis.

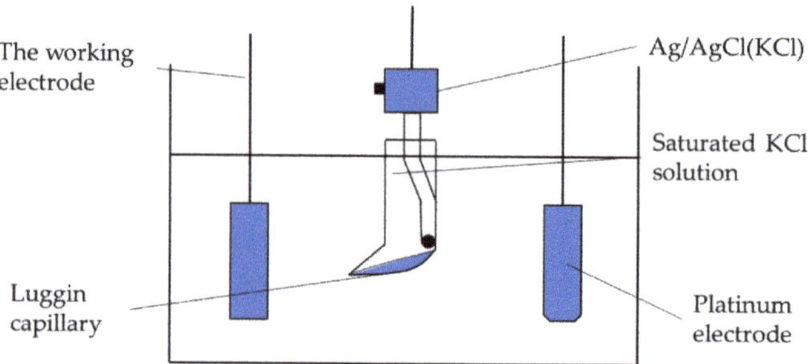

Figure 2. Electrochemical detection device schematic.

3. Results

3.1. Coating Morphology Analysis and Composition

The SEM photographs in Figure 3 (image type is SEI) show the surface topography of the composite coating before corrosion. It can be seen that before the corrosion, the coating structure of different nano-CeO_2 particles is composed of different sizes of cells, the arrangement is tight, and no obvious defects are found. When the concentration of nano-CeO_2 particles is 0 g/L (Figure 3a), the cytoplasm is a spherical hillock-like structure, but the size difference is large, and there are also defects such as pores and protrusions. When a small number of nano-CeO_2 particles is added to the plating solution (Figure 3b), the surface flatness of the coating is improved, but the cell structure has partial protrusions, the boundary is tortuous, and there are some defects such as pores. When the concentration is increased to 1 g/L (Figure 3c), the surface of the coating is dense and flat, the structure is compact, the cells are closely arranged, and the boundary is very blurred, and there are no obvious protrusions and impurity pores. When the concentration of nano-CeO_2 particles in the plating solution reaches 1.5 g/L or more (Figure 3d), the surface morphology of the coating can be seen to have obvious agglomeration, and the surface of the coating is rough and uneven, with protrusions and defects generated. According to the analysis, the scanning jet of the plating solution accelerates the ion transport, increases the limiting current density, and strengthens the cathodic polarization, so that the deposition is performed at a high flow density [5]. The formation process of the compact nickel-based coating is similar to that of soil plant growth, and the nano-CeO_2 particles dispersed in the plating solution are similar to the seeds, and are adsorbed on the surface of the substrate by tiny solid particles, because the rare earth element Ce is the third sub-group element. It has a large effective charge number and exhibits strong adsorption capacity. It can adsorb Ni^{2+}, Fe^{2+}, Co^{2+}, and other ions [5]. As the deposition progresses, the seeds gradually grow, forming a cell structure with many different sizes. When the nano-CeO_2 particles are excessive, they are excessively adsorbed on the surface of the metal substrate, causing the surface-active sites of the matrix to be masked and lose their activity, thereby greatly reducing or even inhibiting the nucleation sites, and uneven nanoparticle agglomerates are deposited on the surface of the plating layer. The formation of larger protrusions affects the quality of the coating, and the advantage of nano-CeO_2 particles is not obvious.

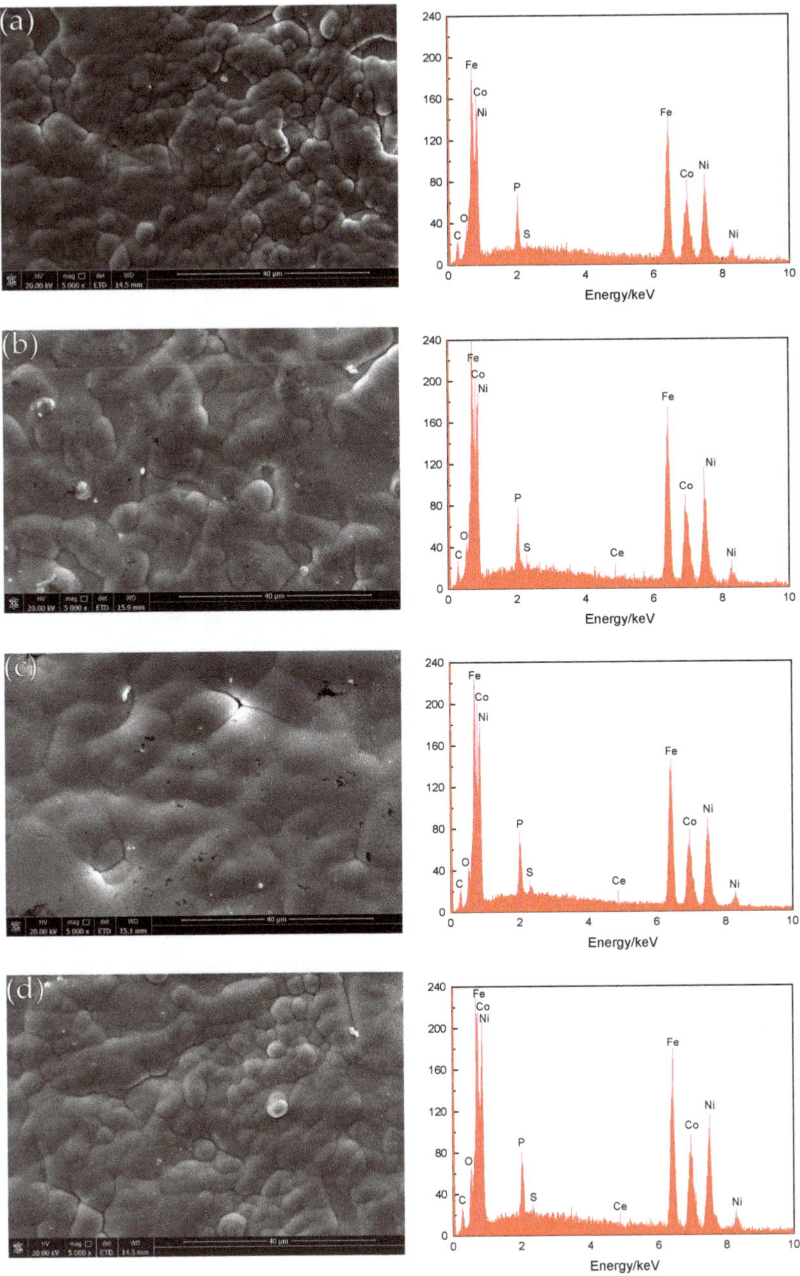

Figure 3. Surface morphology of the coatings before corrosion and EDS spectrum of coatings: (**a**) Ni-Fe-Co-P; (**b**) Ni-Fe-Co-P-0.5g/L CeO_2; (**c**) Ni-Fe-Co-P-1g/L CeO_2; and (**d**) Ni-Fe-Co-P-1.5g/L CeO_2.

After cutting and inlaying the test piece, the cross-section of the test piece is observed by SEM, and the cross-sectional shape of the obtained coating is shown in Figure 4 (image type is SEI). It is obvious that the Ni-Fe-Co-P-CeO_2 composite coating is uniform and dense, and there are no larger defects such as cracks and holes, effectively shielding the corrosion passage of the corrosive medium into the substrate and retarding the corrosion.

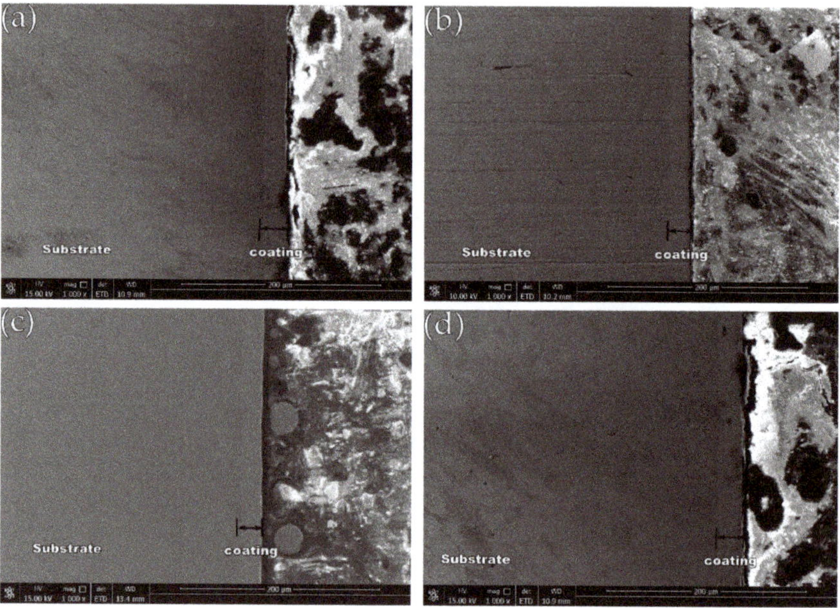

Figure 4. Cross-section morphology of the coatings: (**a**) Ni-Fe-Co-P; (**b**) Ni-Fe-Co-P-0.5 g/L CeO_2; (**c**) Ni-Fe-Co-P-1 g/L CeO_2; and (**d**) Ni-Fe-Co-P-1.5 g/L CeO_2.

Using EDS technology, the EDS spectrum obtained by analyzing the composition of the surface of the coating is shown in Figure 3. Ni, Fe, Co, and P elements are present in all the energy spectra, and an appropriate number of nano-CeO_2 particles are added to the plating solution. The energy spectrum of the surface of the coating shows a slight peak of Ce element (Figure 3b–d), which indicates that the prepared coating is a quaternary Ni-Fe-Co-P alloy coating and Ni-Fe-Co-P-CeO_2 composite coating. Figure 5 shows the mass fraction of P element in the coatings of different nano-CeO_2 particles obtained by EDS analysis. It can be seen that the mass fraction of P element increases first and then decreases with the increase of the concentration of nano-CeO_2 particles, and when the concentration of nano-CeO_2 is 1 g/L, the maximum value is 3.40%. Since P element will be enriched and hydrolyzed on the surface of the electrolyte to form hypophosphite, a phosphorus-rich film is formed between the coating and the interface of the corrosive medium to make the nickel-based coating exhibit high corrosion resistance. Adding an appropriate number of nano-CeO_2 particles to the plating solution increases the P content in the coatings. The increase of the P element content shortens the film formation time of the phosphating film on the surface of the coatings, and also increases the thickness of the phosphating film, which contributes to the improvement of the corrosion resistance of the coatings [19,20].

Figure 6 shows an elemental view of the surface of the Ni-Fe-Co-P-1 g/L CeO_2 composite coating, wherein the Ce element diagram (Figure 6f) represents nano-CeO_2 particles, and it can be seen that the alloying elements and the nano-CeO_2 particles are uniformly distributed on the surface of the plating layer. Studies have shown that the uniform distribution of elements and particles is due to the improved corrosion resistance of the coating.

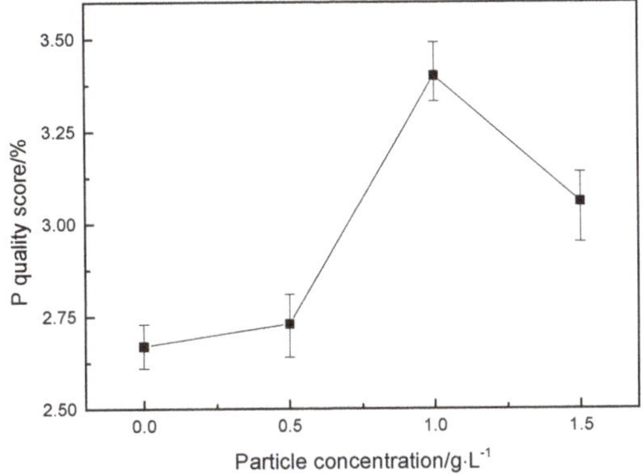

Figure 5. P mass fraction of coatings with different concentration of nano-CeO_2 particles.

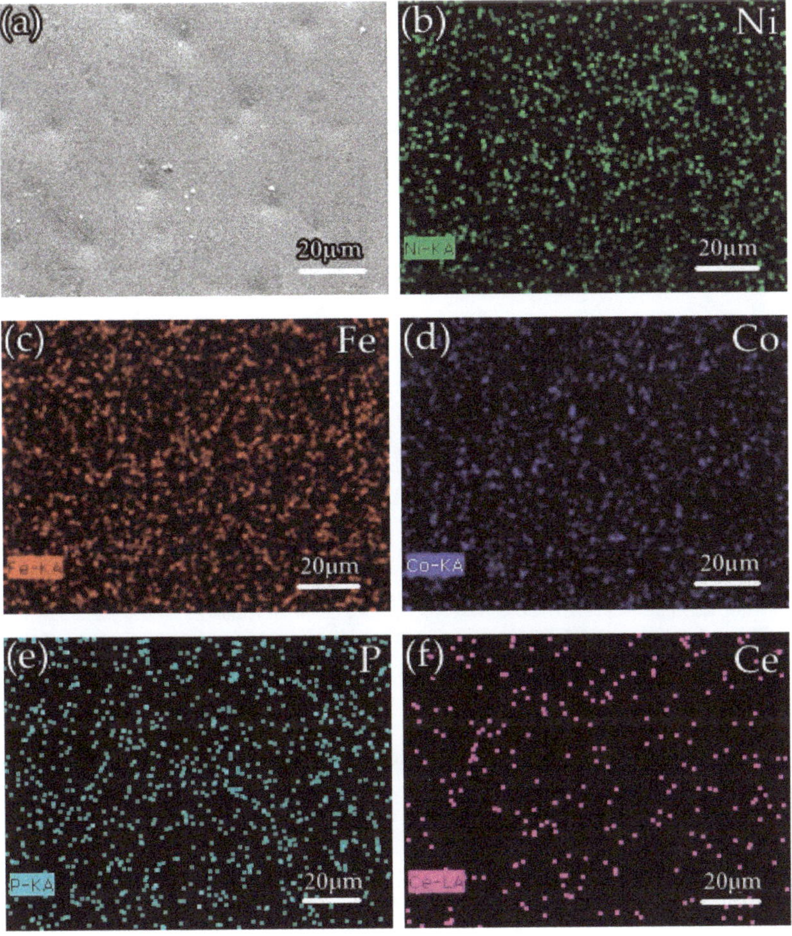

Figure 6. Elemental surface mapping of Ni-Fe-Co-P-1 g/L CeO_2 composite coating:(**a**) the SEM image of the analyzed surface (**b**) Ni content; (**c**) Fe content; (**d**) Co content; (**e**) P content; and (**f**) Ce content.

3.2. Plating Phase Structure

Figure 7 is an XRD pattern of the coating obtained by X-ray diffraction test. It can be seen that the coating is a typical amorphous structure, and there is a significant diffuse scattering broadening peak (Ni (110)) between 42° and 48° in 2θ. The peak width of the diffraction peak of the nanocrystalline alloy coating did not produce obvious changes, and peak intensity changes were not obvious, indicating that the nano-CeO_2 particles did not obviously change in the phase structure of Ni-Fe-Co-P coating. For the nickel-phosphorus coating, the crystal structure depends mainly on the P element content in the coating. The authors of [21] have shown that when P content is lower than 5%, it is usually crystalline structure, and when P content is higher than 6.5%, it becomes amorphous structure. In this test, since the prepared plating layer is a quaternary alloy plating layer, and the atomic structure, size, and electronegativity of Ni, Fe, Co, and P elements are largely different, the amorphous forming ability is enhanced. Therefore, the plating layer is still amorphous when the P content is low. It is generally believed that the amorphous coating has better corrosion resistance due to the absence of local electrochemical potential difference between crystal grains and grain boundaries in the crystalline coating [18].

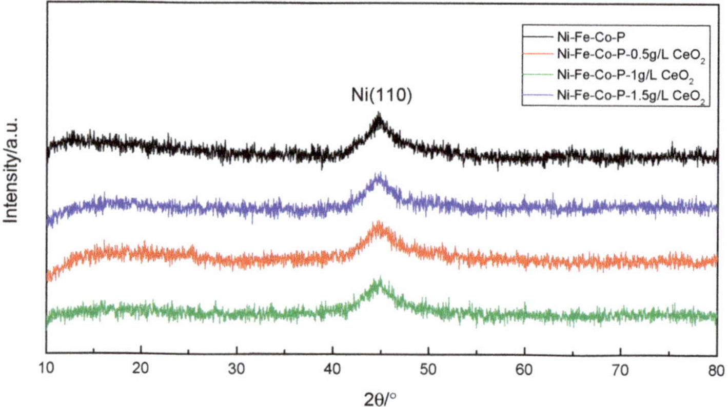

Figure 7. XRD patterns of coatings with different concentration of nanometer CeO_2 particles.

3.3. Tafel Polarization Curve

Figure 8 shows the polarization curves of the composite coatings in the 50 g/L NaCl solution. The corrosion parameters obtained by Cview 2 software and polarization curve epitaxy are shown in Table 4. It can be seen from Figure 8 that the anodic polarization process of the composite coating is hindered and a significant passivation behavior occurs, and the composite coating is obtained when the concentration of nano-CeO_2 particles in the plating solution is 1 g/L. The passivation zone is significantly larger than the remaining composite coating. It can be seen from Figure 8 and Table 4, compared with the polarization curve of pure Ni-Fe-Co-P alloy coating, that the polarization curve of composite coating prepared by co-deposition of a certain number of nano-CeO_2 particles by scanning electrodeposition technology moves up and left as a whole. With the increase of the concentration of nano-CeO_2 particles in the plating solution, the self-corrosion potential is continuously shifted, and the corrosion current density is gradually reduced. When the concentration of nano-CeO_2 particles is 1 g/L, the prepared Ni-Fe-Co-P-CeO_2 composite coating has the most positive self-corrosion potential (−0.19372 V) and the minimum corrosion current density (1.5375×10^{-5} A·cm^{-2}). While continuing to increase the concentration of nano-CeO_2 particles, the corrosion potential is negatively shifted, and the corrosion current density is significantly increased, indicating that corrosion resistance has begun to decline. According to the principle of corrosion electrochemistry, the larger the corrosion potential is, the smaller the corrosion current density is, the smaller the corrosion tendency of the material is, and the better the corrosion resistance is. Therefore, the concentration of nano-CeO_2 particles is 1 g/L. The Ni-Fe-Co-P-CeO_2 composite coating has the best corrosion resistance. Studies have shown that anodic polarization can slow metal corrosion, and the degree of anodic polarization directly affects the speed of the anode process [22]. Compared with the pure Ni-Fe-Co-P coating, the addition of nano-CeO_2 particles increases the hindrance of the corrosion process of the nickel-based coating. The Ba and Bc of the polarization curve of the composite coating are increased compared with the coating of the undoped nano-CeO_2 particles; especially, the blocking effect (Ba) of the anode is more significant. When excessive nano-CeO_2 particles are added to the plating solution, too much rare earth oxide adsorbs on the surface of the substrate, hindering the adsorption of Ni, Co, and Fe element on the surface of the substrate, which hinders the deposition of particles, which is not conducive to the plating. The formation of its corrosion resistance has been weakened.

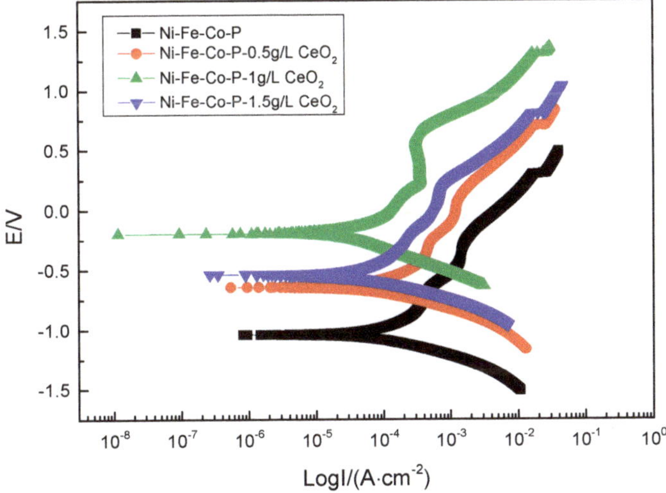

Figure 8. Polarization curves of coatings with different concentrations of nanometer CeO_2 particles.

Table 4. Composition of plating solution.

Sample	Ba (mV)	Bc (mV)	I_{corr} (A·cm^{-2})	E_{corr} (V)	Error (%)
Ni-Fe-Co-P	156.7	155.8	8.0989×10^{-5}	−1.0394	5.96
Ni-Fe-Co-P-0.5 g/LCeO_2	243.09	156.47	6.6569×10^{-5}	−0.64157	6.14
Ni-Fe-Co-P-1 g/LCeO_2	336.01	174.46	1.5375×10^{-5}	−0.19372	7.01
Ni-Fe-Co-P-1.5 g/LCeO_2	246.58	244.66	4.5404×10^{-5}	−0.5361	9.93

3.4. Analysis of Electrochemical Impedance Spectroscopy

In order to further explore the mechanism of electrochemical corrosion of Ni-Fe-Co-P-CeO_2 composite coating, more corrosion kinetic information is obtained. The AC impedance analysis of the composite coating is performed under open circuit potential. The electrochemical impedance spectrum obtained in Figure 9 is shown. The Nyquist diagram of the composite coating (Figure 9a) shows a single capacitive reactance arc characteristic, and the Bode diagram (Figure 9c) has only one peak, indicating that the time constant is 1, and the electrode reaction process is mainly affected by the charge [23]. The transfer effect also indicates that the corrosive medium only contacts the interface of the coating and does not penetrate into the surface of the substrate due to diffusion. It can be seen from the phase angle curve of the Bode diagram (Figure 9c) that the maximum phase angle of the Ni-Fe-Co-P-CeO_2 composite coating is higher than that of the pure nickel-based coating (56.379). From the impedance curve (Figure 9b), the impedance modulus of the composite coating doped with nano-CeO_2 particles is higher than that of the undoped nano-CeO_2 particles throughout the scanning frequency interval. This shows that the corrosion resistance of the Ni-Fe-Co-P alloy coating is effectively improved by co-depositing nano-CeO_2 particles. It can also be seen from the Nyquist diagram of Figure 9 (Figure 9a) that the radius of the capacitive reactance of the Ni-Fe-Co-P-CeO_2 composite coating is much larger than that of the Ni-Fe-Co-P alloy coating. When the concentration of nano-CeO_2 particles is 1 g/L, the radius of the capacitive anti-arc is the largest, and the radius of the capacitive anti-arc is used as the characterization of the corrosion resistance of the coating. The larger the radius, the greater the resistance of charge transfer and the harder the corrosion reaction. This result shows that the Ni-Fe-Co-P-CeO_2 composite coating has better corrosion resistance. The AC impedance spectrum is modeled by the equivalent circuit diagram shown in Figure 10 and fitted by Zview software. The obtained fitting data is shown in Table 5. In the equivalent circuit diagram, Rs is the resistance in the solution. Rp is a charge transfer resistor, CPE is a constant phase angle element, and its impedance is

$$Z = 1/Y_0(j\omega)^{-n}$$

its type has two parameters: constant Y_0, its dimension is $\Omega^{-1}\cdot cm^{-2}\cdot s^{-n}$; parameter n, dimensionless index. When n = 1, the CPE component is the ideal capacitor. When n = 0, the CPE component is pure resistance, and in the actual solution, n is between 0 and 1 [7]. Obviously, with the addition of nano-CeO_2 particles, the charge transfer resistance of the composite coating increases first and then decreases but the charge transfer resistor (R_p) of the doped nano-CeO_2 particles is always larger than that of the pure nickel-based coating, and the corrosion resistance is extremely high great improvement. When the concentration of nano-CeO_2 particles in the plating solution is too large, the nano-CeO_2 particles are easily agglomerated, and the inclusions formed are increased, resulting in loose coating structure, and the strengthening effect of the nano-CeO_2 particles is weakened.

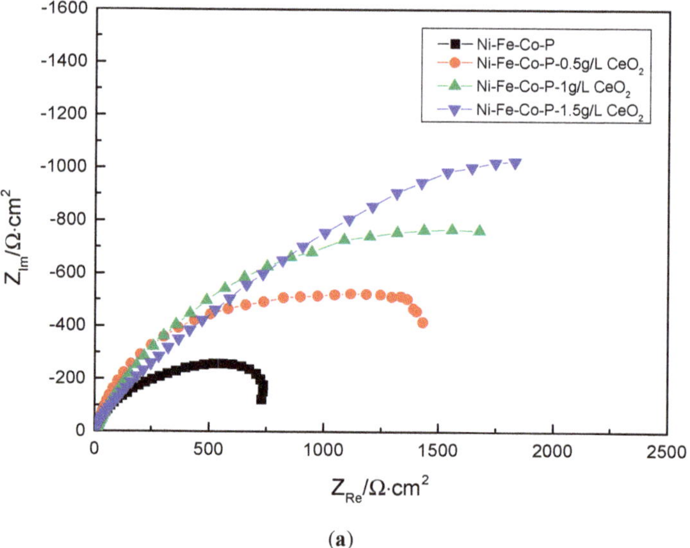

(a)

Figure 9. *Cont.*

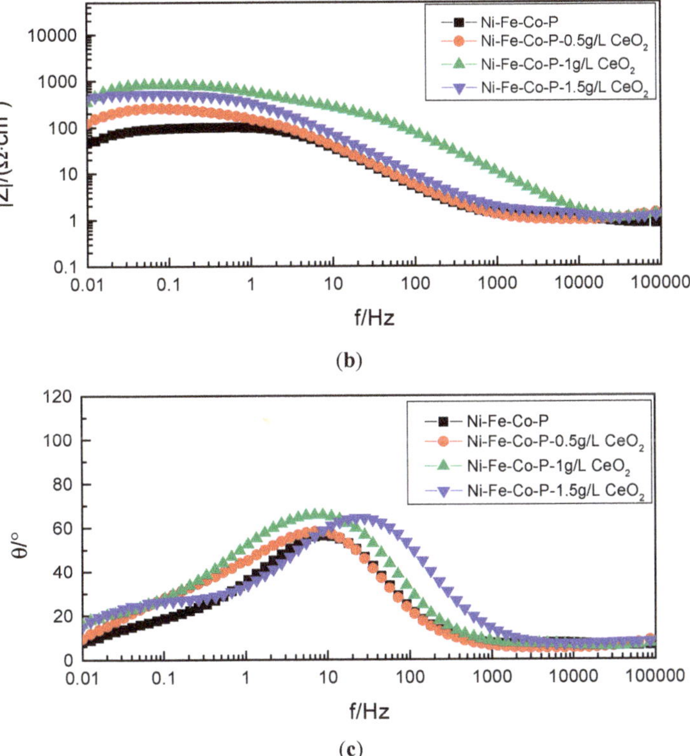

Figure 9. Alternating current impedance method EIS of coatings with of coatings with different concentrations of nanometer CeO_2 particles: (**a**) Nyquist diagram, (**b**) Bode diagram—impedance curve, and (**c**) Bode diagram—phase Angle curve.

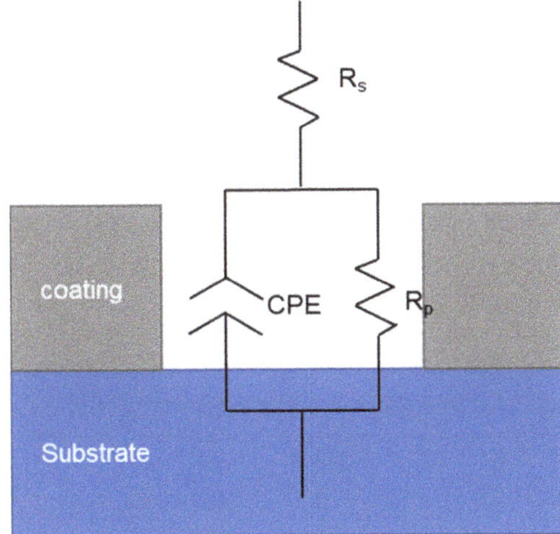

Figure 10. Equivalent circuit diagram.

Table 5. Equivalent circuit diagram parameter value.

Sample	R_s ($\Omega\cdot cm^{-2}$)	CPE-T ($F\cdot cm^{-2}$)	CPE-P	R_p ($\Omega\cdot cm^{-2}$)	Error (%)
Ni-Fe-Co-P	10.47	0.0011242	0.73659	776.1	4.46
Ni-Fe-Co-P-0.5 g/L CeO_2	11.31	0.000526	0.79027	1513	3.12
Ni-Fe-Co-P-1 g/L CeO_2	2.229	0.00054346	0.62269	2941	4.30
Ni-Fe-Co-P-1.5 g/L CeO_2	2.807	0.00084912	0.60494	2631	6.85

3.5. *Surface Morphology after Corrosion of the Coating*

The SEM photographs in Figure 11 (image type is SEI) shows the surface topography of the composite coating after corrosion. It can be seen that after 5 days of etching in 50 g/L NaCl solution, many micro-protrusions of different sizes appear on the surface of the coating, and the surface appears more frequently black corrosion product. A large number of narrow and shallow microcracks extend along the boundaries of the cell structure. The degree of corrosion of the composite coating with different concentrations of nano-CeO_2 particles in the plating solution is different. Among them, the coating of undoped nano-CeO_2 particles (Figure 11a) is most corroded, and a large amount of corrosion product is deposited on the surface. When the concentration of nano-CeO_2 particles is 1 g/L (Figure 11c), the coating is the least corroded and has a stronger retarding effect on corrosive media.

Generally speaking, the corrosion of metal in NaCl solution is mainly due to the presence of Cl^-, the Cl^- radius is small, the penetrating ability is very strong, and the adsorption is unevenly in the vicinity of the boundary and the impurity, so that the local dissolution is dominant, and pitting micropores are formed. Even if the surface has a passivation film formed by the metal, Cl^- can form a soluble compound with the cation of the passivation film, destroying the dynamic balance of dissolution and repair of the passivation film, causing the passivation film to be gradually eliminated and continue to be corroded, resulting in etching. The deepening of the hole can quickly become a corrosion pit. When sprayed electrodeposition is used to prepare Ni-Fe-Co-P alloy coating, it is always accompanied by hydrogen evolution reaction, which retards the discharge deposition of Ni, Co, and Fe elements, which leads to the formation of pinholes or pits on the surface of the coating. In addition, the transition elements are sprayed. During the electrodeposition process, the amount of hydrogen absorption is large. The hydrogen atoms that penetrate into the cell by diffusion will cause distortion of the cytoplasm, forming a large internal stress, and stress corrosion occurs during the corrosion process. After the corrosion, the surface of the plating layer is easily cracked [24]. Since the deposition process of the sprayed electrodeposited Ni-Fe-Co-P alloy coating is a process of uneven reduction and accumulation of Ni, Co, Fe, and P, the atomic size is different and the arrangement is different, which can only be disorderly stacked and reflected to the plating layer [25]. On the top, the cell material is relatively dispersed, which provides conditions for the diffusion of corrosive media, but the corrosion condition after the addition of nano-CeO_2 particles is improved. The reason is: first, the filling effect of the nano-CeO_2 particles between the coating boundaries makes the structure of the composite coating is more uniform and dense, the porosity is greatly reduced, and the rare earth elements have strong affinity with impurity elements such as O and H, and these impurity elements can be wrapped to form a rare earth composite phase, and the surface of the coating is dispersed. The composite phase coverage causes the permeation channel of Cl^- ions to be effectively intercepted, thereby enhancing the corrosion resistance of the composite coating [26]. Secondly, because the potential of nano-CeO_2 particles is in Ni, Co, and Fe metals, it is easy to form microscopic galvanic cells at the interface between nano-CeO_2 particles and nickel-based alloy. Nano-CeO_2 particles are used as the cathode, and Ni, Co, and Fe are the anode. This galvanic reaction changes the coating from local spot corrosion to uniform corrosion, which helps to slow down the corrosion. When the concentration of nano-CeO_2 particles is too large, the metal ions are precipitated in a large amount as a complex, and the amount of precipitation increases, but the actual deposition rate decreases, and the corrosion tendency of the coating increases [27].

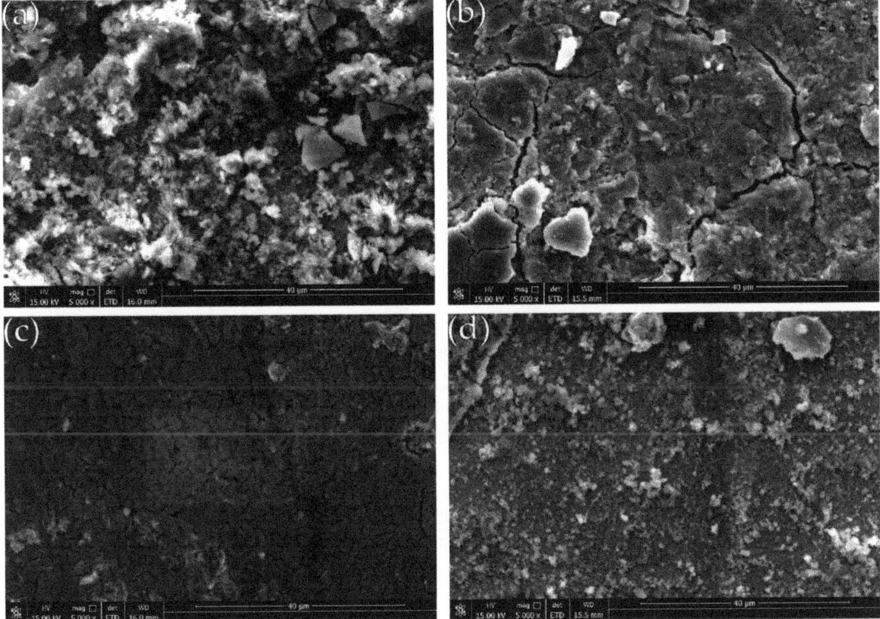

Figure 11. Surface morphology after plating corrosion: (**a**) Ni-Fe-Co-P; (**b**) Ni-Fe-Co-P-0.5 g/L CeO_2; (**c**) Ni-Fe-Co-P-1 g/L CeO_2; and (**d**) Ni-Fe-Co-P-1.5 g/L CeO_2.

4. Conclusions

In this paper, a Ni-Fe-Co-P-CeO_2 composite coating was prepared using the scanning electrodeposition technique. To explore the impact of the concentration of nano-CeO_2 particles in the plating solution on the micro morphology, structure, and composition of the coating, and to study the strengthening mechanism of nano-CeO_2 particles on the electrochemical corrosion behavior of Ni-Fe-Co-P alloy coating, the following conclusions were drawn:

(1) The surface structure of Ni-Fe-Co-P-CeO_2 composite coating is dense, with fewer defects, and the bonding between the coating and the substrate is good. The addition of nano-CeO_2 particles increases the P mass fraction of the coating, which helps slow down corrosion.

(2) The Ni-Fe-Co-P-CeO_2 composite coating is still amorphous in the case of low P mass fraction.

(3) The Ni-Fe-Co-P-1 g/L CeO_2 composite coating has the most positive self-corrosion potential, the lowest self-corrosion current density, and the best corrosion resistance.

(4) With the increase of the concentration of nano-CeO_2 particles in the plating solution, the impedance spectrum of Ni-Fe-Co-P-CeO_2 composite coating is nonlinearly related to the charge-transfer resistance of the equivalent circuit, which increases first and then decreases. Regularly, the Ni-Fe-Co-P-1 g/L CeO_2 composite coating has the largest charge transfer resistance (2941 $\Omega \cdot cm^{-2}$) and the weakest corrosion tendency.

(5) After corrosion, micro-cracks and a large number of corrosion products appear on the surface of the coating. After doping with appropriate number of nano-CeO_2 particles, the alloy coating can inhibit this corrosion, and the corrosion degree of the Ni-Fe-Co-P-1 g/L CeO_2 composite coating is the smallest, showing the best corrosion resistance.

Author Contributions: X.F. and S.D. designed the experiments; W.M., Q.W., and S.D. performed the experiments and analyzed the data; X.F. contributed reagents and materials. S.D. and W.M. wrote the paper; and X.F. and J.L. provided corrections on the original draft.

Funding: This research was funded by the China Postdoctoral Science Foundation (Grant number 2017M621665), the Postdoctoral Science Foundation of Jiangsu Province of China (Grant number 2018K022A), and Postgraduate Research and Practice Innovate Program of Jiangsu Province (Grant number SJCX19_0145).

Conflicts of Interest: The authors declare no conflict of interest.

References

1. Guan, P.P.; Liu, A.M.; Shi, Z.N.; Hu, X.W.; Wang, Z.W. Corrosion Behavior of Fe-Ni-Al Alloy Inert Anode in Cryolite Melts. *Metals* **2019**, *9*, 399. [CrossRef]
2. Melchers, R.E. The effect of corrosion on the structural reliability of steel offshore structures. *Corros. Sci.* **2005**, *47*, 2391–2410. [CrossRef]
3. Karimzadeha, A.; Aliofkhazraei, M.; Walsh, F.C. A review of electrodeposited Ni-Co alloy and composite coatings: Microstructure, properties and applications. *Surf. Coat. Technol.* **2019**, *372*, 463–498. [CrossRef]
4. Tam, J.; Lau, J.C.F.; Erb, U. Thermally Robust Non-Wetting Ni-PTFE Electrodeposited Nanocomposite. *Nanomaterials* **2019**, *9*, 2. [CrossRef] [PubMed]
5. Shen, L.; Fan, M.; Qiu, M.; Jiang, W.; Wang, Z. Superhydrophobic nickel coating fabricated by scanning electrodeposition. *Appl. Surf. Sci.* **2019**, *483*, 706–712. [CrossRef]
6. Shen, L.D.; Xu, M.Y.; Jiang, W.; Qiu, M.B.; Fan, M.Z.; Ji, G.B.; Tian, Z.J. A novel superhydrophobic Ni/Nip coating fabricated by magnetic field induced selective scanning electrodeposition. *Appl. Surf. Sci.* **2019**, *489*, 25–33. [CrossRef]
7. Zhang, W.W.; Li, B.S.; Ji, C.C. Synthesis and characterization of Ni-W/TiN nanocomposite coating with enhanced wear and corrosion resistance deposited by pulse electrodeposition. *Ceram. Int.* **2019**, *45*, 14015–14028. [CrossRef]
8. Chen, J.; Zhao, G.L.; Matsuda, K.; Zou, Y. Microstructure evolution and corrosion resistance of Ni-Cu-P amorphous coating during crystallization process. *Appl. Surf. Sci.* **2019**, *484*, 835–844. [CrossRef]
9. Safavi, M.S.; Babaei, F.; Ansarian, A.; Ahadzadeh, I. Incorporation of Y_2O_3 nanoparticles and glycerol as an appropriate approach for corrosion resistance improvement of Ni-Fe alloy coatings. *Ceram. Int.* **2019**, *45*, 10951–10960. [CrossRef]
10. Hasanpour, P.; Salehikahrizsangi, P.; Raeissi, K.; Santamaria, M.; Calabrese, L.; Proverbio, E. Dual Ni/Ni-Co electrodeposited coatings for improved erosion-corrosion behaviour. *Surf. Coat. Technol.* **2019**, *368*, 147–161. [CrossRef]
11. Abedini, B.; Ahmadi, N.P.; Yazdani, S.; Magagnin, L. Structure and corrosion behavior of Zn-Ni-Mn/Zn-Ni layered alloy coatings electrodeposited under various potential regimes. *Surf. Coat. Technol.* **2019**, *372*, 260–267. [CrossRef]
12. Su, Q.D.; Zhu, S.G.; Bai, Y.F.; Ding, H.; Di, P. Preparation and elevated temperature wear behavior of Ni doped WC-Al2O3 composite. *Int. J. Refract. Met. Hard Mat.* **2019**, *81*, 167–172. [CrossRef]
13. Liu, C.S.; Wei, D.D.; Huang, X.Y.; Mai, Y.J.; Zhang, L.Y.; Jie, X.H. Electrodeposition of Co-Ni-P/graphene oxide composite coating with enhanced wear and corrosion resistance. *J. Mater. Res.* **2019**, *34*, 1726–1733. [CrossRef]
14. Krawiec, H.; Vignal, V.; Krystianiak, A.; Gaillard, Y.; Zimowski, S. Mechanical properties and corrosion behaviour after scratch and tribological tests of electrodeposited Co-Mo/TiO_2 nano-composite coatings. *Appl. Surf. Sci.* **2019**, *475*, 162–174. [CrossRef]
15. Mehr, M.S.; Akbari, A.; Damerchi, E. Electrodeposited Ni-B/SiC micro- and nano-composite coatings: A comparative study. *J. Alloys Compd.* **2019**, *782*, 477–487. [CrossRef]
16. Xu, R.D.; Wang, J.L.; He, L.F.; Guo, Z.C. Study on the characteristics of Ni-W-P composite coatings containing nano-SiO_2 and nano-CeO_2 particles. *Surf. Coat. Technol.* **2008**, *202*, 1574–1579. [CrossRef]
17. Xu, R.D.; Wang, J.L.; Guo, Z.C.; Wang, H. Effects of rare earth on microstructures and properties of Ni-W-P-CeO_2-SiO_2 nano-composite coatings. *J. Rare Earths* **2008**, *26*, 579–583. [CrossRef]
18. Sheng, M.Q.; Weng, W.P.; Wang, Y.; Wu, Q.; Hou, S.Y. Co-W/CeO_2 composite coatings for highly active electrocatalysis of hydrogen evolution reaction. *J. Alloys Compd.* **2018**, *743*, 682–690. [CrossRef]
19. Lelevic, A.; Walsh, F.C. Electrodeposition of Ni-P alloy coatings: A review. *Surf. Coat. Technol.* **2019**, *369*, 198–220. [CrossRef]

20. Lin, J.D.; Chou, C.T. The influence of phosphorus content on the microstructure and specific capacitance of etched electroless Ni-P coatings. *Surf. Coat. Technol.* **2019**, *368*, 126–137. [CrossRef]
21. Badrnezhad, R.; Pourfarzad, H.; Madram, A.R.; Ganjali, M.R. Study of the Corrosion Resistance Properties of Ni-P and Ni-P-C Nanocomposite Coatings in 3.5 wt % NaCl Solution. *Russ. J. Electrochem.* **2019**, *55*, 272–280. [CrossRef]
22. Dehgahi, S.; Amini, R.; Alizadeh, M. Microstructure and corrosion resistance of Ni-Al_2O_3-SiC nanocomposite coatings produced by electrodeposition technique. *J. Alloys Compd.* **2017**, *692*, 622–628. [CrossRef]
23. Wang, C.; Shen, L.D.; Qiu, M.B.; Tian, Z.J.; Jiang, W. Characterizations of Ni-CeO_2 nanocomposite coating by interlaced jet electrodeposition. *J. Alloys Compd.* **2017**, *727*, 269–277. [CrossRef]
24. Ranganatha, S.; Venkatesha, T.V.; Vathsala, K.; Kumar, M.K.P. Electrochemical studies on Zn/nano-CeO_2 electrodeposited composite coatings. *Surf. Coat. Technol.* **2012**, *208*, 64–72. [CrossRef]
25. Ji, X.L.; Yan, C.Y.; Duan, H.; Luo, C.Y. Effect of phosphorous content on the microstructure and erosion-corrosion resistance of electrodeposited Ni-Co-Fe-P coatings. *Surf. Coat. Technol.* **2016**, *302*, 208–214. [CrossRef]
26. Li, B.S.; Zhang, W.W.; Li, D.D.; Wang, J.J. Electrodeposition of Ni-W/ZrO_2 nanocrystalline film reinforced by CeO_2 nanoparticles: Structure, surface properties and corrosion resistance. *Mater. Chem. Phys.* **2019**, *229*, 495–507. [CrossRef]
27. Jin, H.; Wang, Y.Y.; Wang, Y.T.; Yang, H.B. Synthesis and properties of electrodeposited Ni-CeO_2 nano-composite coatings. *Rare Metals* **2018**, *37*, 148–153. [CrossRef]

MDPI
St. Alban-Anlage 66
4052 Basel
Switzerland
Tel. +41 61 683 77 34
Fax +41 61 302 89 18
www.mdpi.com

Materials Editorial Office
E-mail: materials@mdpi.com
www.mdpi.com/journal/materials

www.ingramcontent.com/pod-product-compliance
Lightning Source LLC
LaVergne TN
LVHW071244170726
843515LV00006BA/1327